National Center for Construction Education and Research

AF834669

# *Pipeline Maintenance Level One*

Prentice Hall
Upper Saddle River, NJ
Columbus, OH

contren™ Learning Series

American Petroleum Institute
Pipeliner Training and Assessment Program (PTAP)

This information is general in nature and intended for training purposes only. Actual performance of activities described in this manual requires compliance with all applicable operating, service, maintenance, and safety procedures under the direction of qualified personnel. References in this manual to patented or proprietary devices do not constitute a recommendation of their use.

---



10 9
ISBN 0-13-046677-8

# *Preface*

This volume was developed by the National Center for Construction Education and Research (NCCER) in response to the training needs of the construction, maintenance, and pipeline industries. It is one of many in NCCER's *Contren™ Learning Series*. The program, covering training for close to 40 construction and maintenance areas, and including skills assessments, safety training, and management education, was developed over a period of years by industry and education specialists.

NCCER also maintains a National Registry that provides transcripts, certificates, and wallet cards to individuals who have successfully completed modules of NCCER's *Contren™ Learning Series*, when the training program is delivered by an NCCER Accredited Training Sponsor.

The NCCER is a not-for-profit 501(c)( 3) education foundation established in 1995 by the world's largest and most progressive construction companies and national construction associations. It was founded to address the severe workforce shortage facing the industry and to develop a standardized training process and curricula. Today, NCCER is supported by hundreds of leading construction and maintenance companies, manufacturers, and national associations, including the following partnering organizations:

## PARTNERING ASSOCIATIONS

- American Fire Sprinkler Association
- American Petroleum Institute
- American Society for Training & Development
- American Welding Society
- Associated Builders & Contractors, Inc.
- Association for Career and Technical Education
- Associated General Contractors of America
- Carolinas AGC, Inc.
- Carolinas Electrical Contractors Association
- Citizens Democracy Corps
- Construction Industry Institute
- Construction Users Roundtable
- Design-Build Institute of America
- Merit Contractors Association of Canada
- Metal Building Manufacturers Association
- National Association of Minority Contractors
- National Association of State Supervisors for Trade and Industrial Education
- National Association of Women in Construction
- National Insulation Association
- National Ready Mixed Concrete Association
- National Utility Contractors Association
- National Vocational Technical Honor Society
- North American Crane Bureau
- Painting & Decorating Contractors of America
- Portland Cement Association
- SkillsUSA-VICA
- Steel Erectors Association of America
- Texas Gulf Coast Chapter ABC
- U.S. Army Corps of Engineers
- University of Florida
- Women Construction Owners & Executives, USA

Some features of NCCER's *Contren™ Learning Series* are:

- An industry-proven record of success
- Curricula developed by the industry for the industry
- National standardization providing portability of learned job skills and educational credits
- Credentials for individuals through NCCER's National Registry
- Compliance with Apprenticeship, Training, Employer, and Labor Services (ATELS) requirements for related classroom training (CFR 29:29)
- Well-illustrated, up-to-date, and practical information

# *Acknowledgments*

This module is a product of the farsightedness and leadership of the following sponsors:

Alyeska Pipeline Service Company
Bradford Brothers, Inc.
Ed Walton Construction Co.
Enron Transportation Services
Equistar Pipeline Operations
Green Equipment Company
Koch Pipeline Company, L. P.
Lakehead Pipeline Company
Marathon Ashland Pipe Line LLC
Matrix Service Mid-Continent
Plantation Pipe Line Company
Sun Pipeline
West Shore Pipe Line
Williams Energy Services

This curriculum would not exist were it not for the dedication and unselfish energy of those volunteers who served on the Authoring Team. A sincere thanks is extended to:

Ken Ballon
Lloyd Bell
Cecil Bradford, III
Robert Buchanan
Daniel Cerkoney
Ron Davenport
Vance Davis
Don Gilley
Greg Guy
Roy Harlan
Neal Jones
Bruce Larson
Grady McWhorter
Daniel Pallo
Philip Tideman
Thomas Trauthwein

*A final note:* This book is the result of a collaborative effort involving the production, editorial, and development staff at Prentice-Hall, Inc., and the National Center for Construction Education and Research. Thanks to all of the dedicated people involved in the many stages of this project.

# Contents

Module 66101-02

# *Introduction to the Pipeline Industry*

# Introduction to the Pipeline Industry
# Annotated Instructor's Guide

**Module 66101-02**

## MODULE OVERVIEW

This module provides an overview of the pipeline industry. Topics covered include basic functions and purposes of pipelines and facilities, roles of control personnel, and operations procedures.

## PREREQUISITES

Please refer to the Course Map in the Trainee Module. Prior to training with this module, it is recommended that the trainee shall have successfully completed the following:

Core Curriculum

## OBJECTIVES

Upon completion of this module, the trainee will be able to:

1. Explain the basic functions and purposes of pipelines and facilities and identify the characteristics and hazards of common pipeline products.
2. Identify maps and drawings used to depict pipelines and facilities.
3. Explain the roles of control personnel and equipment in the overall operation of a pipeline.
4. Explain liquid pipeline hydraulics and gas pipeline pneumatics.
5. Explain the types and purposes of pipeline equipment.
6. Explain pipeline electrical power systems and corrosion control.
7. Review operations, maintenance, and emergency procedures and perform documentation required for pipeline operations.

## PERFORMANCE TASKS

This is a knowledge-based module – there is no performance testing.

## NCCER STANDARDIZED CRAFT TRAINING PROGRAM

The National Center for Construction Education and Research (NCCER) provides a standardized national program of accredited craft training. Key features of the program include instructor certification, competency-based training, and performance testing. The program provides trainees, instructors, and companies with a standard form of recognition through a National Craft Training Registry. The program is described in full in the *Guidelines for Accreditation*, published by the NCCER. For more information on standardized craft training, contact the NCCER by writing us at P.O. Box 141104, Gainesville, FL 32614-1104, calling 352-334-0911, or e-mailing info@nccer.org. More information may be found at our Web site, www.nccer.org.

## HOW TO USE THIS ANNOTATED INSTRUCTOR'S GUIDE

Each page presents two sections of information. The larger section displays each page exactly as it appears in the Trainee Module. The narrow column ties suggested trainee and instructor actions to each page and provides icons to call your attention to material, safety, audiovisual, or testing requirements. The bottom of each page includes space for your notes.

**Teaching Tip**

If you see the Teaching Tip icon, that means there is a teaching tip associated with this section. Also refer to any suggested teaching tips at the end of the module.

## SAFETY CONSIDERATIONS

Ensure that the trainees are equipped with appropriate personal protective equipment. Emphasize the importance of following all safety precautions and procedures when working with power tools.

## PREPARATION

Before teaching this module, you should review the Module Outline, Objectives, Performance Tasks, and the Materials and Equipment List. Be sure to allow ample time to prepare your own training or lesson plan and gather all required equipment and materials.

## MATERIALS AND EQUIPMENT LIST

Transparencies
Markers/chalk
Blank acetate sheets
Transparency pens
Pencils and scratch paper
Copies of Quick Quiz*
Module Examinations**
Overhead projector and screen
Whiteboard/chalkboard
Appropriate personal protective equipment
Copies of your company's policy and procedures manual
Samples of P&IDs, blueprints, and strip maps/alignment sheets
Posterboard or other large sheets of paper
Colored markers
Assorted valves
Copies of *API 650* and *API 653*
Copies of *API 510* and *2510*
Assorted fittings
Copies of *49 CFR Part 192* (Gas) and/or *49 CFR Part 195* (Liquid), including *Parts 192.327* and *195.248*
Assorted corrosion coupons
Various examples of documentation, including:
Operations logs
Work orders
Event logs

* Located at the end of this Annotated Instructor's Guide.
** Located in the Test Booklet packaged with this Annotated Instructor's Guide.

## TEACHING TIME FOR THIS MODULE

An outline for use in developing your lesson plan is presented below. Note that each Roman numeral in the outline equates to one session of instruction. Each session has a suggested time period of 2½ hours. This includes 10 minutes at the beginning of each session for administrative tasks and one 10-minute break during the session. Approximately 15 hours are suggested to cover *Introduction to the Pipeline Industry*. You will need to adjust the time required for hands-on activity and testing based on your class size and resources.

| Topic | Planned Time |
|---|---|
| **Session I. Introduction** | |
| A. Origination | ____________ |
| B. Transport | ____________ |
| 1. Laboratory | ____________ |
| C. Storage | ____________ |
| 1. Laboratory | ____________ |
| D. Delivery | ____________ |
| **Session II. Pipeline Products and Map Drawings** | |
| A. Natural Hydrocarbons | ____________ |
| B. Refined Products | ____________ |
| C. Special Products | ____________ |
| D. Protective Measures | ____________ |
| E. Piping and Instrumentation Diagrams | ____________ |
| F. Blueprints | ____________ |
| G. Strip Maps/Alignment Sheets | ____________ |
| **Session III. Operations, Fluids and System Hydraulics, and Gases and System Pneumatics** | |
| A. Scheduling | ____________ |
| B. Pipeline Operations | ____________ |
| C. Control Systems | ____________ |
| D. Communications | ____________ |
| E. Fluids | ____________ |
| F. Basic Hydraulics | ____________ |
| G. The Ideal Gas Laws | ____________ |
| H. Measuring Energy Gases | ____________ |
| I. Commercial Natural Gas Combustible Limits and Specific Gravity | ____________ |
| **Session IV. Pipeline Equipment** | |
| A. Pumps and Motors | ____________ |
| B. Compressors | ____________ |
| C. Valves | ____________ |
| 1. Laboratory | ____________ |
| D. Actuators/Operators | ____________ |
| E. Tanks | ____________ |
| F. Pressure Vessels | ____________ |
| G. Pipe | ____________ |
| H. Fittings | ____________ |
| I. Manifolds | ____________ |
| J. Traps | ____________ |
| K. Measuring/Sampling Equipment | ____________ |

**Session V. Electrical Power Systems and Corrosion Control**

A. Power Generation and Transmission ____________

B. Power Distribution ____________

C. DC Power Distribution ____________

D. Types and Causes of Corrosion ____________

  1. Laboratory ____________

E. Inspection and Testing ____________

F. Maintenance and Protection ____________

**Session VI. Regulations, Procedures, and Documentation**

A. Regulations ____________

B. Abnormal Operating Conditions ____________

C. Company Policies and Procedures ____________

D. Qualifications ____________

E. Documentation ____________

F. Summary

  1. Summarize module ____________

  2. Answer questions ____________

G. Module Examination ____________

  1. Trainees must score 70% or higher to receive recognition from the NCCER.

  2. Record the testing results on Craft Training Report Form 200 and submit the results to the Training Program Sponsor.

Module 66101-02

# *Introduction to the Pipeline Industry*

Instructor's Notes:

Instructor's Notes:

# Introduction to the Pipeline Operator Qualification Rule, API, and NCCER

*It could be said that petroleum products keep America running. Petroleum products like oil and natural gas are the principal sources of energy in the United States and are used in many manufacturing and chemical processes. Petroleum products are primarily transported by pipeline. There are more than a million miles of liquid and natural gas pipeline in this country, ranging in diameter from two inches to four feet. These pipelines bring crude oil and natural gas from production fields to processing plants and refineries—and ultimately to the consumer.*

## Pipeline Operator Qualification

Petroleum products are considered hazardous, so pipeline personnel must take extra precautions and follow specific procedures when working with them. To help ensure the safe handling and transportation of petroleum products, the United States Department of Transportation (DOT) mandates safe practices as federal regulations.

One such regulation, the Operator Qualification (OQ) rule, requires pipeline operators to develop a written qualification plan to evaluate their personnel and their contract personnel (see *49 CFR Part 192 Subpart N—Gas,* and *49 CFR Part 195 Subpart G—Liquid*). The evaluation hinges on the operator's ability to perform certain covered tasks and to be able to recognize and respond to abnormal operating conditions that may be encountered while performing these covered tasks.

## Covered Tasks

A pipeline performance task is deemed *covered* if the activity passes the following four-part test:

1. Is it performed on a pipeline facility?
2. Is it an operations or maintenance task?
3. Is it performed as a requirement of *49 CFR Part 195 (Liquid)* or *49 CFR Part 192 (Gas)*?
4. Does it affect the operation or integrity of the pipeline facility?

If the answer to all four questions is *yes*, then the activity is considered *covered* and must figure into the evaluation process of pipeline personnel.

## American Petroleum Institute

The OQ rule and its related covered tasks provided the catalyst to develop a unified pipeline training and assessment program—so that pipeline operators and owners have ready access to the assessment and training tools they need. This effort was the direct result of the vision and leadership provided by The American Petroleum Institute (API) and API members.

Efforts to unify the industry are nothing new to API. Indeed, the impetus for forming the Institute in 1919 was the need to standardize engineering specifications for drilling and production equipment. Today, the most pressing issues revolve around public perceptions and government policies toward the industry. Speaking with one voice on these issues has become as essential as having interchangeable parts in the field. API's members determine the petroleum industry's positions on public policy as well as the standards that govern its day-to-day operations worldwide.

## Consortium on Operator Qualification

API assembled a focus group whose primary purpose was to identify covered tasks and provide compliance recommendations to pipeline operators.

The Consortium on Operator Qualification (COOQ) became the guiding force for the identification and development of covered tasks for the liquid pipeline industry. Their work product, API publication 1161, *Guidance Document for the Qualification of Liquid Pipeline Personnel,* formed the basis for the resulting Pipeliner Training and Assessment Program.

The covered task list created in API publication 1161 has now evolved to include gas covered tasks and is maintained by NCCER. The most current covered task list can be downloaded at www.nccer.org.

## National Center for Construction Education and Research

With the guidance and support of API, its members, and the Consortium on Operator Qualification, the National Center for Construction Education and Research (NCCER) was able to design and implement the Pipeliner Training and Assessment Program.

NCCER is a not-for-profit 501(c)(3) education foundation established in 1995 by the world's largest and most progressive construction and maintenance contractors and national contractor associations. NCCER was founded to address the

severe workforce shortage facing the construction and maintenance industries and to develop a standardized training process and curricula. Today, NCCER is supported by hundreds of leading construction and maintenance companies, manufacturers, and national associations and developed more than 35 construction and maintenance titles.

NCCER also maintains a National Registry, which contains the records of trainees and instructors who have received NCCER training. All completion information for every trainee is sent to the NCCER and filed within the Registry. The Registry can then confirm training and skills for workers as they move from company to company, state to state, or even within their own company.

### *Standardized Curriculum (NCCER's Contren™ Learning Series)*

NCCER curricula are developed using the expertise of industry subject matter experts (SMEs). The use of SMEs as an integral part of every new development or revision ensures that the curriculum is current, industry-driven, and indicative of industry practices. The development and/or revision process is facilitated from start to finish by NCCER project managers, who work closely with the SMEs and a select team of technical writer(s) on the development and presentation of the curriculum.

### *Pipeline: Five Training Paths*

The pipeline curriculum focuses on essential skills and competencies within five skill areas:

- Corrosion Control
- Electrical and Instrumentation
- Maintenance
- Mechanical
- Operations (Field and Control Center)

SMEs from the pipeline industry have determined the skills and competencies needed based on API's covered tasks and the consensus of the group/industry. Specific company information is generally not incorporated into the curriculum unless the SMEs determine that the information is the industry standard.

### *Training Modules and Covered Tasks*

The smallest unit of instruction in the curriculum is a *module*. Each curriculum is composed of modules (the equivalent of stand-alone chapters) that are logically sequenced into complete *levels* of instruction. Every module is task-oriented and requires the successful completion of a written (knowledge-based) examination and a performance (hands-on) test. Modules may be individually selected and/or combined for efficient task training. Modules in the pipeline curricula, to the extent possible, will align with API-developed covered tasks. While not all modules will relate to covered tasks, all covered tasks will relate to at least one module.

### *Benefits of Training to the Pipeline Marketplace*

Applying NCCER's standardized training process allows pipeline owners and contractors in the United States to establish the benchmark for acceptable skill and competency levels within the pipeline industry. This process will assist training organizations in developing and/or improving standardized pipeline training programs that effectively match their resources and operating conditions while responding to regulatory demands and to the needs of the industry. The process will also recognize training organizations that meet the minimum standards of competence, warranting regulatory, public, and professional confidence. This in turn will foster national unity within the pipeline industry for the mutual benefit of individual technicians, owners, contractors, and ultimately pipeline-users and society.

## Additional Resources

To review the Pipeliner Training and Assessment Program Implementation and Resource Guide and the latest version of pipeline covered tasks, visit NCCER's Web site at www.nccer.org, or contact:

**National Center for Construction Education and Research**
PO Box 141104
Gainesville, FL 32614-1104
(888) 622-3720
Web site: www.nccer.org

For more information on API, please contact:

**American Petroleum Institute**
1220 L Street, Northwest
Washington, D.C. 20005
(202) 682-8000
Web site: www.api.org

To order *API Pub 1161*, contact Global Engineering Documents by calling 1-800-854-7149 or visiting their Web site at www.global.his.com.

Additional sources of information on the pipeline operator qualification regulation include:

**DOT – Office of Pipeline Safety**
www.ops.dot.gov

**Transportation Safety Institute**
www.tsi.dot.gov



Instructor's Notes:

## COURSE MAP

This course map shows each curriculum path in the Pipeliner Training program. The suggested training order begins on the left with the Pipeline Core modules and proceeds to the right. Which curriculum path you follow depends on your areas of specialty and interest. On a given curriculum path, skills levels increase as you advance to the right on the course map, although some are single-level curricula. The local Training Program Sponsor may adjust the training order.

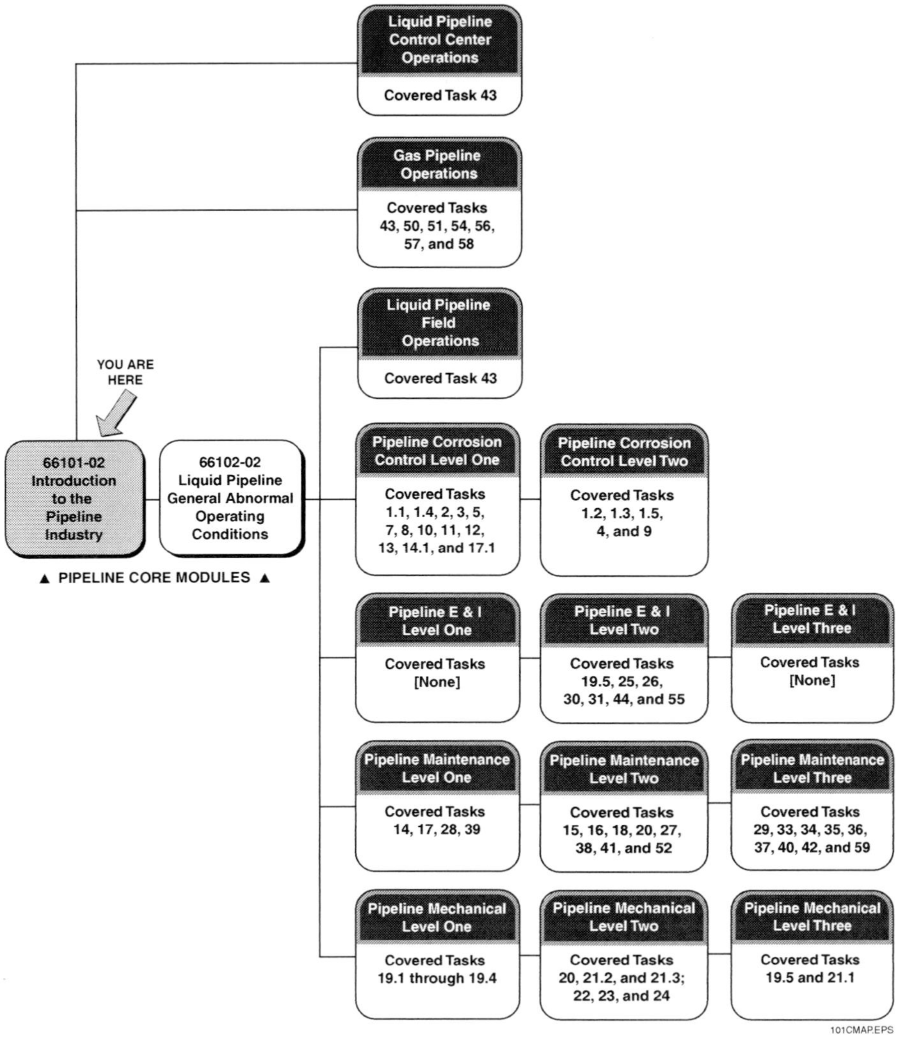






**Homework**

**Assign reading of Section 1.0.0.**

Instructor's Notes:

## MODULE 66101-02 CONTENTS

## MODULE 66101-02 CONTENTS (Continued)





Instructor's Notes:

## Figures





Instructor's Notes:

## Tables

Instructor's Notes:

MODULE 66101-02

# *Introduction to the Pipeline Industry*

## Objectives

When you have completed this module, you will be able to do the following:

1. Explain the basic functions and purposes of pipelines and facilities and identify the characteristics and hazards of common pipeline products.
2. Identify maps and drawings used to depict pipelines and facilities.
3. Explain the roles of control personnel and equipment in the overall operation of a pipeline.
4. Explain liquid pipeline hydraulics and gas pipeline pneumatics.
5. Explain the types and purposes of pipeline equipment.
6. Explain pipeline electrical power systems and corrosion control.
7. Review operations, maintenance, and emergency procedures and perform documentation required for pipeline operations.

## Prerequisites

Before you begin this module, it is recommended that you successfully complete the following: Core Curriculum.

## 1.0.0 ◆ INTRODUCTION

Pipelines are used to transport **petroleum**, petroleum **products**, liquid carbon dioxide, and other liquids and gases over great distances. In spite of the costs of constructing pipelines, they are a relatively economical means of transport. A pipeline system consists of pipes, pump stations, compressor stations, and other facilities to support the safe, continuous movement of liquids and gases. Pipelines carry petroleum from the source, the oil and gas fields, to the **refineries**, to and from gas processing plants, and eventually to consumers. Pipes are most often buried in the ground, but they are also laid under water and sometimes erected above the ground.

Today, a network of gas and liquid pipelines crisscrosses the United States. The first pipeline on a continental scale was completed in the 1930s, but the real boom in cross-country lines came during World War II when the Big Inch and the Little Big Inch were constructed to provide fuel for wartime industry and the military. The Big Inch began operating in 1943, was 24 inches in diameter, and stretched 1,341 miles from Longview, Texas, to Phoenixville, Pennsylvania, and Linden, New Jersey. This made it the largest **crude oil** pipeline in the world. The Little Inch, opened in 1944, was 20 inches in diameter and 1,475 miles long. It carried **refined products** from Beaumont, Texas, to Linden, New Jersey.

The most famous crude oil pipeline in the U.S. crosses 800 miles of permanently frozen ground in Alaska. The Trans-Alaska Pipeline, which began operations in 1977, is 48 inches in diameter and carries crude oil from the Prudhoe Bay field to the shipping terminal at Valdez, Alaska. Along the way, it crosses 34 major rivers and streams and three major mountain ranges, going from sea level to as high as 4,800 feet and back to sea level. Protection of the delicate tundra, permafrost, wilderness environment, and wildlife in its path were and still are critical concerns.

All pipeline operators are concerned with safety and environmental impact, particularly those carrying petroleum products that are hazardous to the environment as well as to the people who live near the pipeline. Good construction



**Materials**

**Ensure that you have everything required to teach the course. Check the Materials and Equipment List at the front of this module.**

**Classroom**

**Explain that terms shown in bold are defined in the Glossary at the back of this module.**

**Audiovisual**

**Show Transparency 1, Course Objectives.**

**Teaching Tip**

**See the general Teaching Tip at the end of this module.**

**Classroom**

**Review the history of pipeline systems. Emphasize the importance of protecting people and the environment.**

**Classroom**

**Ask a trainee to name the four functions of a petroleum pipeline system.**

**Review how oil and gas are formed. Refer to Figure 1. Explain the four types of traps.**

**Discuss drilling and completion. Ask trainees why a well may not be completed.**

practices, good maintenance and operations practices, continuous monitoring, and rapid response are necessary to prevent or to minimize the effects of pipeline leaks. Operations of a pipeline must be conducted in a fashion that will ensure the safety of employees and the general public and protect the environment.

The petroleum pipeline system can be divided into the following functions:

- Origination
- Transport
- Storage
- Delivery

### 1.1.0 Origination

All petroleum products produced or shipped originate from the domestic and foreign crude oil or natural gas from the oil and gas fields. Exploration teams from petroleum companies use geological and geophysical methods to find possible deposits of oil and gas deep under the layers of the earth. Drilling confirms and accesses the find; the well is completed and production begins.

#### *1.1.1 Exploration*

Oil and gas are formed in the earth from decaying organic matter that was buried long ago. The combination of time, pressure, and heat causes the matter to undergo a chemical change into hydrocarbons. These fluids and gases then tend to move, or migrate, upward in the earth until they either escape at the surface or are trapped in the earth's layers. It is this trapped fluid and gas that is sought by exploration.

The oil/gas-bearing formation is basically the layer of earth in which the hydrocarbon deposit is trapped until it is drilled. The formation must have the porosity to hold the fluid and gas and have the structure to keep it from moving further upward. Traps can have a variety of structures, and some of the most typical are the anticline trap, stratigraphic trap, fault trap, and salt dome trap, as shown in *Figure 1*.

#### *1.1.2 Drilling and Completion*

When a possible deposit is indicated, a well must be drilled to try to find it. Many wells have been drilled without finding the sought-after oil and gas. There may not be enough to warrant production costs or it cannot be drawn out economically. However, when a workable deposit is found, the well is completed for production. Completion includes casing the well with concrete, perforating the casing with

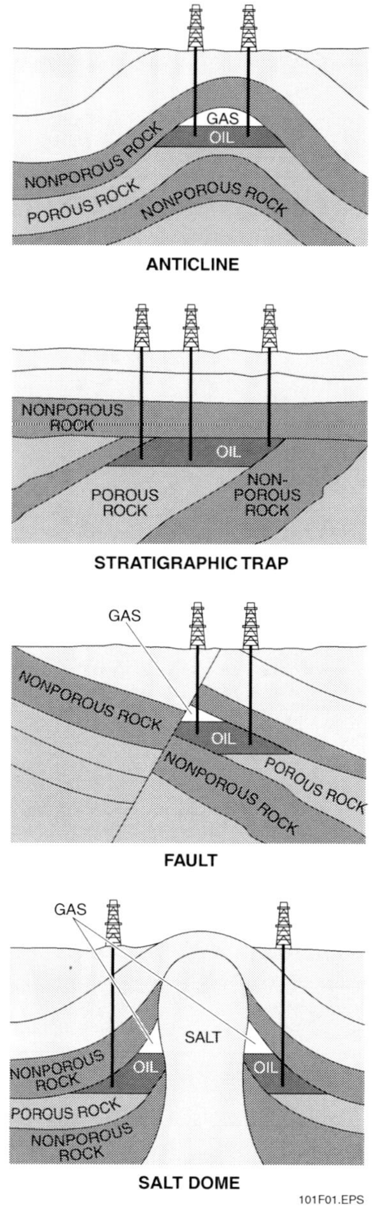


*Figure 1* ◈ Types of traps.



Instructor's Notes:

holes for the product to flow through tubing within the casing, and installing pumps or valves on the top of the hole to control the flow. Tests of the producing well may show a large enough deposit to warrant other wells being drilled and completed in the area. All of these wells comprise a field. It is from these wells that the oil or gas is gathered to transport through the pipeline.

Many crude oil reservoirs have a gas cap. The oil produced from these reservoirs often contains various amounts of associated gas. Some formations, however, trap only gas. This trapped gas forms a gas or gas-condensate reservoir, which produces only gas or gas with varying amounts of liquid gas condensate. Drilling gas wells is very similar to drilling and completing crude oil wells. The primary difference between crude oil and gas wells is in the type of surface facility equipment needed to process the oil or gas after it is produced.

### 1.1.3 Production

Different types of pumps or valves, such as a Christmas tree and a beam pump, are used to cap the well depending on the properties of the product being withdrawn and the formation producing it (*Figure 2*). Oil and gas are often found in the same formation along with some amount of water. The amount of each may vary in relation to each other, and this may affect the production of the well. Since gas is lighter than oil, and oil is lighter than water, they are layered in that order: gas on the top, then oil, then water on the bottom.

Because the formation had to be closed off in some way for the product to accumulate, pressure is usually also trapped in the formation. When it is drilled, that pressure can push the oil, gas, and water into the hole with some force. When it is controlled, this pressure is good because it can push the oil and gas up to the surface with little or no pumping. Generally a newer well will have more pressure than an older one, so in time, pumping power may be needed to pull the product out. However, some wells start with only a little pressure and must be pumped from the beginning. Control of the flow is important since producing a well too fast can affect the life of the well. These conditions, along with other properties, will determine how the well is capped and produced. A Christmas tree of valves may be installed to control the flow, and a beam pump may be used to pull the product out. Gas wells are completed with only a Christmas tree.

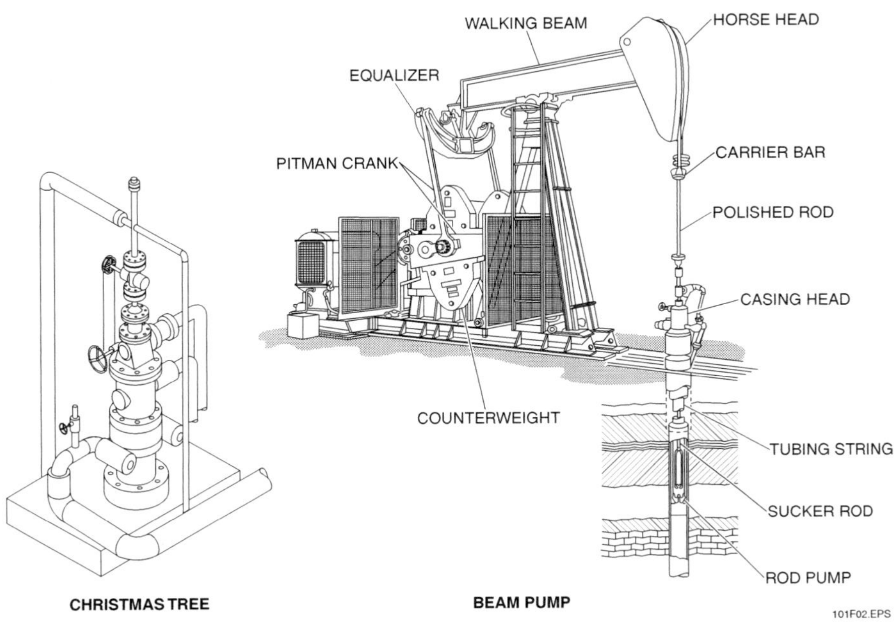


*Figure 2* ◆ Christmas tree and beam pump.



**Classroom**

**Ask trainees to explain the primary difference between crude oil and gas wells.**

**Discuss the effects of pressure on production and how to compensate for too little or too much pressure. Answer any questions trainees may have.**

Laboratory

**Divide trainees into pairs. Have each pair write down the steps in the process of moving oil and gas from the field to the consumer. Review and discuss the answers.**

Classroom

**On the whiteboard/ chalkboard, have trainees list the characteristics of the different types of pipe used in pipelines.**

### 1.2.0 Transport

Lines from the oil and gas fields are called *gathering lines* or *flow lines*. These gathering lines bring the oil and gas together into a facility for separation and processing. The vapor gas must travel in a different pipeline from the crude oil, so the gas and oil are separated in huge high-pressure **tanks**, called *gas/oil separators*. *Figure 3* shows the high-pressure separators.

From the separation or gas processing facility, the crude oil and gas is directed into a large **main line**, or trunk line, for the longest part of the journey to a refinery or a shipping terminal. The gas may be stored in tanks or routed through gas pipelines to market. Gas main lines transport gas to a gas distribution center for distribution to consumers, industrial facilities, or storage caverns. From the refinery, a large main line, or trunk line, will carry products across country with distribution lines branching off the main line along the way. A large distribution line may carry the liquid to a central distribution point, such as a regional or city station, from which smaller lines or vehicles carry it to consumers. Pipelines are constructed to meet *ASME Piping Standard, B31.4-1998, Pipeline Transportation Systems for Liquid Hydrocarbons, and other Liquids*. *Figure 4* shows the transportation chain.

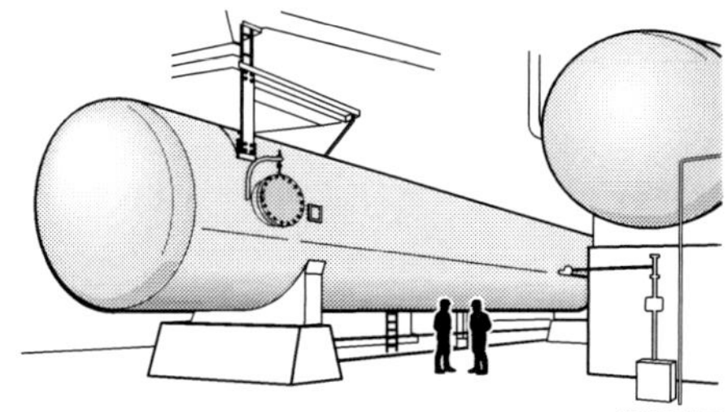

*Figure 3* ◆ High-pressure separators.

101F03.EPS

#### *1.2.1 Pipe*

The size of pipe used in pipelines varies from 2-inch diameter pipe for gathering lines or small distribution lines, to 48-inch pipe for main lines or trunk lines. Most modern pipe is made of steel, either seamless or with a welded longitudinal seam, that is produced in lengths that are 40-60 feet long. These are often welded into 80-foot sections before being delivered to the construction site. Steel pipe is vulnerable to **corrosion**, which can cause damage to the pipe and reduce wall thickness, compromising the integrity of the pipe. Pipe coating and wrapping are the primary

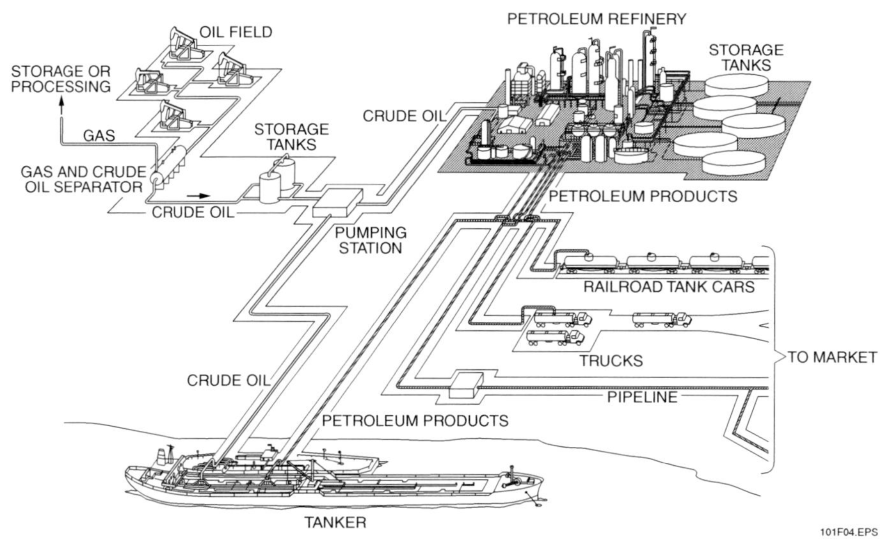


101F04.EPS

*Figure 4* ◆ Transportation chain.



Instructor's Notes:

means of protecting the pipe. **Cathodic protection** is used after the pipe is buried to inhibit a corrosive environment.

### *1.2.2 Construction*

At the time of construction, permits are obtained for a specified right-of-way for the pipeline. Earth-moving equipment clears the land, an access road is built along the route, and trenches are excavated for the pipe. Trenches require a minimum depth of coverage as stated in *Pipeline Safety Regulations, Part 195.248* for liquid and *Part 192.327* for gas.

Pipe is hauled to the right-of-way and welded in strings by teams of welders. After welding, radiographic inspection is used to check the welds. Sometimes, pipe must be bent by bending machines to fit the terrain or to go around obstacles. To inhibit corrosion, the pipe must be coated and wrapped before burial. This can be done at the mill or at the site.

The trench is dug and then prepared for the pipe with a layer of protective bedding. The pipe is strung or laid in the trench by sideboom tractors. Long sections are tied-in or welded together to form a continuous conduit. The trench is then filled with backfill to cover the pipe (see *Figure 5*). The pipe must be **hydrostatically tested** or checked for leaks. This involves filling the line with water and pressurizing it to see if the system will maintain the pressure intended.

When piping crosses lakes and rivers, it may be laid and weighted with concrete or steel anchors to keep it on the bottom. *Figure 6* shows an add-on weight installation. In most rivers, the crossings are drilled under the river bed without disturbing the bottom.

Pipelines are usually buried except at pump and compressor stations, some valve stations, and terminals. However, pipe is sometimes erected aboveground by using H-shaped supports, called bents, to hold the pipe. These bents are located at specified intervals along the line. The uprights on the bents are vertical support members (VSMs). In areas like Alaska, the VSMs have to be self-cooling to avoid heating the permafrost. This is not the normal installation, but it is necessary in frozen environments. *Figure 7* shows a bent with self-cooling VSMs.

Supports have sliding plates, or shoes, to allow the pipe to contract or expand with weather or to move with **seismic activity**. The shoes allow the pipe to move as much as 20 feet back and forth and 10 feet up and down. Aboveground lines are also constructed in a zigzag pattern instead of a straight line to allow for movement. *Figure 8* shows a common elevated pipeline section.

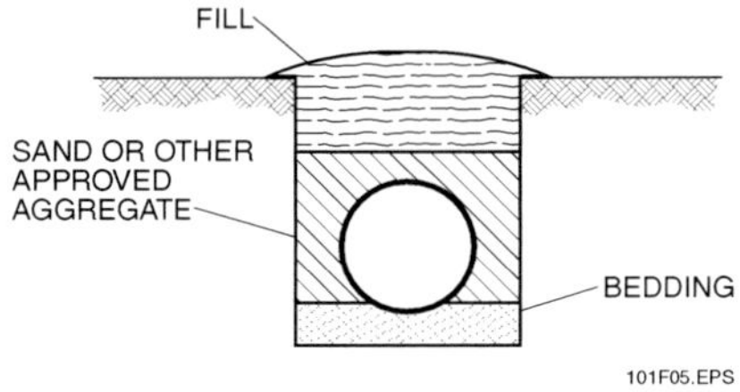


*Figure 5* ◈ Pipe burial.

*Figure 6* ◈ Add-on weight installation.

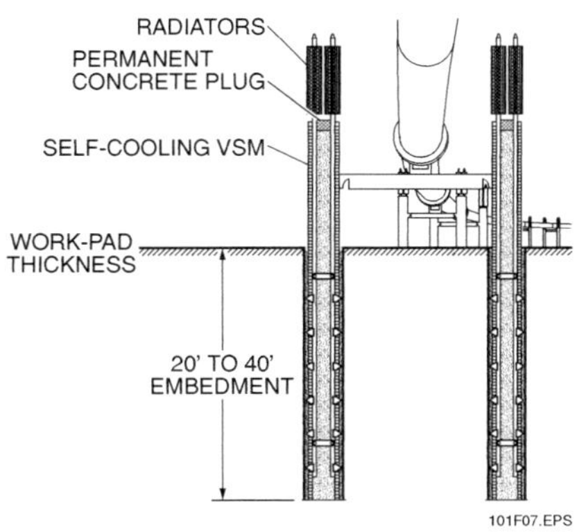


*Figure 7* ◈ Bent with self-cooling VSMs.

**Hand out copies of *49 CFR Parts 195.248* and *192.327*. Ask trainees to identify the minimum depth requirements.**

**Review the process of preparing and laying pipelines.**

**Show Transparency 2 (Figure 5). Discuss various types of approved aggregates.**

## Classroom

**Referring trainees to Figure 9, trace the path of product through a liquid pipeline pump station.**

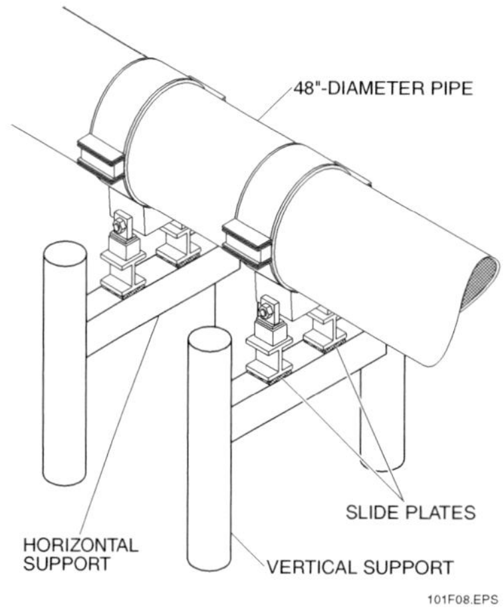


*Figure 8* ◆ Common elevated pipeline section.

Pump and compressor stations are spaced strategically along the pipeline at intervals that are dependent on ground elevation, terrain, and other pipeline design factors to maintain the desired flow rate in the line. The pressure that causes the fluid or gas to flow is lost mostly through friction of the product but also by changes in elevation. This pressure must be changed by pumps and compressors to maintain flow. Drag-reducing agents (DRA) can also be added to the product in the line to reduce the friction loss.

A pump or compressor station facility includes pumps or compressors for moving great volumes of liquid or gas, methods of controlling the flow and pressure, means of measuring and accounting for the commodities being moved, storage facilities, and other devices for quality control. *Figure 9* shows an example of a liquid pipeline pump station layout. *Figure 10* shows an example of a gas pipeline compressor station layout.

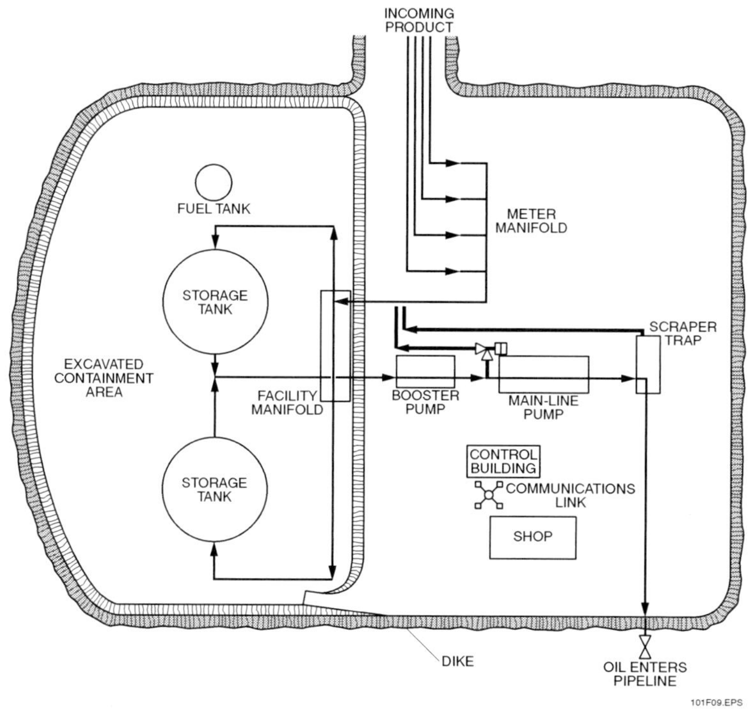


*Figure 9* ◆ Example of a pump station layout.



Instructor's Notes:

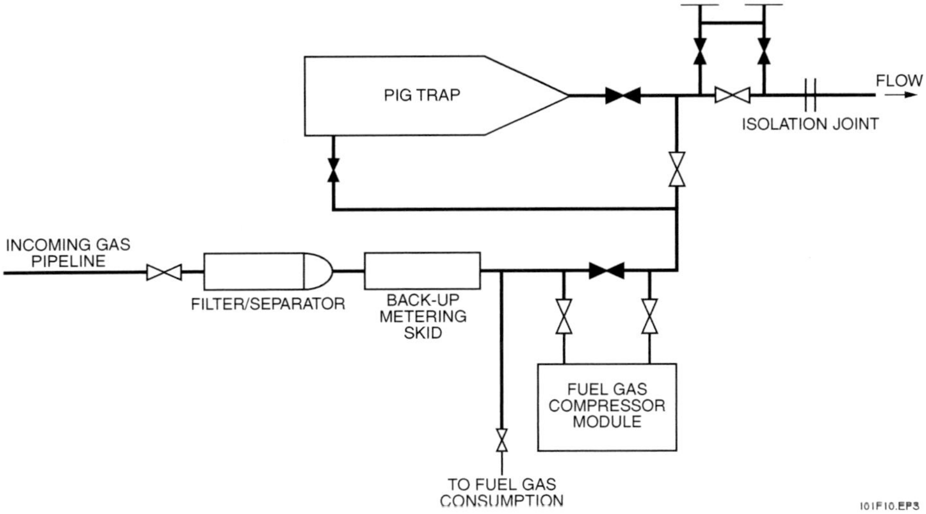


*Figure 10* ◈ Example of a gas pipeline compressor station layout.

### 1.2.3 Operation and Maintenance of the Pipeline

The pipeline is monitored closely and inspected for potential equipment failure and line integrity. Crews identify problems and dig up the pipe to perform repairs. Small **anomalies** may be repaired temporarily while large anomalies may require replacing the pipe. In this case, a bypass line may be constructed so that flow continues while the damaged line is repaired.

The inside of the pipe must also be cleaned periodically to remove wax, scale, and debris. A scraper **pig** is inserted into the line at one point and flows with the liquid or gas to another point, where it is removed. The pig scrapes the inside walls using the hydraulic force of the flow behind it. An instrument pig, also called a smart pig, is inserted into the line periodically to check for anomalies.

### 1.2.4 Petroleum Products

Different petroleum products can be shipped consecutively from the point of origin in the same pipeline. This is called batching. To maintain quality, batches of each product are arranged in the line and pumped under pressure so that little or no mixing occurs during transit.

## 1.3.0 Storage

Storage facilities may be located along the pipeline route, at the origination sites, at pump stations, and at terminals. Storage is necessary to meet seasonal needs and other demand fluctuations. Storage is also necessary for the efficient operation of production equipment, pipelines, tankers, refineries, and for emergency use. Large volumes of natural gas are typically stored in underground storage fields throughout the United States.

The trend towards giant tankers created a need for large storage facilities at the loading and unloading ends of tanker runs. A large crude oil terminal typically has several tanks, each with capacities up to half a million **barrels**. A small pump station might have only one tank with a capacity of about 50,000 barrels. Stations along a product line may need separate tanks for storage of each product coming through. These are sometimes called relief tanks or surge tanks. *Figure 11* shows some storage tanks.

## 1.4.0 Delivery

A wide range of delivery methods is used for gas and petroleum products depending on their destination, intended use, and transport properties.



**Audiovisual**

**Show Transparency 3 (Figure 10). Trace the path of product through a gas pipeline compressor station.**

**Classroom**

**Ask a trainee to explain how a pipeline is maintained. Answer any questions trainees may have.**

**Define batching. Ask a trainee how quality is maintained during batching.**

**Laboratory**

**Divide trainees into pairs. Have each pair list five things they know about storage. Then ask trainees to share their lists with the class.**

**Audiovisual**

**Show Transparency 4 (Table 1). Ask trainees which types of transport are used for ocean routes and which for rivers and canals.**

**Homework**

**Assign the reading of Sections 2.0.0–3.0.0 for the next session.**

**Materials**

**Ensure that you have everything required for demonstrations during this session.**

**Classroom**

**Have trainees write down the three basic stages involved in producing petroleum products from crude oil. Review and discuss the answers.**

**Referring to Figure 12, explain the refinery process.**

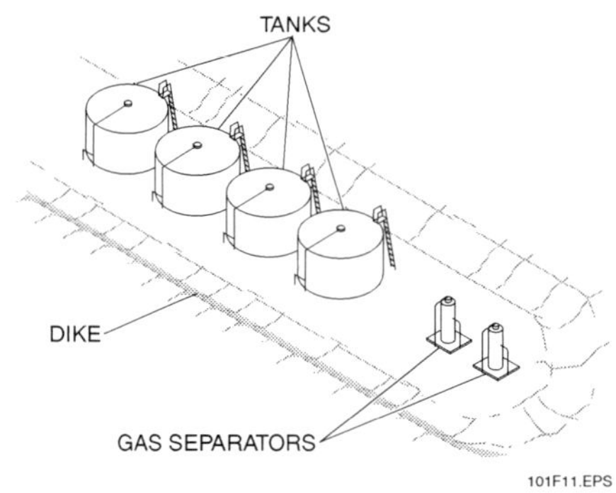


*Figure 11* ◆ Storage tanks.

Petroleum is carried chiefly by pipeline, tanker, barge, tank truck, and railroad tank car. Nearly all product moves through pipelines for at least part of the route. Some products are delivered to city or regional stations by pipeline. From there they may be piped or delivered to local consumers or distribution sites. Gasoline, diesel, and kerosene are delivered by tank truck to gas stations for purchase by users. Industrial products may be shipped to companies by pipeline, truck, or railroad. Supertankers are used to move product on ocean routes, while barges are used on rivers and canals. While liquid petroleum products are typically measured in barrels, gas is typically measured using cubic feet. Depending on the volume being measured, gas quantities can be measured using billion cubic feet (BCF), thousand cubic feet (MCF), or hundred cubic feet (CCF). *Table 1* shows typical capacities that can be moved by different means of transport.

**Table 1** Means of Transport

| Transport | Capacity |
|---|---|
| Tanker Truck | 200 barrels |
| Railroad Tanker | 1,500 barrels |
| Barge | 15,000 barrels |
| Supertanker | 1 million barrels+ |
| Large Liquid Pipeline | 2 million barrels+/day |
| Large Gas Pipeline | 1 BCF/day |

101T01.EPS

## 2.0.0 ◆ PIPELINE PRODUCTS

A pipeline can carry a wide range of petroleum and gas products. Each of these products has its own characteristics and rules for handling. Each of these products also presents its own set of hazards for the pipeline workers, for the public, and for the environment. An understanding of the products in the pipeline, the hazards involved, and protective measures to be taken is essential to safe and secure operations.

Petroleum products are produced from crude oil in three basic stages. The first stage is separation and consists of separating the oil into its various fractions, or parts. The main process in this stage is fractional distillation, which separates light, medium, and heavy fractions. The second stage is conversion, in which less useful fractions are converted into more valuable ones. The third stage is treatment, which improves the quality and performance of petroleum products. *Figure 12* shows a diagram of the refinery process.

Products in a typical pipeline are discussed in terms of their characteristics and associated hazards. These products may include:

- Natural hydrocarbons
  - Crude oil
  - Liquefied petroleum gas
  - Wellhead natural gas
- Refined products
  - Gasoline
  - Diesel fuel
  - Fuel oil
  - Jet fuel
  - Consumer-ready natural gas
- Special products
  - Chlorine gas
  - Hydrogen chloride gas
  - Propylene
  - Acetone
  - Benzene
  - Ethane and ethylene
  - Anhydrous ammonia
  - Methane gas
  - Carbon dioxide

### 2.1.0 Natural Hydrocarbons

Natural hydrocarbons include crude oil, liquefied petroleum gas, and natural gas.

#### *2.1.1 Crude Oil*

Crude oil is unrefined petroleum—the source of the myriad of petroleum products. Its specific gravity ranges from 9 to 55 as compared with the specific gravity of water, which is 10. A light crude is one with a higher gravity, and a heavy crude has a lower gravity.

Because the petroleum may have different ingredients or fractions, its color can range from yellow



Instructor's Notes:

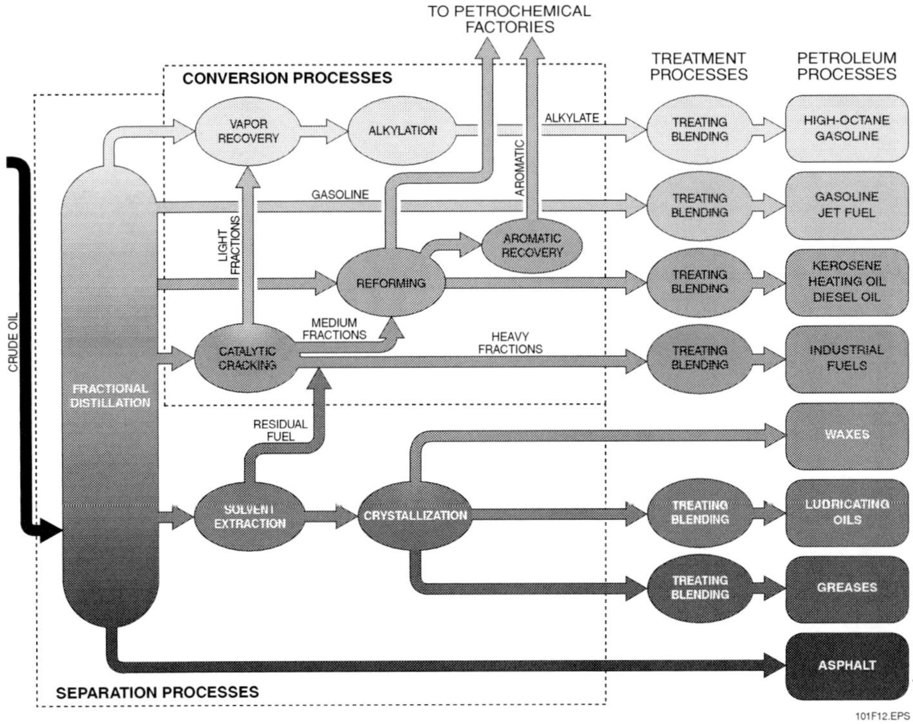


*Figure 12* ◆ Diagram of the refinery process.

to black. It may have a paraffin, asphalt, or mixed base. Crude oil also contains substances other than hydrocarbons. These impurities include metallic compounds and sulfur and may make up as much as 10% of the oil. Oil containing a large amount of sulfur compound is called a sour crude. A sweet crude contains little or no sulfur.

### *2.1.2 Liquefied Petroleum Gas*

Liquefied petroleum gas (LPG) is a mixture of heavier, gaseous, paraffinic hydrocarbons, principally butane and propane. These gases, easily liquefied at moderate pressures, may be transported as liquids and converted to gases on release of the pressure. Thus, LPG is a portable source of thermal energy that finds wide application in areas in which it is impractical to distribute natural gas. LPG is used in industry for cutting and welding metals and on farms for operating various kinds of machinery. The principal sources are natural and refinery gas, from which the liquefied petroleum gases are separated by fractionation.

Butane/butylene mix, a form of LPG, is extremely flammable. It releases vapors at well below ambient temperatures and readily forms flammable mixtures with air. If ignited, it will burn in open air or explode in confined spaces. Its vapors are heavier than air and may travel long distances to a point of ignition and then flash back. Alkane/chlorine gas mixtures have produced explosions. Closed systems will form a white frost at the point of the release. Liquid spills will vaporize, forming cold, dense vapor clouds that should be avoided even with respiratory equipment.

### *2.1.3 Wellhead Natural Gas*

Natural gas is the name given to the naturally occurring hydrocarbon mixtures found in many underground wells. Natural gas is an unrefined



**Classroom**

**Have a trainee explain the difference between light crude and heavy crude. Ask trainees to name the various ingredients and impurities crude oil may have.**

**Stress the dangers of various gas mixtures. From your own experience or readings, share a story of a major gas explosion and the damage it caused.**

**Safety**

**Stress the danger of exposure to petroleum products. Remind trainees to wear personal protective equipment and to follow safety procedures.**

**Teaching Tip**

**See the Teaching Tip at the end of this module.**

product that is a naturally occurring gaseous mixture of hydrocarbons that may also contain nitrogen, helium, carbon dioxide, hydrogen sulfide, and water vapor. Because the content or makeup of natural gas can vary from well to well and region to region, it is normally referred to as wellhead natural gas. Natural gas is processed to recover the **natural gas liquids (NGLs)**, the heavier hydrocarbon components that include butane, propane, and natural gasoline.

> **WARNING!**
> Exposure to petroleum products can cause eye, skin, and respiratory tract irritation. Always wear proper personal protective equipment (PPE) and refer to company procedures and the product material safety data sheet (MSDS) for specific details.

Natural gas is highly compressible, highly expandable, and has a low specific gravity. Although gaseous at normal temperatures and pressures, the gases making up the mixture are variable in form and may be found as gases or as liquids under varying temperatures and pressures.

## 2.2.0 Refined Products

Refined petroleum and gas products include gasoline, jet fuel, diesel fuel, heating oil, and kerosene. Other refined products may include propane, butane, consumer-ready natural gas, and blending components.

### *2.2.1 Gasoline*

Gasoline is a volatile, flammable liquid hydrocarbon refined from crude oil and used universally as a fuel for internal-combustion, spark-ignition engines. Natural gasoline, or casinghead gasoline, is the liquid hydrocarbon recovered from wet natural gas. Gasoline is classified as regular, premium, and aviation grades according to how smoothly it burns in an engine.

### *2.2.2 Diesel Fuel*

Diesel fuel is a light hydrocarbon mixture for diesel engines, similar to furnace fuel oil. It has a boiling range just above that of kerosene. Diesel fuel requires less refining than gasoline.

### *2.2.3 Fuel Oil*

Refined fuel oil comes in two grades. The specifications for both are defined by *American Society for Testing and Materials (ASTM) Specification D 396*.

No. 1 distillate, such as kerosene, is a light fuel. No. 2 fuel oil diesel is a distillate fuel oil prepared for use in atomizing-type burners or for smaller industrial burner units. Lesser grades of fuel oil, called *residual fuel oils*, are made from the topped crude of refinery operations. Residual fuel oil is used for the production of electric power, space heating, and various industrial purposes.

### *2.2.4 Jet Fuel*

Jet fuel is a highly refined kerosene petroleum distillate. Jet fuel has good flow properties in low-temperature conditions, and it burns efficiently and cleanly. It must be delivered pure and free of rust, water, and sediment.

Military jet fuel is a highly refined kerosene distillate that meets stricter military specifications.

### *2.2.5 Consumer-Ready Natural Gas*

Wellhead natural gas is normally run through a refining process to remove any undesired contents, remove contents that can be used for other commercial purposes, and achieve a more consistent safe product and heat content. The final product is known as commercial natural gas or consumer-ready natural gas.

Consumer-ready natural gas is still a mixture of different gases but is made up primarily of methane. Other gases, such as ethane, propane, nitrogen, and carbon dioxide can also be found in trace quantities. The composition of the consumer-ready natural gas remains fairly consistent over time. The heat value range of typical consumer-ready natural gas is 1,020 to 1,045 Btus per cubic foot.

## 2.3.0 Special Products

Special products include other by-products that are produced and carried through pipelines.

### *2.3.1 Chlorine Gas*

Chlorine gas is a product used for water purification, chemical processes, and the manufacture of plastic. Chlorine gas is a strong oxidizing agent; it can react explosively with organic products and it is poisonous. Liquid and gas chlorine under pressure can be hazardous. Chlorine is nonflammable and nonexplosive, but like oxygen, it can support combustion of certain substances.



Instructor's Notes:

It can react violently or form explosive compounds with many chemicals, such as acetylene, turpentine, ether, ammonia gas, and hydrogen, and finely divided metals.

### 2.3.2 Hydrogen Chloride Gas

Hydrogen chloride gas is a fuming, poisonous gas used to make hydrochloric acid, an important industrial chemical used in many chemical compounds, in metallurgy, in food processing, and as a cleaning agent. Hydrogen chloride is nonflammable and noncombustible. However, if dissolved in water, it produces hydrochloric acid, which attacks most metals.

### 2.3.3 Hydrogen Sulfide

Hydrogen sulfide is a flammable, colorless, gaseous compound of hydrogen and sulfur, which in small amounts has the odor of rotten eggs. It is formed naturally by the decomposition of organic animal and vegetable matter by bacterial action. Hydrogen sulfide is found in crude oil, natural gas, sewers, and swamps and is created by a number of industrial processes. It is extremely corrosive and poisonous. Hydrogen sulfide causes damage to the eyes, skin, breathing passages, and lungs, attacks and paralyzes the central nervous system, especially the parts that control the lungs and heart. In large amounts, it deadens the sense of smell. Hydrogen sulfide is an extremely dangerous compound, and exposure is deadly.

### 2.3.4 Propylene

Propylene is used in plastics manufacturing. It is extremely flammable and will readily ignite at ambient temperatures. The flammable vapors, being heavier than air, may travel long distances along the ground to a source of ignition and then flash back. Propylene presents the danger of rocketing cylinders, ruptured containers, and explosion. Reignition or explosion can occur if the flames are extinguished without stopping the source flow. An electrostatic charge may build up during handling if the equipment is not grounded.

### 2.3.5 Acetone

Acetone is a clear, colorless liquid with a sweet, pungent odor that is used as a solvent. It is extremely flammable and may cause flash fire. It is incompatible with a variety of materials, including inorganic acids, caustic, amines, alkanolamines, oxygen, halogens, aldehydes, ammonia, oxidizing agents, and chlorinated compounds. Contact with these should be avoided during transport, handling, and storage.

### 2.3.6 Benzene

Benzene is an extremely flammable liquid, readily igniting at ambient temperatures. The vapor forms an explosive mixture with air. Flammable vapors, being heavier than air, may travel long distances to a point of ignition and flash back. Incomplete combustion can produce carbon monoxide.

### 2.3.7 Ethane and Ethylene

Ethane and ethylene are simple asphyxiants. They are dangerous when exposed to heat or flame and can react violently with oxidizing materials. Both are flammable over a wide range in air, with an even broader range in pure oxygen.

### 2.3.8 Anhydrous Ammonia

Anhydrous ammonia is the liquid form of pure ammonia gas. It is used as a fertilizer because of its high nitrogen content. It is also used as an industrial refrigerant in cold-storage and ice manufacturing plants, since it absorbs a large amount of heat when going from a liquid back to a gas. Anhydrous ammonia is made by compressing dry ammonia gas.

Ammonia is a colorless alkaline gas made up of one part nitrogen and three parts hydrogen. It is lighter than air and has a sharp, stinging odor. Ammonia can be safely inhaled if it is greatly diluted in air, but concentrated ammonia gas can cause suffocation and death. It does not burn in air, but it burns in oxygen with a weak yellow flame. Ammonia is made by combining hydrogen and nitrogen, and as a by-product in the production of coal and coke gas.

### 2.3.9 Methane Gas

Methane is a member of the alkaline or paraffin series of hydrocarbons. It is called marsh gas because it forms by anaerobic bacterial decomposition of vegetable matter in swampy land. Coal miners know it as *firedamp* because it is combustible when mixed with air. It is a major component of natural gas and of coal gas, and forms in large amounts in sewage disposal processes.

In addition to its use as a fuel, methane is important as a source of organic chemicals and of hydrogen. In the presence of steam and catalysts, it yields carbon monoxide and hydrogen, which can be converted to liquid hydrocarbons or to methanol and other alcohols. The incomplete combustion of methane with air produces carbon black, which is used in rubber and ink. Chlorination of methane yields various chlorides and chloroform.



**Teaching Tip**

**See the Teaching Tip at the end of this module.**

**Classroom**

**Have trainees identify the special products that are extremely flammable (propylene, acetone, and benzene).**

**Classroom**

**Ask trainees to list the information contained on an MSDS.**

**Stress the importance of handling and containing spills carefully. Review Figure 13.**

### *2.3.10 Carbon Dioxide*

Carbon dioxide ($CO_2$) is a colorless, odorless, tasteless gas which is about 15 times heavier than air. Under normal conditions, it is stable, inert, and nontoxic. The decay of all organic matter and the burning of hydrocarbon fuels produces carbon dioxide. It is obtained commercially from four sources: naturally occurring underground $CO_2$ wells, fermentation, combustion of carbonaceous fuels, and as a by-product of chemical processing.

Carbon dioxide gas may be liquid or solid based on compression and lowered temperatures. It may be used as a refrigerant, inerting medium, chemical reactant, neutralizing agent for alkalies, and pressurizing agent. Liquid carbon dioxide is an effective fire-extinguishing agent because it rapidly reduces the temperature of burning materials below the ignition point. The gas, being heavy and inert, displaces air, which is the source of the fire's oxygen, and blankets the flames. Oil tankers, barges, storage vessels, and pipelines are quickly and safely made inert with carbon dioxide to allow for welding and other repairs.

## 2.4.0 Protective Measures

**Material safety data sheets (MSDSs)** are good sources of information on these products and must be made available to all employees of the company. There is an MSDS for each hazardous substance, listing its properties and hazards, including a hazard rating for health, fire, reactivity, and other hazards. MSDSs also detail safe handling, personal protective equipment requirements, and first aid/emergency procedures.

All products pose hazards to the environment if spilled or disposed of improperly. Areas of possible spills should have containment structures and/or procedures. These include dikes around storage and transfer facilities and containment booms around tankers being filled. The marine environment is particularly sensitive to oil spills because the tides and currents spread the oil and complex marine life is so vulnerable. *Figure 13* shows temporary containment of a spill from a tank.

Oil may be harmful on land, too. It can seep into the groundwater or streams where it contaminates plant and animal life and eventually human

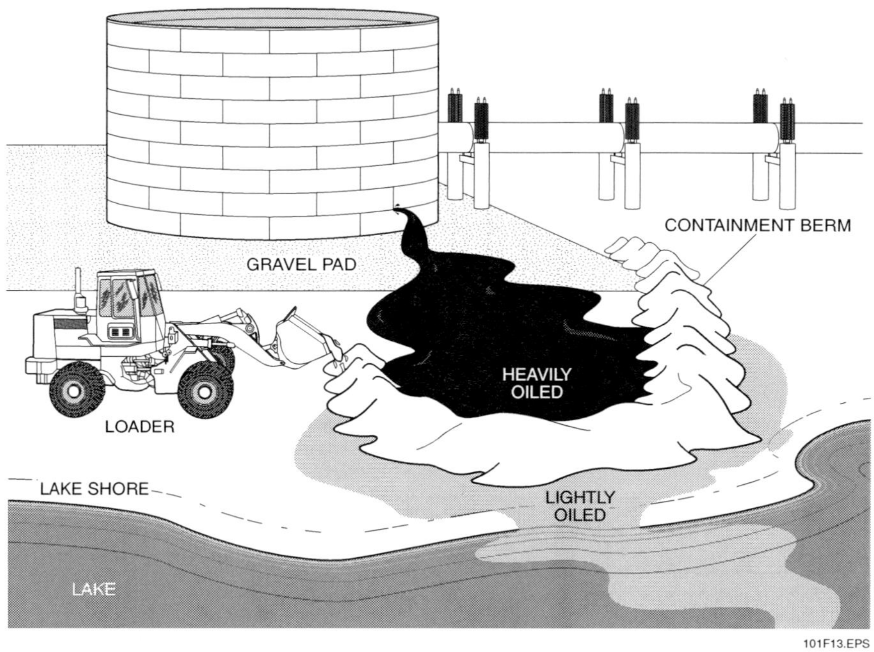


*Figure 13* ◆ Temporary containment of a tank spill.



Instructor's Notes:

life. Handling and disposal of hazardous chemicals are addressed in company procedures and MSDSs. Detailed procedures exist for handling all petroleum products, and the observance of these procedures is strictly regulated and enforced by federal, state, and local agencies.

## 3.0.0 ◆ MAPS AND DRAWINGS

Maps and drawings are an important means of communicating information about pipeline equipment, flow, and processes. Some of these are piping and instrumentation diagrams (P&IDs), strip maps, blueprints, and alignment sheets.

### 3.1.0 Piping and Instrumentation Diagrams (P&IDs)

A P&ID depicts the functional layout of a piping system and its components, valves, and instrumentation as clearly and accurately as possible. It is accurate to the extent that all components are connected to each other as shown in relation to flow-path orientation. A P&ID does not represent the actual physical layout of equipment; for example, a valve that may appear to be right at the discharge of a pump can actually be located some distance away or even on another floor. Many times, a P&ID will use a broken line around groups of equipment to indicate that they are all in the same building (see *Figure 14*).

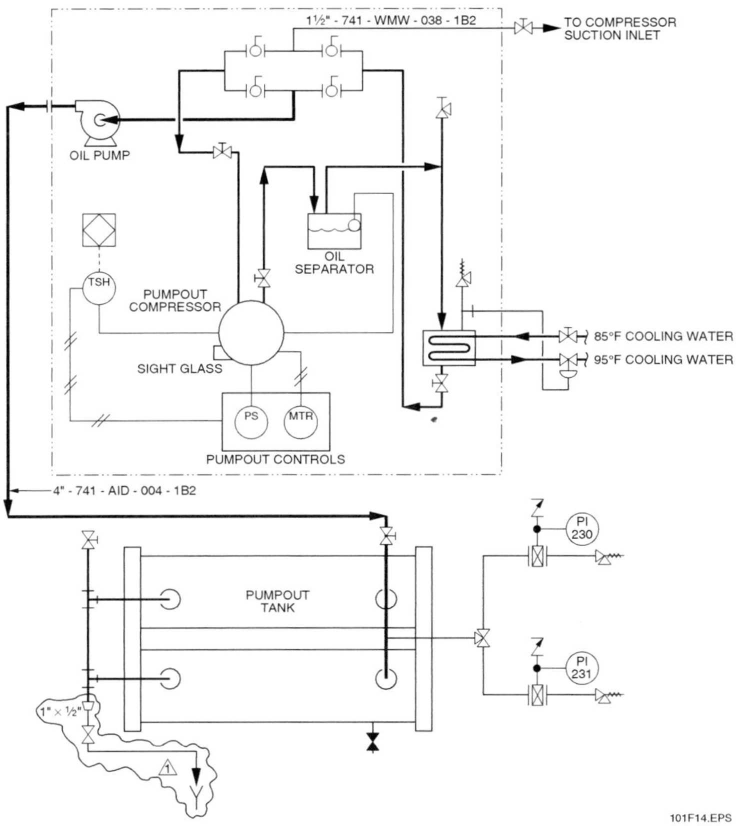


*Figure 14* ◆ Sample P&ID.



**Demonstration**

**Hand out samples of P&IDs, blueprints, and strip maps/alignment sheets. Have trainees familiarize themselves with each.**

**Classroom**

**Referring to Figure 14, point out the broken line and arrows. Ask trainees what these show.**

**Classroom**

**Ask trainees to name the information in a blueprint. Write their answers on the board.**

**Distinguish between strip maps and alignment sheets. Answer any questions trainees may have.**

The P&ID depicts all components of a particular system, including pipe sizes, flange sizes, valve sizes, flow direction, and references to other diagrams. Standardized symbols are used for these components instead of pictures. Industry standards for these symbols have been developed and adopted by organizations, such as the American National Standard Institute (ANSI), the Instrument Society of America (ISA), and the American Society for Mechanical Engineers (ASME), to aid interpretation. The P&ID title block and legend should be examined closely for explanations of the diagram.

### 3.2.0 Blueprints

Blueprints are duplicates, or copies, of plans for buildings, assemblies, systems, and other constructions. These are working drawings, including specifications for a project that usually includes site plans, floor plans, elevations of buildings, cross sections, and detail drawings. They were originally made by a photographic process that produced white lines against a blue background, hence, the name. However, they are now produced by computer-aided design (CAD) software and generally have black or blue lines on a white background. The blueprint shows the size and location of all structures and equipment in the project to scale and how they are physically put together. *Figure 15* shows a site-plan blueprint.

### 3.3.0 Strip Maps/Alignment Sheets

Strip maps (see *Figure 16*) are combinations of different types of maps or drawings over the same area such as the pipeline route. The strips of each type of map or drawing are laid out side by side

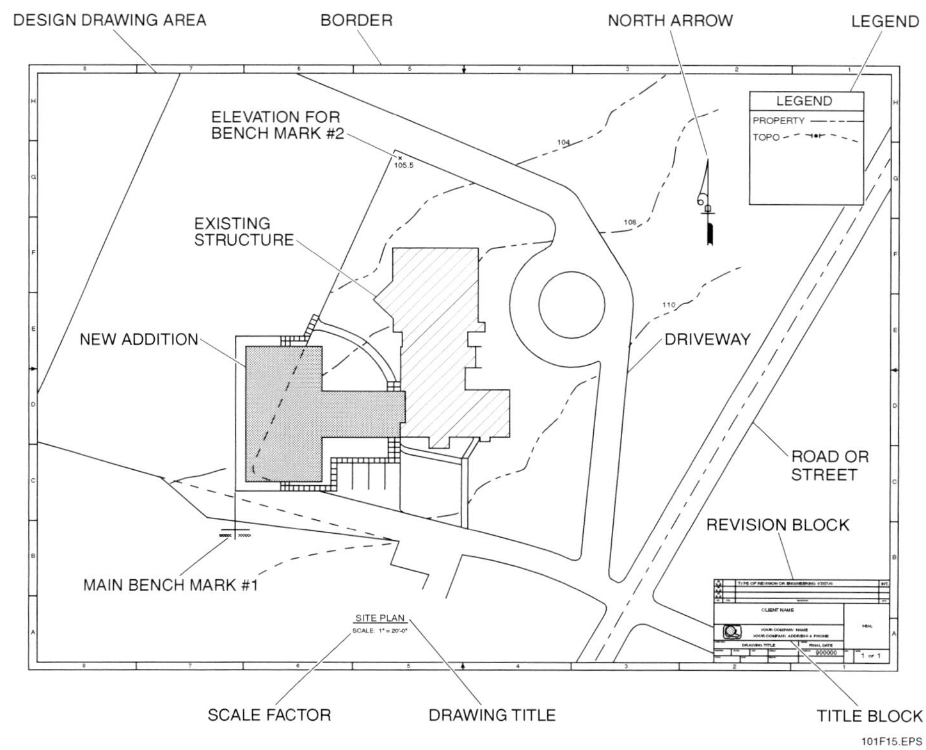


*Figure 15* ◆ Site-plan blueprint.



Instructor's Notes:

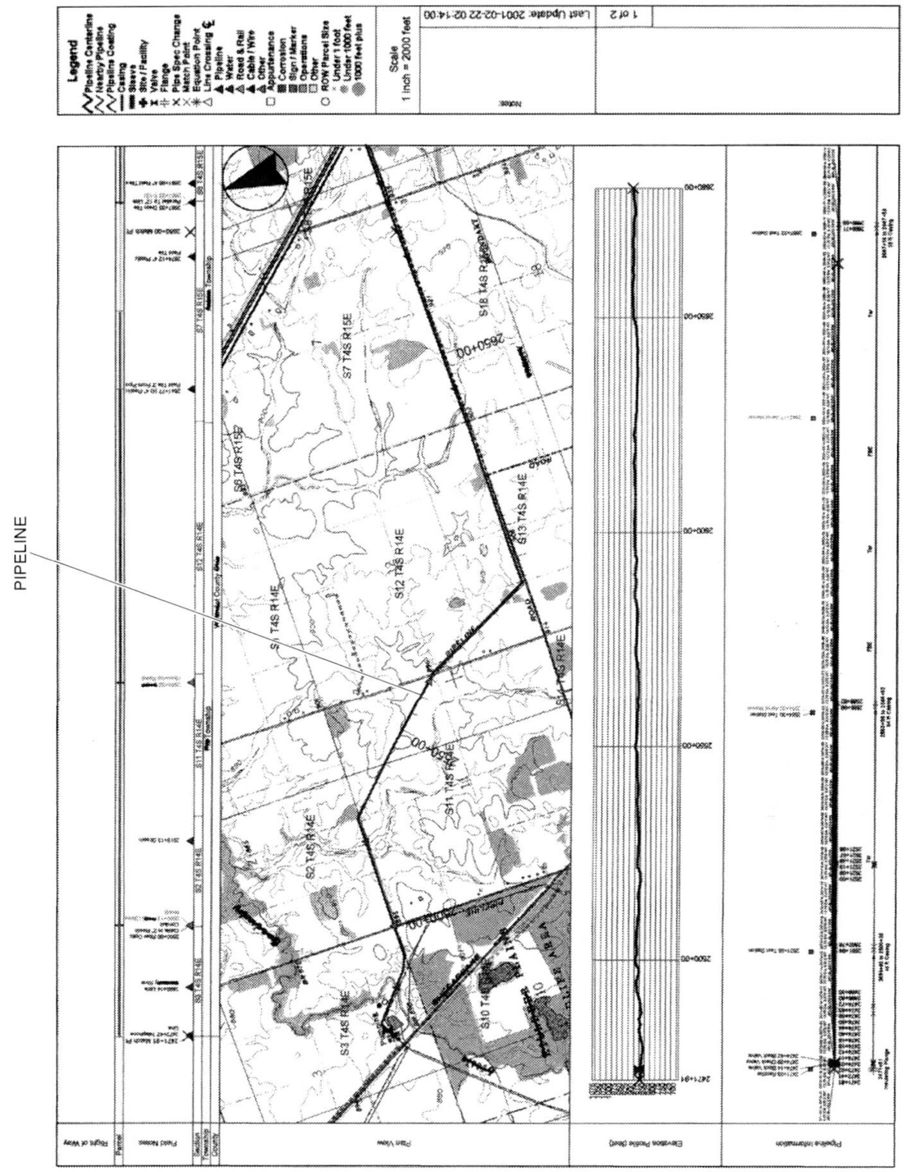


101F16.EPS

*Figure 16* ◆ Sample strip map.

**Homework**

**Assign the reading of Sections 4.0.0–6.0.0 for the next session.**

**Materials**

**Ensure that you have everything required for demonstrations during this session.**

**Classroom**

**Have trainees list the tasks often performed by a scheduler. Write their answers on the board.**

on the same sheet so that they line up along the route. There may be views shown of the ground from an aerial, or plan view, and side, or elevation view, which shows the ground profile along a specific line. The pipeline will be superimposed on all views. Strip maps help to coordinate surface features with topography as well as with other pipeline data at any point on the route.

Alignment sheets are various types of maps, drawings, and diagrams that show different features along the pipeline route. The alignment sheets will show the relationship of the pipeline to true north, geographic features, political boundaries, fences, rivers, property lines, and other pipelines. *Figure 17* shows one type of alignment sheet.

Strip maps and alignment sheets vary greatly from company to company. These names may be used interchangeably or may mean very different things.

## 4.0.0 ◆ OPERATIONS

Controlling the fast-paced, continuous, and complex operation of the pipeline is a critical job done largely by the scheduler, controller, and dispatcher with the help of other personnel and systems like Supervisory Control and Data Acquisition (SCADA), programmable logic controllers (PLCs), and remote terminal units (RTUs). All control functions are documented by filling out forms or printing out reports.

### 4.1.0 Scheduling

Before a shipment of product reaches the pipeline and comes under the control of the dispatching office, it has to be accumulated and scheduled with the pipeline scheduler. The scheduler at a pipeline facility typically performs the following functions:

- Takes nominations from shippers
- Creates pumping orders for pipelines to meet shippers' needs
- Validates and reviews operating information for schedule adjustments
- Receives delivery information in time for shipments
- Validates all shipping information for billing
- Utilizes power optimization to save costs and maximize throughput efficiency

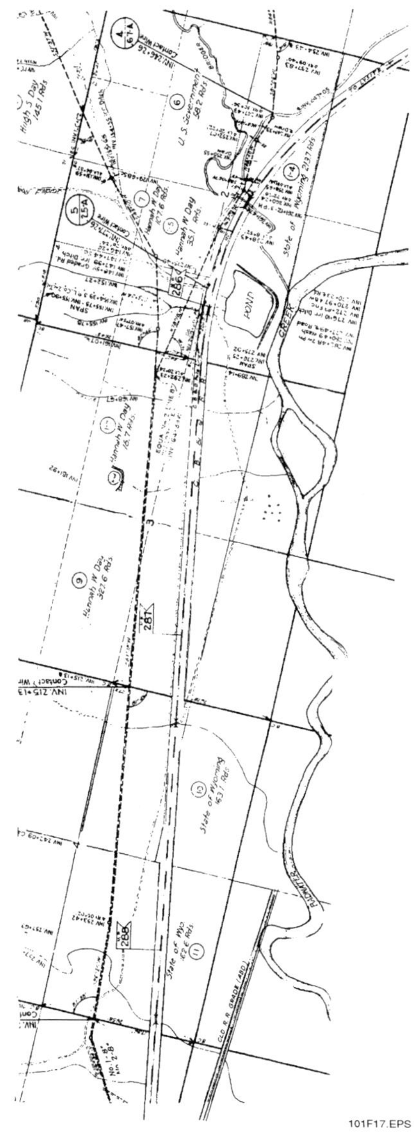

101F17.EPS

*Figure 17* ◆ One type of alignment sheet.



Instructor's Notes:

### 4.2.0 Pipeline Operations

The pipeline is monitored closely and inspected for potential equipment failures and line integrity. These problems can arise from bad welds, corrosion spots, faulty valves, third party damage, or shifting of the pipe. When aboveground, the pipe may also be damaged by accidents or sabotage. In areas with faults or history of earthquakes, seismic activity must also be closely monitored. Flow can also be interrupted by obstructions in the pipe, such as stuck pigging devices, foreign objects, and debris.

Computers and automation are used to control and monitor the pipeline operation. Communication is maintained between all stations regarding the status of the operation. Controllers have the ability to shut down the pipeline flow if a problem is detected and to pinpoint the location of the problem. *Figure 18* shows a typical control center layout.

### 4.3.0 Control Systems

SCADA is the software used to monitor and control the flow of products through the pipeline every day. Wireless telemetry and SCADA systems provide remote alarm, control, and data collection for many types of pipeline operations.

SCADA is installed with the main computer system and hooked up through a server to RTUs and PLCs in the field. Sensors along the pipelines precisely measure operating conditions, flow rates, and the status of control devices, such as flow and pressure regulators and valves. The RTUs collect vital information in the field and transmit the data via telephone, microwave, or satellite circuits to the control center.

### 4.4.0 Communications

Communications are the systems that keep the pipeline operating. All along the pipeline, everyone must have open communications at all times because everything that occurs on the pipeline affects the entire system.

A variety of communications equipment and devices may be used in a pipeline company's operations. These are set up for continuity during normal or emergency conditions. It is important that communications be reliable and always available. Most companies have redundant, or backup, systems to ensure that this is true. Some of the equipment used includes:

- Dedicated satellites or microwave-based radio communications systems that provide linkage between the control room personnel and various pump stations and valve locations
- Hardwire communications that link the control room and all offices and all remotely monitored and operated facilities, such as pump stations and RTUs
- Portable cellular phones or vehicle phones provided to all key and supervisory personnel
- Locally activated pagers or nationwide pagers provided to key personnel for use in routine operating or emergency events

101F18.EPS

*Figure 18* ◆ Typical control center layout.



**Classroom**

**Have trainees name some problems that can arise with pipelines. Point out that computers and automation are used to prevent problems from occurring.**

**Discuss the setup of control systems. Ask trainees what SCADA and RTUs do.**

**Explain the equipment used to facilitate communications. Ask if trainees have any questions.**

Classroom

**Define hydraulics and fluid. Explain the different properties of fluids.**

- Fixed repeaters installed at points within the system that provide internal long distance and local radio communications

In the event of a power loss, backup systems are provided, which may include standby generators for powering communications equipment, batteries, or uninterruptible power systems (UPS). Portable self-contained mobile repeaters may be used at an incident scene and utilized to augment existing radio systems or replace any that may be inoperable. A radio link (base station) may be used with portable units to allow for emergency communications between facilities in the event that telephone and/or microwave transmissions become inoperable. Redundant communications and power systems should be installed at key facilities to allow for control in the event of loss of primary systems.

## 5.0.0 ◆ FLUIDS AND SYSTEM HYDRAULICS

The operation of any liquid pipeline is built on the principles of hydraulics, or the science of the behavior of fluids at rest or in motion. These principles are in turn built on the properties and characteristics of fluids.

### 5.1.0 Fluids

A fluid is a substance that flows, and some flow more easily than others. A fluid is also a substance that changes form with a slight force or pressure; it is elastic, however—meaning it will try to return to its previous form when the pressure is removed. Some fluids are more elastic than others. Elasticity is a measure of the fluid's resistance to change in shape.

Fluids include all liquids and gases. Water at ordinary temperature is a liquid fluid; at high temperatures, a gas fluid. Air is a gas fluid. All fluids can be greatly affected by changes in temperature and pressure.

No fluid is perfect and frictionless, and different fluids have different levels of friction, or resistance to flow. This resistance to flow is called *viscosity*.

Some fluids are thicker than others. This is called *density*. Density can be a large factor in the viscosity of a fluid. It is a measure of how tightly packed the fluid molecules are or how much of the fluid's matter exists within a given unit of volume.

Measuring density in a liquid is readily done with a hydrometer because they are not usually compressible. Measuring density in a gas is more difficult because gases can easily be compressed. Compressibility is the ability to be squeezed, or forced into a smaller volume.

Density helps determine the compressibility of a fluid. When a fluid is compressed, it becomes denser. The specific gravity of a substance is derived from its density. Specific gravity is the ratio of the density or mass of a given volume of the substance to the density of the same volume of a standard substance, such as water. The water and the liquid substance must be at the same temperature. For gases, the gas must be at the same temperature and pressure.

American Petroleum Institute (API) gravity is a unit of measure developed specifically for liquid petroleum products. The heavier the substance, the lower the number of its API gravity. Temperature will affect API gravity and should always be corrected to 60°F.

The properties of fluids that are used to describe the fluids are density, viscosity, compressibility, inertia, and elasticity. *Table 2* shows some fluid properties.

**Table 2** Fluid Properties

| Fluid | API Gravity @ 60°F | Specific Gravity @ 60°F | Vapor Pressure @ 100°F | Absolute Viscosity | Compressibility @ 1000 psi @ 60°F |
|---|---|---|---|---|---|
| EP Mix 80/20 | 214 | 0.41 | 550 psi | 0.11 cP | 7.3% |
| NGLS | 146 | 0.51 | 325 psi | 0.15 cP | 3.8% |
| Raw Feed Propane | 146 | 0.51 | 150 psi | 0.15 cP | 3.8% |
| Butane | 112 | 0.58 | 88 psi | 0.18 cP | 2.0% |
| Anhydrous Ammonia | 97 | 0.62 | 200 psi | 0.25 cP | 1.6% |
| Gasoline | 63 | 0.66 | 2 psi | 0.70 cP | 0.8% |
| Jet Fuel #1 | 42 | 0.82 | 2 psi | 1.4 cP | 0.8% |
| Diesel Fuel #2 | 33 | 0.86 | 1 psi | 4.7 cP | 0.5% |
| Crude Oil | 50 and below | 0.78–0.98 | > 1 psi | 6–400 cP | > 1.0% |
| Water | 10 | 1 | | 1.12 cP | 0.3% |

101T02.EPS



Instructor's Notes:

Other properties of fluids, particularly those found in pipelines, include volatility and flammability. Volatility is the tendency of a fluid to readily vaporize. Sudden vaporization can cause a violent reaction or explosion. Flammability is a measure of how readily a fluid catches fire when in contact with heat or sparks. The volatility and flammability of specific pipeline products is described in another section of this module.

### 5.2.0 Basic Hydraulics

Some laws of hydraulics apply to gases as well as liquids but only under certain conditions. In general, the following sections pertain only to liquid fluids.

Hydraulics is usually divided into two branches: hydrostatics for liquids at rest and hydrodynamics for liquids in motion.

#### *5.2.1 Hydrostatics*

Hydrostatic principles are fundamental to the design of many machines and instruments. The design of hydraulic presses and the operation of instruments, such as manometers and barometers, depend on these principles.

A simple hydraulic press consists of a liquid-filled cylinder that contains two pistons, one smaller than the other. When a force is applied to the smaller piston, the force travels through the liquid to the larger piston, producing a force larger than the initial one. The increase in force is proportional to the ratio of the larger piston to the smaller one. *Figure 19* shows a simple hydraulic press.

The laws of hydrostatics are the following:

1. The amount of pressure caused by the weight of a column of liquid is determined by the height of the column.
2. An object placed in a liquid loses an amount of weight equal to the weight of the liquid it displaces.
3. A fluid in a container transmits pressure equally in all directions.

#### *5.2.2 Hydrodynamics*

Hydrodynamic principles are basic to hydraulic engineering, to the design of certain machines, and to pipeline engineering. Hydrodynamics define the plans for water systems, canals, and irrigation systems, as well as airplane design and hydraulic machines like water turbines.

Pipelines for water and other fluids have to take into account the pipe size and the fluid's friction on the pipe to determine fluid velocity through the pipe. Liquid flow can be steady or unsteady. An unsteady flow results from changes in the velocity, temperature, or pressure of a liquid or its movements around obstacles. The three basic laws of hydrodynamics apply only to steady-flowing liquids. The laws of hydrodynamics are the following:

1. The velocity of a liquid flowing through a pipe increases as the area of the pipe decreases and vice versa.
2. The pressure of a liquid increases as its velocity decreases and vice versa.
3. The velocity with which a liquid flows through an opening in a container equals the velocity of a body falling from the surface of the liquid to the opening.

*Figure 20* shows liquid velocity and pressure relationships.

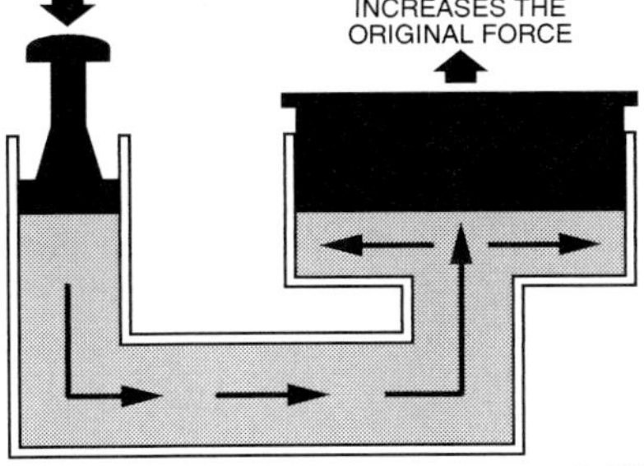


*Figure 19* ◆ Simple hydraulic press.

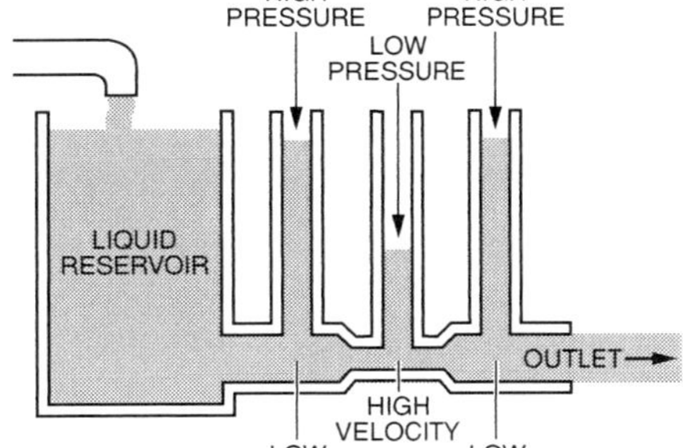


*Figure 20* ◆ Liquid velocity and pressure relationships.



**Audiovisual**

**Show Transparency 5 (Figure 19). Have a trainee explain how a hydraulic press works.**

**Classroom**

**Explain the three laws of hydrostatics.**

**List the three laws of hydrodynamics on the board and discuss each law. Note that the laws apply only to steady-flowing liquids.**

**Audiovisual**

**Show Transparency 6 (Figure 20). Explain the relationships of liquid velocity to area and of pressure to velocity.**

**Classroom**

**Ask trainees to list on the board the primary factors in pipeline design. Explain MOP and MAOP.**

**Define pneumatics. Ask trainees what the primary goal of gas compression is.**

**Ask trainees to name the three variables that have to be calculated and controlled when compressing gas.**

**Teaching Tip**

**See the Teaching Tip at the end of this module.**

### 5.3.0 Pipeline Hydraulics

Fluid properties are major factors which affect the design of the pipeline. The primary factors in pipeline design are:

- Fluid characteristics
- Desired flow rate
- Maximum operating pressure (MOP)
- Maximum allowable operating pressure (MAOP) (for gas)
- Elevation profile/terrain
- Fittings/valves
- Installation costs
- Environmental concerns

Pressure is referenced in several ways in the pipeline industry. The maximum operating pressure (MOP) is defined as the maximum internal pressure allowed liquid during the operation of a pipeline. The term MOP is typically used more in the liquid pipeline industry and the regulations are referenced in *49 CFR 195*. The MAOP is defined as the maximum internal pressure permitted during the operation of a gas pipeline. The term MAOP is typically used more in the gas pipeline industry and the regulations can be referenced in *49 CFR 192*.

Although fluids are defined by their ability to flow, something has to start the flow and overcome their inertia and friction. A force, such as difference in pressure, gravity, or an external force physically pushing it, can do this. Usually, this is done by pumps. Stopping the flow once started will require another force, such as resistance (friction), or a stationary object, such as a closed valve.

The resistance or friction of the fluid against the pipe will be constantly trying to slow it down. Most pumps work by providing a difference in pressure and must continue providing that force all along the line to overcome the friction. There are drastic differences in the amount of pressure required to deliver the proper volume of different types of products. But the pressure in the system must be kept under the maximum limits.

Friction is also affected by the roughness of the pipe interior, the pipe diameter and length, the product density and viscosity, the flow rate, and obstacles, such as valves and bends. Pipeline design can help reduce some of these, but higher fluid viscosity and density will increase the friction.

When the fluid is traveling uphill, pumps must be used to overcome the fluid's inertia and resistance and the force of gravity. When the flow in the pipe is downhill, gravity may be all that is needed.

Another factor in the flow of the pipeline is the temperature. The higher the temperature, the lower the viscosity and vice versa. When the product is hot, it flows better. Sometimes heaters are included in the pipeline equipment to keep the product at the optimum temperature.

## 6.0.0 ◆ GASES AND SYSTEM PNEUMATICS

Pneumatics, as it relates to the pipeline industry, is the study of the storage, compression, and transmission of gases. The compression of gas has one primary goal, to deliver the gas at a pressure higher than the original pressure. The original pressure may vary from very low absolute pressures to several thousand pounds. The pressure rise may vary from a few ounces to many thousand pounds, and the volumes handled may vary from a few cubic feet per minute to hundreds or thousands of cubic feet.

There are three variables that have to be calculated and controlled when compressing and moving gas in a pipeline. These variables are pressure (p), volume (v), and temperature (k). These three characteristics of gases have a close relationship with one another and are explained in the ideal gas laws.

### 6.1.0 The Ideal Gas Laws

Pressure, volume, and temperature of gases have a close relationship with one another. By keeping one of them constant, the other two will have either a direct or inverse proportion. The relationships between pressure, volume, and temperature are explained using the ideal gas laws. These ideal gas laws are Boyle's law, Amonton's (Gay-Lussac's) law, and Charles' law.

#### *6.1.1 Boyle's Law*

Boyle's law states that when temperature is kept constant and volume is decreased, pressure will increase. His law gives the relationship between pressure and volume if temperature and amount are held constant.

- If the volume of a container is increased, the pressure decreases.
- If the volume of a container is decreased, the pressure increases.

If the volume is increased, the gas molecules have farther to travel and they will impact the container walls less often. This means the gas pressure will be less because there are fewer molecule impacts per unit of time.

If the volume is decreased, the gas molecules have a shorter distance to travel, thus striking the walls more often. This results in pressure being increased because there are more molecule impacts per unit of time.



Instructor's Notes:

*6.1.2 Amonton's (Gay-Lussac's) Law*

Amonton's law states that when volume is kept constant and temperature is increased, pressure also increases. Amonton's law was discovered by Joseph Louis Gay-Lussac in the early 1800s. His law gives the relationship between pressure and temperature when volume and amount are held constant.

- If the temperature of a container is increased, the pressure increases.
- If the temperature of a container is decreased, the pressure decreases.

If the temperature is increased, the gas molecules will move faster and they will impact the container walls more often. This means the gas pressure inside the container will increase. Gay-Lussac's law is a direct mathematical relationship. This means there are two connected values and when one goes up, the other also increases.

*6.1.3 Charles' Law*

Charles' law states that when pressure is kept constant and volume is increased, temperature also increases. The pressure that a gas exerts on the walls of its container is determined by the momentum of the atoms and molecules of the gas, which in turn is determined by the temperature. As the temperature increases, the atoms and molecules move faster, and so exert a greater pressure on the walls.

This law gives the relationship between volume and temperature if pressure and amount are held constant.

- If the volume of a container is increased, the temperature increases.
- If the volume of a container is decreased, the temperature decreases.

If the temperature is increased, the gas molecules will move faster and they will impact the container walls more often. This means the gas pressure inside the container will increase, but only for an instant. The greater pressure on the inside of the container walls will push them outward, thus increasing the volume. When this happens, the gas molecules will now have farther to go, thereby lowering the number of impacts and dropping the pressure back to its constant value.

### 6.2.0 Measuring Energy Gases

Quantities of natural gas are measured in volume units. A cubic foot of natural gas at a temperature of 60°F and an atmospheric pressure of 14.7 pounds per square inch gauge (psig) is the common unit of measure. Gas production from wells and supplies to power plants are measured in thousands or millions of cubic feet (MCF and MMCF). Resources and reserves are calculated in billions of cubic feet (Bcf) or trillions of cubic feet (Tcf). To put this in perspective, a trillion cubic feet is enough to fill a cube with sides 2 miles long.

The amount of energy that is obtained from the burning of a unit volume of natural gas is measured in British thermal units (Btus). One Btu is the amount of heat required to raise the temperature of 1 pound of water from 60° to 61°F at normal atmospheric pressure (14.7 psig). At sea level, it would take about 75 Btus to heat water enough for a cup of tea. A cubic foot of natural gas on the average gives off 1,000 Btus, but the range of values is 500 to 1,500 Btus. Therefore, 1 cubic foot of some natural gas may make only 7 cups of tea, while another would make as many as 20 cups of tea.

Energy content of natural gas varies because natural gas accumulations vary in the amount and types of energy gases they contain: the more noncombustible gases in a natural gas, the lower the Btu value. In addition, how much of any energy gas that is present in a natural gas accumulation—the mix of combustible gases—also influences the Btu value of natural gas. The more carbon atoms in a hydrocarbon gas, the higher its Btu value. To illustrate, methane typically represents more than 80% of energy gases. Methane contains one carbon atom per molecule; burning 1 cubic foot of methane gives off 1,012 Btus. Butane, possessing four carbon atoms, has a Btu value more than three times larger than that of methane. Molecular hydrogen, on the other hand, though combustible, contains no carbon atoms; its Btu value is three times smaller than that of methane.

### 6.3.0 Commercial Natural Gas Combustible Limits

Commercial natural gas is very flammable when sufficient amounts of oxygen (the oxygen present in air is more than adequate) are present. Mixtures of natural gas and air are explosive only when the mixture is less than 15% or more than 4% natural gas. If the natural gas is less than 4%, it will be too lean a mixture to burn, and over 15%, too rich a mixture to burn. The fact that only a small amount of gas is necessary to create an explosive mixture makes even small leaks in confined spaces very serious.

As mentioned earlier, an odorant is added to commercial natural gas so that it can be easily smelled if there should be a leak. However, it is possible for odorants to be filtered out or partially



**Classroom**

**Explain the ways of measuring natural gas.**

**Explain in greater detail that the danger of explosion is greatest when the mixture of gas and air is between 4 and 15 percent. Ask a trainee why this is true.**

**Classroom**

**Define specific gravity. Ask a trainee whether a gas with a specific gravity of .5 will float or sink. What about a gas with a specific gravity of 1.5?**

**Homework**

**Have trainees read Section 7.0.0 for the next session.**

**Materials**

**Ensure that you have everything required for demonstrations and testing during this session.**

**Classroom**

**Have trainees take turns listing pieces of major pipeline equipment on the board.**

**Audiovisual**

**Show Transparency 7 (Figure 21). Explain how a centrifugal pump works. Discuss how and why pumps are used.**

filtered out of commercial natural gas. This could happen when the natural gas passes through a filtering substance (soil or other). The presence of natural gas should be determined by testing the surrounding air with available instruments such as a combustible gas indicator or another suitable instrument.

### 6.4.0 Commercial Natural Gas Specific Gravity (Weight)

*Specific gravity* is a term used in the comparison of the weight of a gas compared to the weight of the same amount of air measured under the same temperature and pressure. Some gases may or may not be heavier than air. Air is judged to be the standard weight of one (1) with commercial natural gas weighing only .6 or 6/10 when compared to air. This means that natural gas weighs less than air so natural gas will rise up or actually float from the ground and fade away into the atmosphere when a leak occurs. Also, commercial natural gas will dissipate rapidly unless confined in a building, vault, or other structure. However, propane has a specific gravity of 1.5, is heavier than air, and will sink or fall to low places in buildings or in the ground. Butane is even heavier, weighing 2.0 or twice the weight of air, and will cling to the ground. It can be extremely dangerous when leaks occur.

Commercial natural gas is lighter than air and will normally rise and travel the path of least resistance. This means that a leak might travel uphill underground until it reaches a blockage and then travel back down the hill because that is the easiest path to follow.

## 7.0.0 ◈ PIPELINE EQUIPMENT

The pipeline system includes all the pipes, equipment, stations, terminals, and other facilities necessary for operation. The major pipeline equipment consists of:

- Pumps and motors
- Compressors
- Valves
- Actuators/operators
- Tanks
- Piping
- **Manifolds**
- Traps
- Measurement/sampling equipment

### 7.1.0 Pumps and Motors

Pumps provide the pressure and force to move product. Various types of pumps may be used in the pipeline, including centrifugal, positive displacement, hydraulic, and rotary pumps. Each is used in applications requiring different capacities or other pump characteristics. Power is supplied from many different sources, such as turbines, hydraulics, diesel engines, and electric motors. One pump station may use positive-displacement pumps driven by diesel engines, whereas another station may use centrifugal pumps driven by electric motors.

#### *7.1.1 Centrifugal Pumps*

Centrifugal pumps work by rotating and forcing product outward as product flows into the center. Revolving blades or vanes create the suction that pulls product into the pump. The product takes up the space between the vanes and rotates with them. It is then pushed outward until it leaves through the discharge, or exit, pipe. *Figure 21* shows how centrifugal pumps work.

#### *7.1.2 Positive-Displacement Pumps*

In a typical positive-displacement pump (see *Figure 22*), a piston slides back and forth in a cylinder. With each movement, it traps part of the product to be moved. This product is then moved toward the discharge end of the pump, where the pressure rises as the product is squeezed by the piston. The pressure forces the product out. There are several different types of positive-displacement pumps, including lift, or suction, pumps and force pumps.

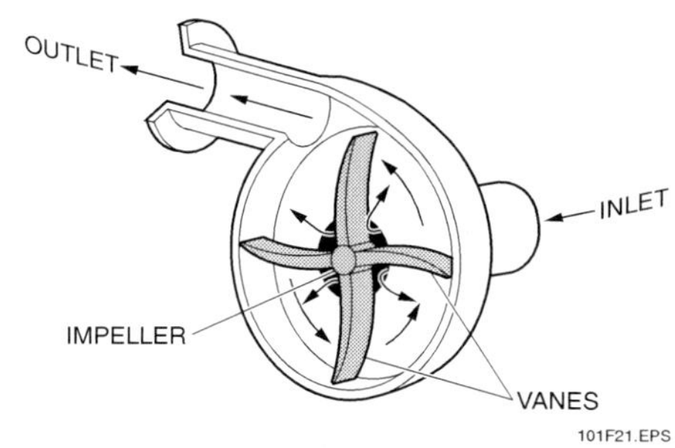


*Figure 21* ◈ How the centrifugal pump works.



Instructor's Notes:

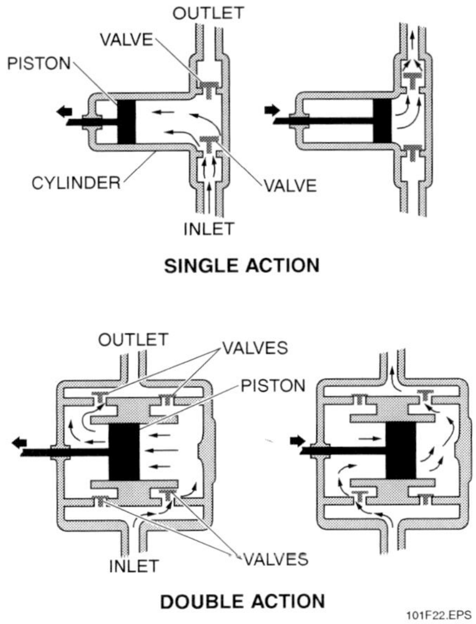


*Figure 22* ◆ Two positive-displacement pumps.

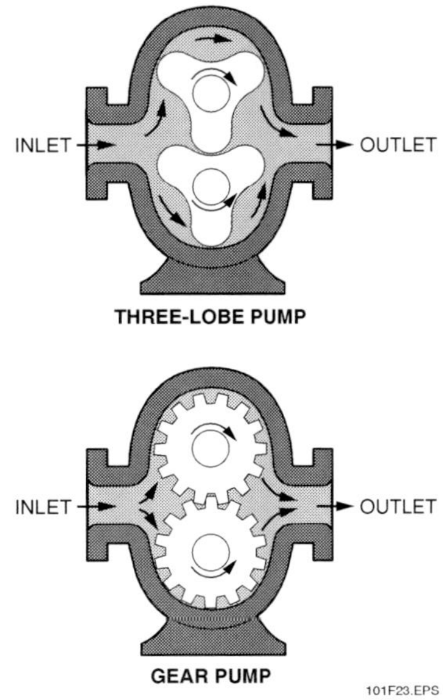


*Figure 23* ◆ Two rotary pumps.

### *7.1.3 Rotary Pumps*

Rotary pumps pump air as well as fluids. They have propeller-like blades that rotate on a single shaft. The rotation at the inlet side speeds up the fluid flow in a direction parallel to the shaft, building up pressure between the rows of blades. The pressure moves the fluid to the discharge side of the pump as the blades continue rotating.

A gear pump, which uses gear teeth instead of a piston, is used to pump thick products. The gears rotate in a housing, trapping product between the gear and the casing, moving it to the outlet side. The pressure in the outlet side builds up until it is high enough to discharge the product. Many types of drivers are used to drive these pumps. *Figure 23* shows two rotary pumps.

### *7.1.4 Drivers*

Crude oil, refined liquid products, and gas are pushed through the pipeline grid by powerful pumps and compressors. The size of this machinery depends on the type of pipeline and the products and volumes being transported. Liquids are usually moved through the pipelines by huge pumps, while gases are moved by compressors. These compressors use pressure to push the gas through the pipeline. The pump and compressor drivers may be electric motors with fixed-speed or variable-frequency drive (VFD), diesel or gas engines, or natural-gas-fueled turbines. The choice of driver depends on the power sources available. The most common drive is a fixed-speed electric motor, although variable-frequency drives are becoming more popular. In remote locations, diesel-powered equipment is often used. Diesels are also used as backups during power failures.

## 7.2.0 Compressors

Compressor stations in a gas line act like pump stations in a fluid line. Compressors boost and maintain the pressure in the line, which keeps the gas flowing. In any compression process, the pressure of gas is increased from one level to another by the addition of energy to the gas being processed.

The centrifugal compressor is classified as dynamic, as opposed to a positive-displacement machine. In a reciprocating compressor, a quantity of gas is drawn into a cylinder, a piston moves, and the trapped gas is compressed by direct volume reduction or positive displacement. By this process, mechanical energy is transformed into pressure energy. *Figure 24* shows a gas compressor.



**Audiovisual**

**Show Transparency 8 (Figure 22). Explain how each type of positive-displacement pump works.**

**Show Transparency 9 (Figure 23). Discuss how the two types of rotary pumps shown operate. Answer any questions trainees may have.**

**Classroom**

**Discuss the various types of drivers and how they work.**

**Referring trainees to Figure 24, trace the compression process.**

Ask a trainee how the compressor flow can be controlled. Ask another trainee to explain what the impeller does.

Place an assortment of valves on a table. Have trainees work in pairs to examine the valves and write down the basic types of valves and applications for each.

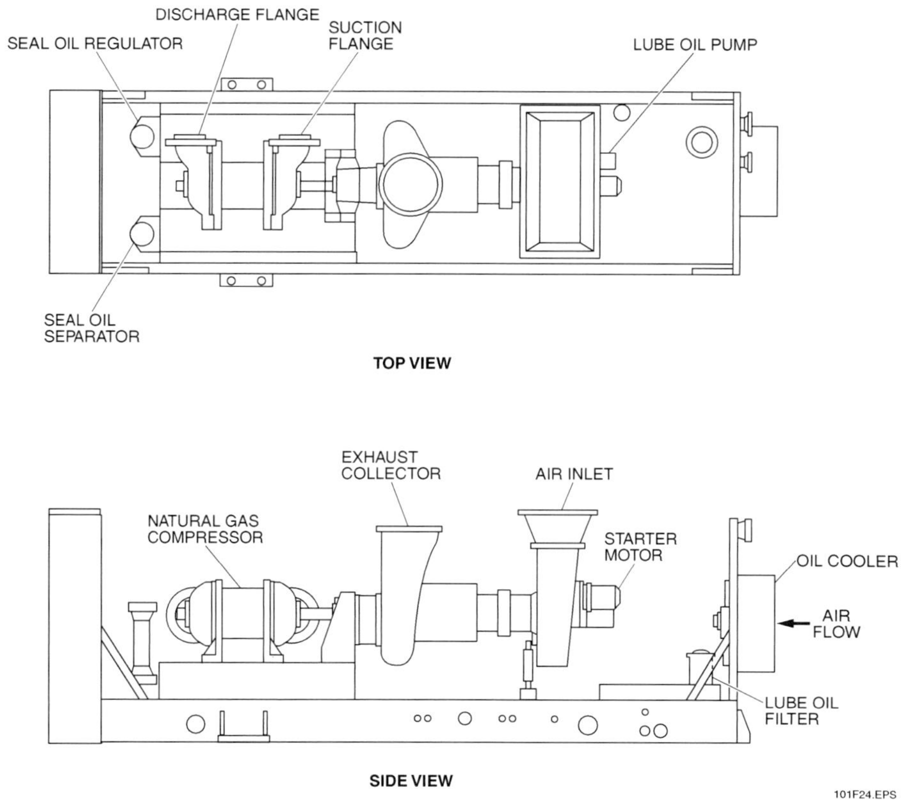


*Figure 24* ◈ Gas compressor.

The combination of inlet vane, impeller, and diffuser or diaphragm is referred to as a stage. A single centrifugal stage has a relatively low pressure capability. When larger ratios are desired, additional stages are added in series. The compressor flow can be controlled and varied by changing the pressure differential across the compressor or by changing the power level. Centrifugal compressors apply inertial forces to the gas. The inertial forces are transmitted by an impeller which, by dynamic centrifugal motion, adds velocity energy through acceleration of the gas. The gas flows from the impeller into the diffuser, where the gas decelerates, and velocity energy is transformed to pressure energy.

## 7.3.0 Valves

Valves are a critical part of pipeline systems. They are used to control the rate of flow in a line, to open or shut off a line completely, or to serve as automatic safety devices. Valves may be identified by type or by function, and can be equipped with locking devices to prevent accidental operation of the valve. Valves may be equipped for manual operation, but most frequently operated valves in the pipeline are equipped with electric-, pneumatic-, or hydraulic-power-driven mechanical operators. Those used extensively include gate, plug, ball, butterfly, check, control, and relief valves.



Instructor's Notes:

### 7.3.1 Gate Valves

In a simple gate valve, a wedge-shaped disc or gate is pressed between inclined seats to close the valve. Operating on the same principle but with different configurations of gates and seats, there are other more complex gate valves. Conduit gate valves have an elongated gate and a port in the lower part for a fluid conduit. *Figure 25* shows a simple gate valve.

### 7.3.2 Plug Valves

Plug valves (see *Figure 26*) are similar in design to the stopcocks used to control flow from containers. The plug usually has an elongated port that may be smaller in cross-sectional area than the pipe. The valve is operated by giving the plug a quarter turn. Operation may be by a lever fitted to the stem, by a gear system, or by power-driven mechanical operators. Plug valves are used where reliability and fast operation are important. There is considerable sliding contact area between the plug and the valve body, and as size increases, the operating force required also increases. Many plugs have interior channels and grooves in which lubricants can be forced into the contact areas to improve turning and sealing. Plug valves are generally run in pipe in which scrapers are not run.

### 7.3.3 Ball Valves

Ball valves are an adaptation of a plug valve, using a ball for the plug. The sealing contact area of the ball valve is that of two seal rings. These seals are made practical by use of tough, low-friction, oil-resistant elastomers.

### 7.3.4 Butterfly Valves

Butterfly valves (see *Figure 27*) control product flow by use of a disc mounted on a stem that can be turned to block or allow flow. This simple lightweight valve is used in parts of the pipeline. Butterfly valves have two valuable features. They are quick-operating and take up minimal space.

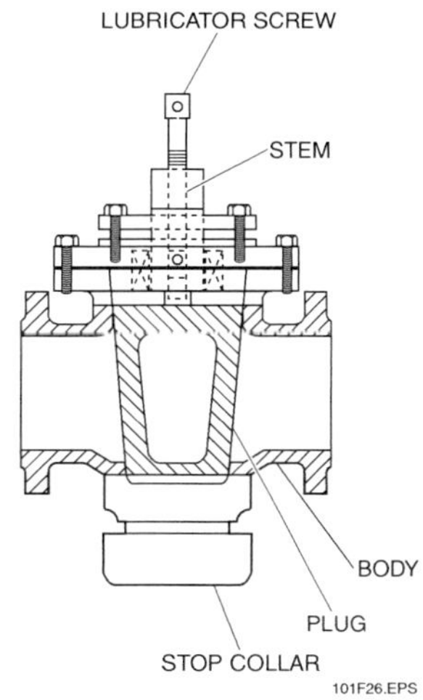


*Figure 26* ◆ Plug valve.

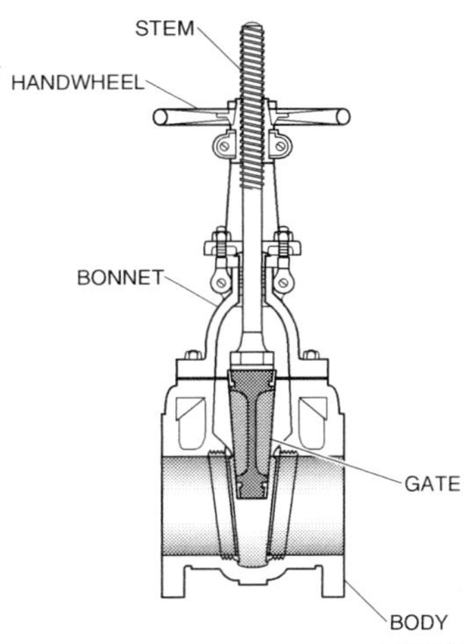


*Figure 25* ◆ Simple gate valve.

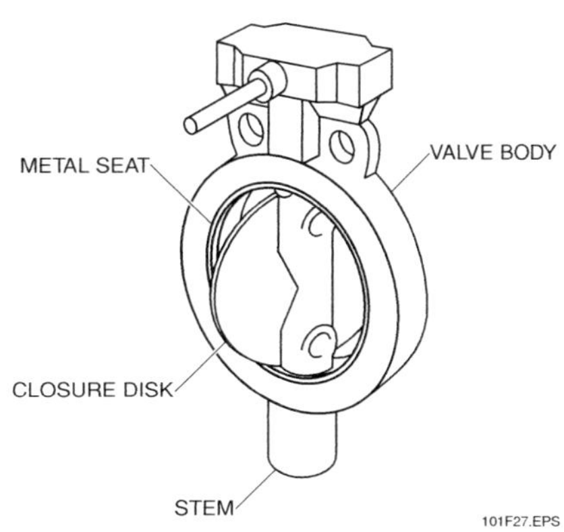


*Figure 27* ◆ Butterfly valve.



**Demonstration**

**When discussing each type of valve, show and pass around an assortment of actual valves. Have trainees compare them with the figures. Explain any differences.**

**Audiovisual**

**Show Transparency 10 (Figure 25). Point out the parts of the simple gate valve and explain how a simple gate valve works.**

**Show Transparency 11 (Figure 26). Explain how plug valves work.**

**Show Transparency 12 (Figure 27). Explain the operation of a butterfly valve. Note the disc mounted on the stem that can be turned to block or allow flow.**

**Audiovisual**

**Show Transparency 13 (Figure 28). Ask a trainee what makes a check valve close.**

**Show Transparency 14 (Figure 29). Have trainees explain how relief valves work.**

**Show Transparency 15 (Figure 30). Note when globe valves are used.**

**Classroom**

**Distinguish between how manual and automatic valves work.**

### *7.3.5 Check Valves*

Check valves (see *Figure 28*) are considered shutoff valves since they permit full flow or prevent flow in one direction. They close automatically when pressure in the flow direction is removed. Check valves can be designed to allow passage of scrapers and are most often used in gathering lines and in station piping. They are often used in conjunction with other shutoff valves.

### *7.3.6 Control Valves*

Automatic control of flow rates and pressures is provided by control valves. They may or may not provide for complete shutoff because flow control is their major function. Port sizes and flow patterns through these valves must be continuously adjusted to provide control by throttling action, increasing their tendency toward damaging vibration, erosion, and corrosion. These valves are used principally at pump stations, although they may be needed at other points in the line.

### *7.3.7 Relief Valves*

Relief valves (see *Figure 29*) may be called safety valves. Their function is to remain closed until pressures rise above design levels. Pressure is released to prevent pressure damage when released products are piped to safe discharge points.

When not opened at set pressure levels, the valves are kept closed by spring pressure, a nitrogen charge, a weight load, or other means. They normally close automatically when the pressure is relieved but may be designed to stay open until reset. They may be operated and monitored remotely.

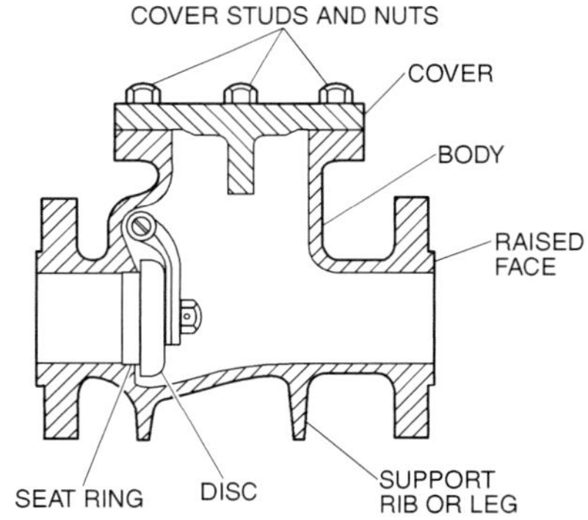


*Figure 28* ◆ Check valve.

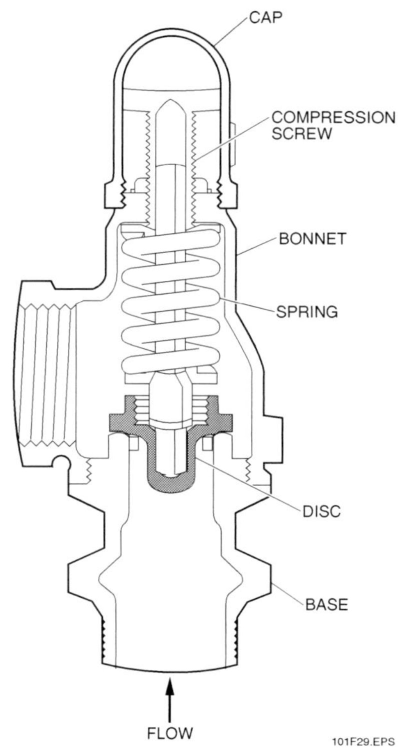


*Figure 29* ◆ Relief valve.

### *7.3.8 Globe Valves*

Other valves in the pipeline include globe valves (see *Figure 30*). Globe valves are used for shutoff and throttling, usually in auxiliary systems, such as steam and water systems.

### *7.3.9 Needle Valves*

Needle valves are a type of small, simple, rugged device for throttling and shutting off extremely high pressures compared with other valves. They are used mostly in small auxiliary piping systems.

## 7.4.0 Actuators/Operators

A valve actuator, commonly called an operator, is a device that opens and closes a valve either manually or automatically. Manual actuators come in the form of handwheels, levers, and



Instructor's Notes:

wrenches that are attached to the stem. Larger valves that require great effort to turn must be equipped with some type of powered mechanism, including electric, pneumatic, or hydraulic units, mounted on the stem. Some valve actuators use a gearbox with wide reduction ratios. Some manual actuators employ high and low gear ratios to enable personnel to turn the valve.

Valves with automatic actuators are controlled from a console in a control center by remote control, sometimes from many miles away. Such automatic actuators are used when the valve is remote from the main work area, when the frequency of operation would require unreasonable human effort, or if rapid opening and closing of the valve is required. Maintenance crews often carry portable valve actuators to work in outlying installations.

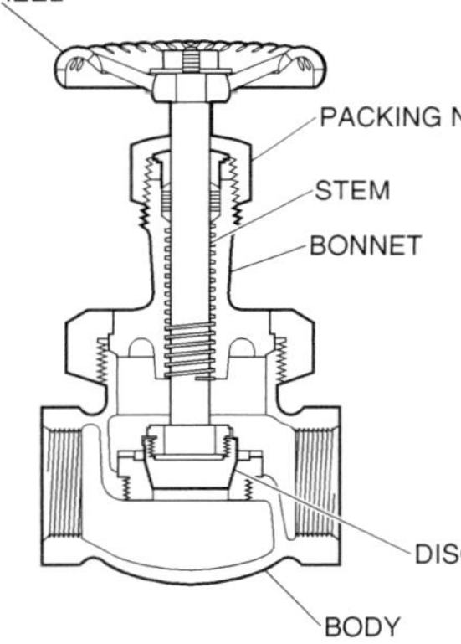


*Figure 30* ◆ Globe valve.

### 7.5.0 Tanks

Tanks are containers used for liquid storage, relief, and breakout. Tanks may exist along the length of the pipeline, at the production site, at pump stations, at terminals, and at delivery points. Although the types and sizes of tanks at each of these sites may vary, they are usually round and made of steel. All atmospheric storage tanks are constructed per *API 650* and repaired per *API 653*.

Tanks must be calibrated for capacity. The dimensions are strapped, or measured, to be used in strapping tables. From these tables, the volume of liquid contained in the tank can be determined. Tank gauges show the tank number and gauge reading from which corresponding volume in barrels can be determined.

Cylindrical tanks may have flat floating roofs, fixed cone-shaped roofs, or both. The first can be used for gasolines and aviation-grade kerosenes, while the second can be used for fuel oils. Some tanks have fixed geodesic domes covering a floating pan. Spheroid tanks are used for volatile products because of their high vapor-pressure capacity.

The floating roof tank (see *Figure 31*) is designed to reduce losses through evaporation of volatile components of crude oil and to reduce the corrosion from sour crude by eliminating the vapor space between the oil and the roof. Seals around the perimeter of the roof prevent vapor loss and rain contamination. The wind girder on the outside rim of the tank adds strength to the shell.

Tanks must also be maintained and repaired per industry standards. Although a mixer helps to keep sediment in suspension in the product, sludge forms in the bottom and must be cleaned out periodically. The tanks are repainted inside and out at specific intervals. Structural inspections and preventive maintenance guard against internal and external damage.

### 7.6.0 Pressure Vessels

Just as tanks are used for the storage of liquid products, pressure vessels are used along the gas pipeline for the storage of gas products. Liquid vessels are typically cylindrical, but like tanks, can vary in size and shape from site to site, depending on application, capacity, and pressure ratings.

Cylinders and pressure vessels can be fabricated of steel, aluminum, or composite materials. They must be manufactured, inspected, marked, tested, and retested in accordance with United

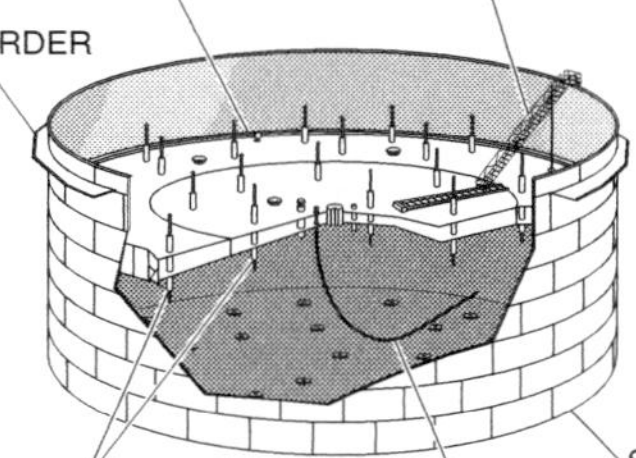


*Figure 31* ◆ Open floating roof tank.



**Classroom**

**Review *API 650* and *API 653*. Familiarize trainees with the construction and repair standards for atmospheric storage tanks.**

**On the whiteboard/ chalkboard, write *Cylindrical Tank with Flat-Floating Roof*, *Cylindrical Tank with Cone-Shaped Roof*, and *Spheroid Tank*. Have trainees contribute facts about each.**

**Review *API Standards 510* and *2510*. Discuss what the regulations cover. Ask trainees what materials can be used to construct pressure vessels.**

**Classroom**

**On the whiteboard/chalkboard, write the four types of pipe mentioned in the text. Have trainees take turns writing a fact about each on the board.**

**Audiovisual**

**Show Transparency 16 (Figure 33). Review the types and identifying characteristics of the fittings.**

**Teaching Tip**

**See the Teaching Tip at the end of this module.**

States **Department of Transportation (DOT)** regulations and exemptions for the type of service that they are used for. Pressure vessels must also be designed and repaired in compliance with *API Standards 510 and 2510*. These regulations cover the manufacture, marking, maintenance, repair, reconditioning, and testing of seamless pressure vessels, and they prescribe the requirements for transportation of hazardous materials within vessels. *Figure 32* shows a pressure vessel.

### 7.7.0 Pipe

Line pipe is made of steel through various methods. The main types of pipe used in the oil and gas industry are seamless, electrical resistance welded (ERW), and double submerged arc welded (DSAW). Other piping, less commonly used now, is welded steel pipe.

Seamless pipe is made from a solid length of metal pierced lengthwise by a mandrel with a rounded nose. The manufactured steel tubes and pipes are used in water, gas, and crude oil transmission and distribution lines and as casing pipes in the petroleum industry.

ERW is the most common form of manufacture for pipe in sizes from 2 to 22 inches outside diameter (OD). While seamless pipe is traditionally stronger and more expensive than comparable ERW pipe, ERW technology is improving and being used more and more frequently.

DSAW pipe is pipe that has longitudinal or spiral butt joints. The joints are produced by at least two passes, including at least one on the inside and one on the outside of the pipe. Pressure is not used, and filler metal for the inside and outside welds is obtained from the electrode or electrodes. This pipe is primarily intended for use in the gas and oil transmission industry.

Welded steel pipe is made by bending strips of steel into the form of a tube and welding the longitudinal seam either by electric resistance, by fusion welding, or by heating the tube and pressing the edges together. Two types of older piping that are still found on pipelines are furnace lap-welded and furnace butt-welded piping.

### 7.8.0 Fittings

Fittings include elbows, tees, reducers, flanges, plugs, and unions. Elbows are connections for directional changes. Tees are used to join three pipe ends. A reducer is used to connect pipes of different sizes. Flanges provide a means of opening and closing piping and attaching pipe ends to mating flanges on valves, tanks, pumps, and other equipment. Plugs and caps close the end of a pipe. A union provides a means of opening and closing piping and permits closing of circuitous piping systems. *Figure 33* shows types of fittings.

Fittings are made in a range of materials based on their service requirements. *Table 3* shows a partial list of materials and their pressure ranges.

**Table 3** Materials and Pressures

| Material | Pressure |
|---|---|
| Cast Iron | 150 psig |
| Malleable Iron | 250 psig |
| Cast Steel | 150–2,500 psig |
| Forged Steel | 1,000–6,000 psig |

101T03.EPS

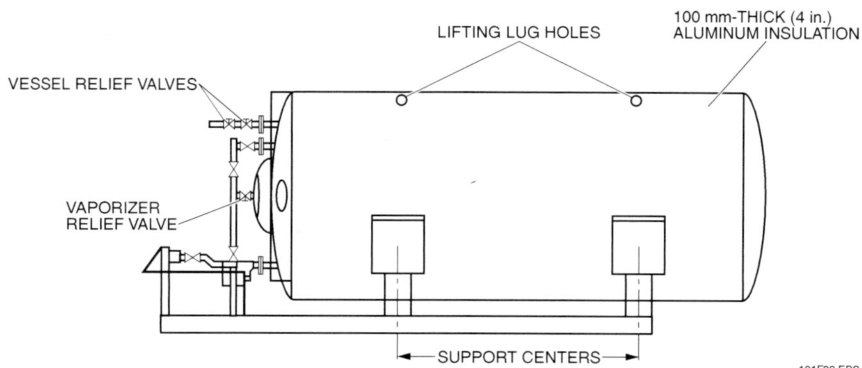


*Figure 32* ◆ Pressure vessel.



Instructor's Notes:

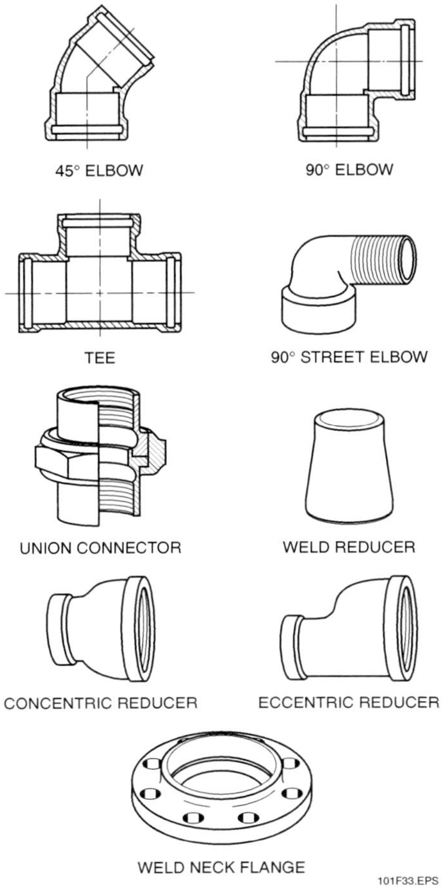


*Figure 33* ◈ Types of fittings.

### 7.9.0 Manifolds

The pipeline uses manifolds to control the flow path of oil, gas, and other petroleum products. Manifolds, with valves and meters, branch off of the line to divert specific volumes of certain products to different locations or other pipelines. *Figure 34* shows a delivery manifold.

The complexity of manifold piping ranges from very simple to very complicated. A booster station manifold may merely allow bypassing of the station. A more extensive manifold at an originating station may connect the main line, field-gathering lines, tanks, and pumps. This manifold may perform any or all of the following operations:

- Pumping product through the main line with tanks inactive
- Receiving product from the field into any tank
- Delivering from the main line into any tank
- Transferring from one tank to another
- Pumping product from a tank into the main line
- Isolating tanks from main-line flow
- Injecting product from any tank into the main line

### 7.10.0 Traps

Traps in the pipeline are designed to launch or catch scrapers and other pigs that are inserted in the pipeline. Valves and/or manifolds are used to isolate the trap for scraper insertion or removal and then to realign the trap to the pipeline flow. *Figure 35* shows incoming and outgoing scraper/pig traps.

Automation of operation has been extended to the launching of scrapers in some instances. Some traps are arranged so that one or more scrapers may be loaded and then launched at any desired time by remote-controlled valve changes.

### 7.11.0 Measuring/Sampling Equipment

All volumes entering, leaving, and passing through the pipeline system must be measured and accounted for. Accurate measurement is required for custody transfer, task purposes, leak detection, and delivery schedules.

The primary method of measurement in the pipeline is metering. Meters, whose accuracy is documented with meter provers, measure and record the volume of product passing through. Factors affecting volume are the temperature, pressure, and density of the product being measured. Various instruments are used to measure these factors concurrently with the volume measurement. Product characteristics are also checked periodically with automatic and manual sampling.

#### *7.11.1 Meters*

Various meters are used along the pipeline to monitor the volume of fluid or gas. Most product measurements are made by turbine meters or positive-displacement meters, although coriolis-type mass flowmeters, vortex, and ultrasonic meters are also used. Meter readings taken before



**Classroom**

**Discuss the types of operations a manifold may perform.**

**Explain the use of traps.**

*Figure 34* ◆ Delivery manifold.

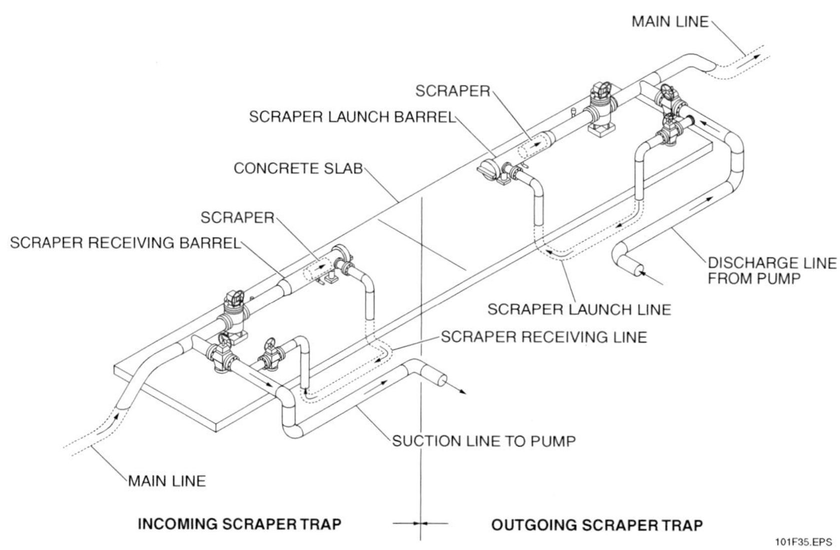


*Figure 35* ◆ Incoming and outgoing scraper, or pig, traps.



Instructor's Notes:

and after deliveries are used to determine the volume delivered, or **throughput**. Meters are typically equipped with transmitting, remote-reading, and print-out accessories to facilitate the delivery of correct product volumes and for preparation of delivery tickets. Data is often transmitted to remote locations. *Figure 36* shows two types of meters.

Gas is typically measured by orifice, turbine, positive-displacement, ultrasonic, and rotary vane meters. These meters are generally automated to correct gas volumes for moisture content and temperature and to provide remote read-out and printing capability. Strainers are installed upstream of meters to remove any foreign material from the stream before it goes through the meter. The wire mesh in strainers is generally more coarse than filter materials are.

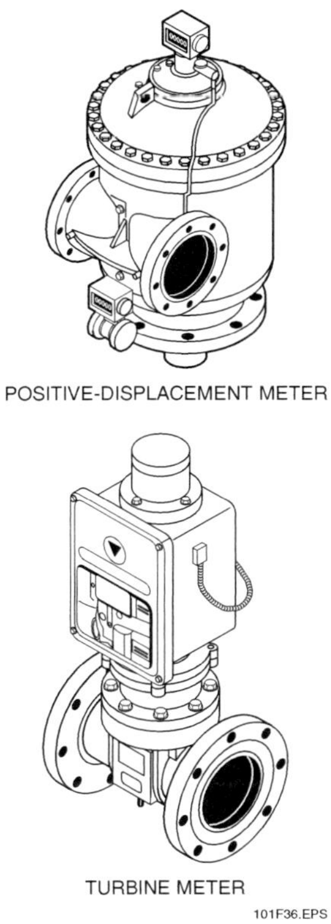


*Figure 36* ◆ Two types of meters.

The orifice plate is the vital component of the orifice meter. It is the part that restricts the flow to develop the difference in pressure that is measured to calculate actual flow. The accuracy of orifice meters is dependent on the quality of the meter that is installed. Expert installation and maintenance are necessary to ensure the continued accuracy of the meter.

Orifice plates are secured between halves of flanges of machined carriers or in the body of a special orifice meter housing constructed so that the plate can be removed for inspection. Some meters allow the plate to be removed without stopping the flow. The orifice meter is made up of the carrier housing or flanges, the orifice plate, gaskets, O-rings, devices for removal, and pressure fittings. Pressure taps are drilled at a location exactly 1 inch on center from either side of the orifice plate. *Figure 37* shows some common orifice plate holders designed for easy removal.

Visual inspection of the housing unit involves close examination of the unit looking for any signs of rust, corrosion, cracks, or any buildup of foreign material around the outside housing joints, suggesting the possibility of a leak. Pressure fittings and sensor connections must also be inspected for leaks or damage of any kind. Plates vary from 2.5 inches to 30 inches in diameter.

### *7.11.2 Provers*

A meter prover is used to calibrate meters for specific conditions of flow and product type. Meters are proved by passing a volume of product through the meter at a known rate and comparing this metered volume with the known volume prover. Two types of provers frequently used are the ball prover and the piston prover. A prover loop is a length of pipe or a calibrated U-shaped pipe with a known volume used to check or prove the accuracy of a meter.

## 8.0.0 ◆ ELECTRICAL POWER SYSTEMS

The electrical power system used by any pipeline company may be quite large or quite small. Power may be self-generated, or partially so, or may be purchased from another source. For the purposes of an introduction, a basic power system will be described from generation to delivery.

An electrical power system is an assembly of equipment and circuits for generating, transmitting, and distributing electrical energy. The elements of a power system form complex networks that require energy control centers to monitor and regulate their operation. The



**Audiovisual**

**Show Transparency 17 (Figure 36). Explain the various types of meters and what they measure.**

**Classroom**

**Ask a trainee to explain what provers are used for. Have another trainee name the two types of provers that are frequently used.**

**Homework**

**Assign the reading of Sections 8.0.0–9.0.0 for the next session.**

**Materials**

**Ensure that you have everything required for demonstrations during this session.**

Classroom

**Discuss the principal elements of an electrical power system.**

**Explain the process of power generation and transmission. Answer any questions trainees may have.**

reliability of the system is critical to pipeline operation. *Figure 38* shows an electrical power system. The principle elements of the system can be divided into the following:

- Power generation and transmission
- Power distribution
- Direct current (DC) power distribution
- Utilization equipment

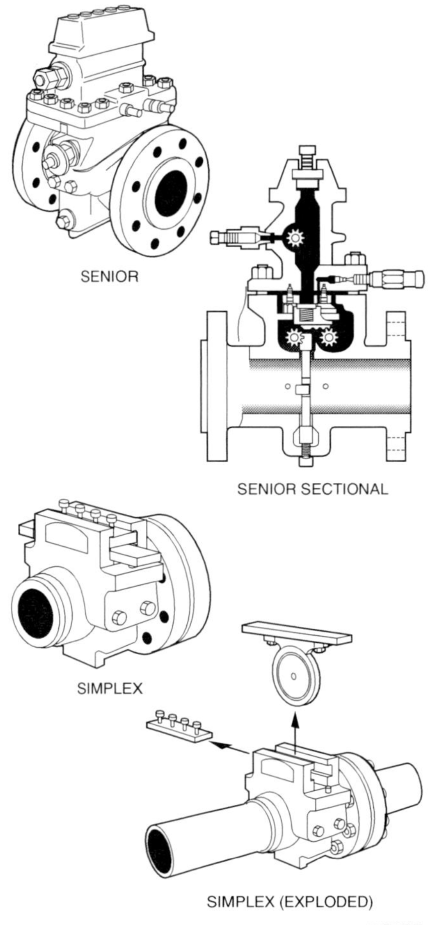


*Figure 37* ◆ Common orifice plate holders.

### 8.1.0 Power Generation and Transmission

Electrical systems require generating stations to produce electric power, and transmission systems to carry it to areas of consumption. Along the way, the voltage of the power delivered must be adapted to the specific needs of consumers. The myriad of uses that consume this energy is the power system load.

#### *8.1.1 Generation*

The large quantities of electricity required to supply a system are produced in generating stations, commonly called power plants. These are actually conversion facilities where the chemical or nuclear energy of fuel, such as coal, oil, gas, and uranium, or the kinetic energy of falling water is converted to electricity.

The generators, which supply the electricity, can be driven by steam turbines, hydraulic turbines, gas turbines, or internal combustion engines. Coal, natural gas, and heavy fuel oil are commonly used fuel for steam turbines.

Three-phase 60 hertz (Hz) AC systems are used almost exclusively in the U.S. Consequently, power system generators are wound for three-phase output at a voltage usually limited by design features. The output of modern generating stations is usually stepped up by transformers to the voltage level of transmission circuits used to deliver power to distant load areas.

#### *8.1.2 Transmission*

The transmission system carries electric power efficiently and in large amounts from generating stations to consumption areas. Such transmission is also used to interconnect adjacent power systems for mutual assistance in case of emergency.

### 8.2.0 Power Distribution

Power is reduced and distributed through a series of substations to provide the voltage required in specific applications. A substation is an assembly of equipment through which electrical energy is passed for transmission, distribution, interconnection, transformation, conversion, or switching. Power distribution involves:

- Substations
- Disconnection means
- Transformers
- Cables/conduits
- Switchgear
- Motor control centers



Instructor's Notes:

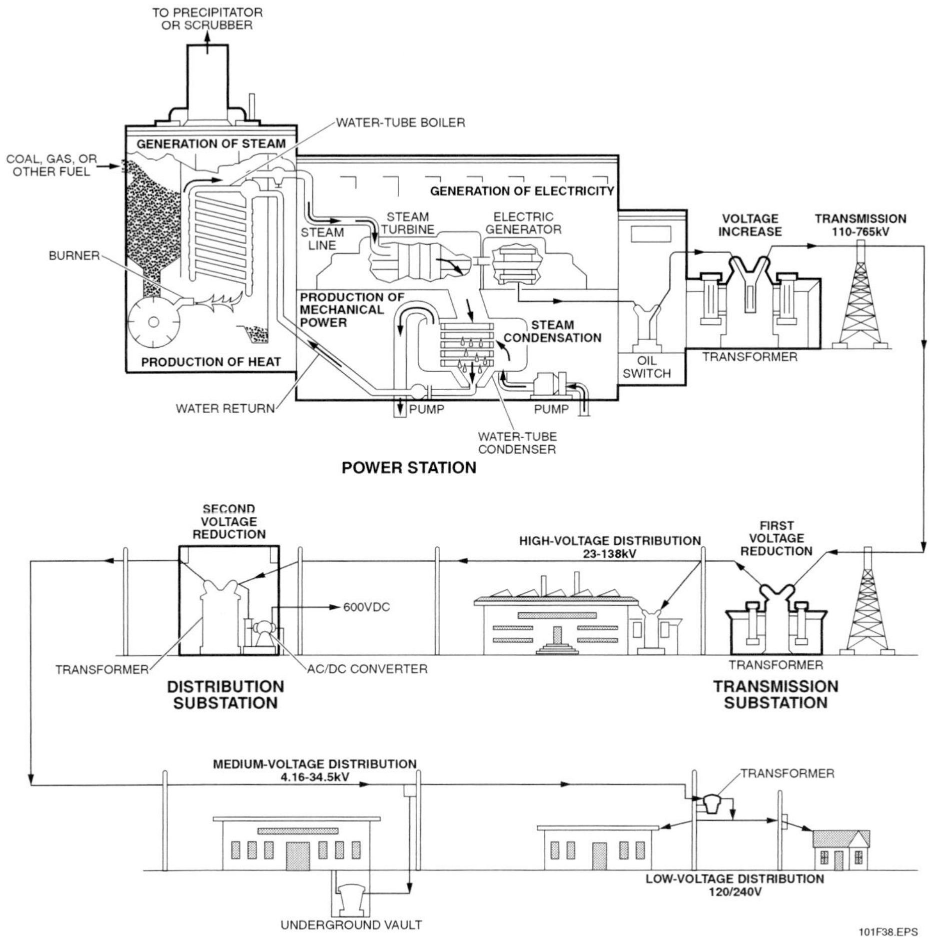


*Figure 38* ◆ Electrical power system.

### *8.2.1 Substations*

Substations are used for some or all of the following purposes:

- Connecting generators, transmission or distribution lines, and loads to each other
- Transforming power from one voltage level to another
- Interconnecting alternate sources of power
- Switching for alternate connections and isolating failed or overloaded lines and equipment
- Controlling system voltage and power flow
- Compensating for reactive power
- Suppressing overvoltage
- Detecting faults, monitoring, and recording information, power measurements, and remote communications



Classroom

**Review the common purposes of substations. Provide examples, if relevant.**

Classroom

**Detail the components of a power distribution system. Ask trainees if they have any questions.**

**Distinguish between raceway and conduit.**

**Ask trainees to list the items an MCC cubicle may contain.**

Minor distribution or transmission equipment installation is not referred to as a substation. Substations are referred to by the main duty they perform. Broadly speaking, they are classified as transmission substations when associated with high-voltage levels, and distribution substations when associated with low-voltage levels. However, they are also referred to in other ways, such as:

- Transformer substations, when the equipment includes transformers
- Switching substations, when the equipment is mainly for connections and interconnections
- Converter stations, when the main function is conversion from alternating current (AC) to DC and vice versa

### 8.2.2 Disconnection Means

Disconnect switches are commonly present so that sections or components can be isolated for maintenance. These have isolation and connection capability but may not be capable of interrupting load.

### 8.2.3 Transformers

Transformers use magnetic core and windings to transform power from one side of the transformer to the other at different voltages. Substation transformers range from small sizes of 1MVA to large sizes of 2,000MVA. Most are insulated and cooled by oil, and adequate precautions must be taken against fire hazard. These include adequate distance from other equipment, fire walls, fire extinguishing means, and pits and drains for containing leaked oil.

Current and potential transformers are for measuring currents and voltages. They are also for providing proportionately low-level currents and voltages for control and protection.

### 8.2.4 Cables and Conduits

Cables are two or more insulated wires and a grounding wire covered by an outer jacket, or sheath. Cable is usually classified by the type of covering it has, either nonmetallic or metallic. Cable may also be classified according to where it can be used. The *National Electrical Code* has definite guidelines on cables and their allowable uses. Because water can cause power loss or short circuits, cables are classified for either dry, damp, or wet locations. They are also classified for exposure to sunlight or rough use.

All cables are marked to show important properties and uses. These markings show the wire size, number of conductors, cable type, and voltage rating. The information is printed on nonmetallic sheathing and on tags attached to metal sheathing.

*Raceway* is a general term that refers to a wide variety of circularly and rectangularly enclosed channels used to house electrical wiring. Conduit is a raceway with a circular cross section similar to pipe that contains cables or wires. Conduit is used to provide protection for conductors and route them from one place to another. In addition, conduit makes it easier to replace or add wires to existing structures and, if it is metal, provides a permanent electrical path to ground.

### 8.2.5 Switchgear

Switchgear includes a wide variety of enclosed circuit breakers, disconnects, busways, and associated items that distribute power from a common point. An appropriate switching arrangement for connections of generators, transformers, lines, and other major equipment is basic to any substation design. Switchgears are use to provide this arrangement.

### 8.2.6 Motor Control Centers

The motor control center (MCC) contains the isolators for all the electrical equipment associated with a main line pump. Each MCC **cubicle** contains some or all of the following:

- Breaker
- Insulator
- Disconnect
- Isolator
- Reset button
- Hand/auto switch
- Run and stop light indicators
- Opened, closed, and stop light indicators

The isolator is the main isolator for the identified equipment and may be padlocked in the OFF position. The run and stop light indicators display the status of the associated equipment. Opened, closed, and stop light indicators are associated with motor-operated isolation valves.



Instructor's Notes:

### 8.3.0 DC Power Distribution

The DC power system provides the controls, monitor, and indicators of power required for control systems. Normal shutdown and automatic malfunction shutdown functions are also provided in the electrical system.

The DC power supply consists of batteries and chargers. A source of DC power is externally supplied by batteries and a battery charger. The charger and certain other equipment require an external power source. The battery charger will automatically charge the batteries. It does so by changing the batteries to full charge and then provides a float charge that will preserve the charge.

The charger will carry continuous and intermittent DC loads up to 80% of the maximum rated output. The charger has limiting characteristics to avoid exceeding its maximum output rating, and the output is further protected by fuses for short-circuit conditions.

## 9.0.0 ◈ CORROSION CONTROL

Corrosion is a natural process that, given the right conditions, can attack any metal or alloy. Most pipeline technicians are familiar with corrosion but do not have specific knowledge of the various types of corrosion. Everyone is surely familiar with rust, the most common form of corrosion which results in overall thinning of a metal. Most corrosion failures, however, are not caused by rust or general corrosion but by localized attack—pitting, crevice corrosion, intergranular corrosion, or stress corrosion cracking.

Each year unchecked corrosion can cost the pipeline industry millions of dollars in material damage and downtime. Corrosion can cause catastrophic failures with potentially severe consequences for the environment and serious injury or even death to personnel. It is necessary today to be able to identify potential problems and risks through a corrosion management program. An entire industry has built up around this need.

The largest organization in the world committed to the study of corrosion is the National Association of Corrosion Engineers (NACE). NACE was established in 1943 by eleven corrosion engineers in the pipeline industry. Today, NACE International has more than 50 years of experience in developing corrosion prevention and control standards. NACE is an excellent source of information for further study.

Federal guidelines have been implemented to support safety of both personnel and the environment. The complete text of the corrosion control requirements can be found in *49 CFR Part 192, Subpart I* and *49 CFR Part 195, Subpart H*. Companies are required to have written guidelines and procedures for most corrosion-related activities.

### 9.1.0 Types and Causes of Corrosion

Many factors contribute to metal corrosion; dissimilar metals that corrode at different rates and are sometimes incompatible, dissimilar soils, dissimilar aeration, bacteria, and stray current are factors that contribute to corrosion.

Pipeline corrosion is most prevalent when the failure of coatings, inhibitors, or cathodic protection occurs in a corrosive environment. These factors do not affect the pipeline equally at all locations, and corrosion does not grow at the same rate throughout a pipeline.

Internal corrosion of oil and gas pipelines is a growing concern to pipeline operators. The causes of internal corrosion of pipelines are chloride, carbon dioxide, hydrogen sulfide, oxygen, and microbiological activity. A recent severe case of internal corrosion caused several failures on a 20-inch oil line after only 2 years of service. The cause of the failure was high carbon dioxide concentrations (9%) in combination with high levels of brine.

Piping and pipeline systems are often great places to grow microbes that can produce corrosive conditions. Microbiologically influenced corrosion (MIC) often works in combination with other forms of corrosion, such as crevice corrosion and under-deposit corrosion, on in the pipe. It can also accelerate corrosion produced by impurities in the system environment, such as hydrogen sulfide and/or carbon dioxide.

For example, MIC can occur in the crevices produced by disbonded coatings on the outside of the pipe that are shielded from cathodic protection. Internally, sites for MIC are where water, debris, and other nutrients accumulate on the pipe wall. Both sites provide the local conditions that promote colonization of microbes, which can make existing corrosion problems worse by producing organic acids or acid-producing gases. There is a long list of microbes that can participate in the corrosion reaction; these include aerobic or anaerobic types (depending on whether they require oxygen or lack of oxygen to survive). They can also make use of nutrients in the environment (sulfates, nitrates, iron) to produce corrodants.



**Classroom**

**Explain the components of a DC power system.**

**Laboratory**

**Divide trainees into small groups. Have trainees brainstorm a variety of possible consequences of corrosion. Ask each group to share their ideas with the class.**

**Classroom**

**Review *49 CFR Part 192, Subpart I* and *Part 195, Subpart H*. Locate and discuss the corrosion control requirements.**

**Explain the factors that may cause corrosion. Emphasize that there are other types of corrosion besides rust.**

**Classroom**

**Discuss the various methods available to detect MIC. Ask a trainee whether sessile or planktonic surveys are better.**

**Explain why injection and sampling systems are used. Answer any questions the trainees may have.**

**Audiovisual**

**Show Transparency 18 (Figure 39). Define corrosion coupons and explain how they are used.**

**Demonstration**

**Pass around several different coupons, both new and used if possible. Have trainees take turns examining each of them.**

**Classroom**

**Ask a trainee to explain what a smart pig can detect.**

**Stress why monitoring for corrosion is essential. Point out the importance of computers in the monitoring process.**

### 9.2.0 Inspection and Testing

Pipeline integrity programs use tools, technologies, and strategies to ensure that pipeline systems are in a condition to perform safely and reliably. Increasingly sophisticated technology is available. Constant monitoring and inspection is essential to any pipeline integrity program.

#### *9.2.1 Microbiologically Influenced Corrosion (MIC)*

A variety of methods is available to detect MIC. Sampling and culturing can be performed for identification of the size of the infection and the active strains of organism present. This information is then used to identify preventative or remedial treatments. Chemical analysis and pH of corrosion products can also give important information relative to the nature of the problem. Evidence taken from the metal surface, called sessile surveys, has a better chance of properly assessing the situation versus sampling of the liquids, called planktonic survey.

#### *9.2.2 Sampling*

Injection and sampling systems are fundamental to corrosion control and process control programs. They are applicable to a large variety of processes in the petroleum, chemical, and water treatment industries. Injection systems are used to inject a wide range of chemicals into processes. Such chemicals include biocides, demulsifiers, corrosion inhibitors, oxygen scavengers, glycol and monoethylene glycol, dewaxers, methanol, odorizers, and product additives.

Sampling systems are used to take samples of the process product. Such samples are then analyzed in the laboratory for inhibitor concentration levels, the presence of metal ions, oxygen levels, scale forming compounds, and a wide range of other substances.

#### *9.2.3 Corrosion Coupons*

Coupons are specimens of materials that are exposed to tests or real environments to assess the effect of degradation on the material. They can be used inside a pipeline or outside if the pipe is underground. Corrosion coupons represent a widely used and relatively simple form of corrosion monitoring. Usually a coupon is exposed to a corrosive environment to represent an actual component/structure. After exposure for a specified amount of time, the coupons can be weighed and examined microscopically to determine and evaluate the type and average rate of corrosion damage. *Figure 39* shows typical coupons.

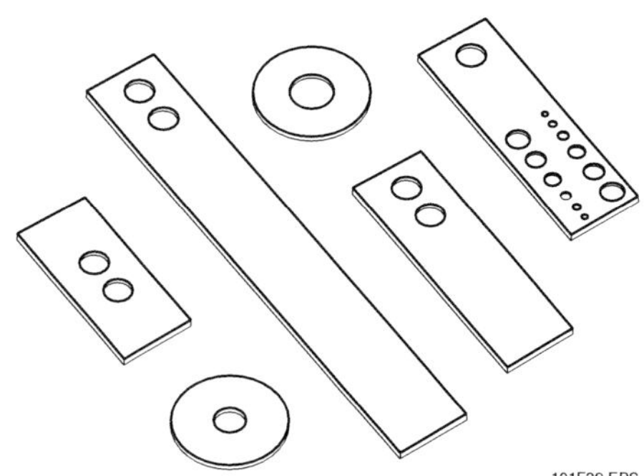

101F39.EPS

*Figure 39* ◆ Typical coupons.

#### *9.2.4 In-Line Inspection, or Smart, Pigs*

Smart pigs are high-tech inspection tools covered with sensors that gather and record information in onboard computers as they pass through the pipeline. They check for various problems, such as thinning, pitting, and general corrosion. Some smart pigs can detect metal loss in areas around welds. *Figure 40* shows an ultrascan in-line inspection (ILI) pig.

### 9.3.0 Maintenance and Protection

Monitoring a pipeline for corrosion is essential in preventing failures. If an operator can identify active or growing corrosion defects, then the rate of corrosion can be individually predicted for each defect, allowing an opportunity for corrective measures before the failure occurs.

COURTESY OF PII GROUP LTD

101F40.EPS

*Figure 40* ◆ Ultrascan in-line inspection pig.



Instructor's Notes:

As soon as production begins in a line, companies begin maintenance programs to reduce the risk of pipeline leaks. Sophisticated electronic control systems monitor changes in pressure, volume, and product flow. A variety of highly technical tools collect information about both the interior and exterior condition of pipelines. In-line inspection (ILI) tools, commonly called smart pigs, assess metal loss, strain, and cracking. This information becomes part of a computer database that is used by experts to identify and plan long-term maintenance programs.

To safeguard the outside surface of the pipe, protective coatings are applied when lines are built. In all cases, as a backup to the protective coating, cathodic protection is installed to help control external corrosion.

### *9.3.1 Inhibitors*

A corrosion inhibitor is a chemical substance that, when added in small concentration to an environment, effectively decreases the corrosion rate. Inhibitors are chemicals that react with a metallic surface or with the environment the surface is exposed to, giving the surface a certain level of protection. Inhibitors often work by adsorbing themselves on the metallic surface, protecting the metallic surface by forming a film. Inhibitors are normally distributed in a solution and injected into a pipeline through a sampling/injection tool. Some are included in a protective coating formulation. Inhibitors slow corrosion processes by one of the following methods:

- Increasing the anodic or cathodic polarization behavior
- Reducing the movement or diffusion of ions to the metallic surface
- Increasing the electrical resistance of the metallic surface

A vapor corrosion inhibitor stops corrosion by releasing protective ions that are attracted to metal. These ions condense on the metal surface, forming a water-repellent layer that contains both positively and negatively charged ions. The layer is maintained and repaired by the constant condensation of new vapor.

Corrosion inhibitors used in deep-water treating of oil and gas wells and deep-water risers differ from the use of land-based inhibitors in several ways, including delivery of the chemical to the point of injection, selection of inhibitor, and monitoring for results.

Each situation must be carefully examined before the proper inhibitor is selected.

### *9.3.2 Cathodic Protection*

Cathodic protection effectively protects underground or submerged metallic structures through the use of a negative potential applied by an external source to the pipeline or structure. Usually, once the pipeline has been made sufficiently negative, it will resist environmental (soil or moisture) corrosion. The method is typically applied to underground pipelines, storage tanks, the interior of water storage tanks, ocean pilings, and electrical transmission towers. Buried steel structures will revert back to their natural state as an iron oxide without proper protection.

Cathodic protection is the most widely applied electrochemical corrosion control technique that reduces the corrosion of a metal surface by making that surface the cathode of an electrochemical cell. Cathodic protection is accomplished by creating and controlling the action of large corrosion cells by connecting the negative terminals of DC power sources to the pipeline and the positive terminals to groundbed anodes (see *Figure 41*).

## 10.0.0 ◆ REGULATIONS, PROCEDURES, AND DOCUMENTATION

Each pipeline company has published operations and maintenance (O&M) and emergency procedures. These procedures are based on federal, state, and local regulations and on industry standards. Regulations have the effect of laws and are enforced by government agencies. Standards are generally accepted practices gained through experience. All pipeline employees are expected to become familiar with all procedures that are relevant to their job.

### 10.1.0 Regulations

The U.S. Department of Transportation (DOT) Office of Pipeline Safety also handles pipeline safety and the quality of pipeline design, construction, and operation. This office administers two sets of regulations, *Code of Federal Regulations (CFR) 49 Part 192* and *CFR 49 Part 195*, that address transportation of natural gas and other gases and liquids by pipeline. The pipeline industry is greatly affected by federal regulations. The Federal Energy Regulatory Commission (FERC) controls the natural gas and crude oil and products lines in addition to the regulations of other federal agencies. The Occupational Safety and Health Administration (OSHA) oversees every facet of worker safety and health, from the wearing of personal protective equipment to provision of hazard information to availability of emergency equipment.



**Classroom**

**Have trainees write down the three methods by which inhibitors slow corrosion processes.**

**Audiovisual**

**Show Transparency 19 (Figure 41). Explain how cathodic protection works. Have trainees name typical applications of cathodic protection.**

**Homework**

**Assign the reading of Section 10.0.0 and have trainees complete the Review Questions for the next session.**

**Materials**

**Ensure that you have everything required for demonstrations and testing during this session.**

**Classroom**

**Clarify what the DOT Office of Pipeline Safety, FERC, and OSHA oversee and/or regulate.**

Classroom

**Have trainees list the AOCs that can occur in a pipeline system.**

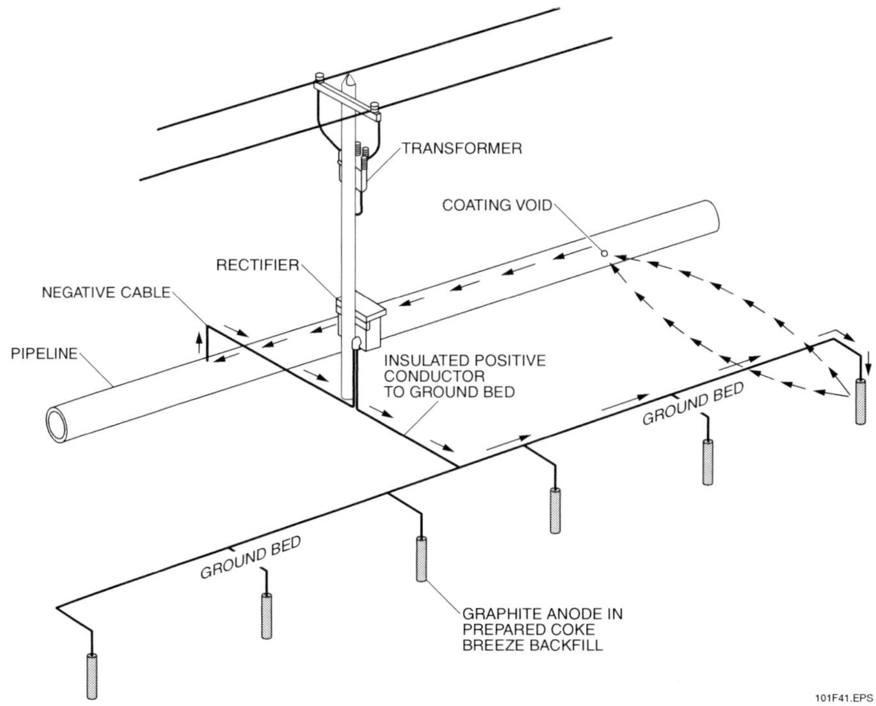


*Figure 41* ◆ Cathodic protection arrangement.

State agencies can adopt federal regulations or enforce more stringent regulations than federal because each state has developed their own form of regulation regarding pipelines. Pipelines operate on an interstate or intrastate basis but not both. Interstate pipelines are located through more than one state and fall under federal jurisdiction and classification. Intrastate pipelines are located in a single state, and different states may classify the pipeline differently. Procedures should be relevant to the states involved.

### 10.2.0 Abnormal Operating Conditions

Abnormal operating conditions (AOCs) are specified by federal law in *49 CFR Part 195.503 and CFR 192.803*. The Operator Qualification Rule requires that an individual recognize and respond to any AOC that could happen while performing a task. An AOC refers to any condition identified by the individual that may indicate failure of a component, or deviation from normal operations that may indicate a condition exceeding design limits or that may result in a hazard to persons, property, or the environment. The following list identifies the AOCs that can occur when operating a pipeline system:

- Abnormal facility condition
- Activation of a safety device
- Communications failure
- Control system failure
- Power interruption
- Fire, explosion, or natural disaster
- Pipeline system damage
- Unexpected hazardous liquid or carbon dioxide encountered
- Unexplained flow rate deviation
- Unexplained pressure deviations
- Unexplained status change
- Other company specific circumstances



Instructor's Notes:

## 10.3.0 Company Policies and Procedures

Company policy and procedure manuals typically include:

- Operating procedures manual
- Maintenance procedures manual
- Safety procedures manual
- Emergency procedures manual

Each procedure will typically include procedure details, purpose, responsibilities, and other supporting information. There are many different requirements depending on the company and the location of the pipeline.

## 10.4.0 Qualifications

With the introduction of the new Operator Qualification Rule in *49 CFR 195.503*, pipeline operators are now required to have a written qualification program. Required elements of the qualification program include:

- Identification of covered tasks
- Qualification in the performance of covered tasks through operator evaluations
- Evaluation measures for individuals following a DOT-reportable accident or for other reasonable causes
- Communication processes
- Identification of intervals at which reevaluations will occur

## 10.5.0 Documentation

Documentation is an important part of the pipeline operation from origin to end point. The DOT requires that many operations and maintenance procedures be documented to record their correct and timely performance. Problems must be documented to track down the cause and to prevent recurrence.

Some documents are forms that must be completed by operations personnel and technicians. Others are forms, reports, and summaries that are printed out periodically through the control system. Some of the documentation required for operations includes:

- Accident/incident reports
- Operations logs
- Work orders
- Daily events/event logs
- Personnel qualifications
- Inspection and repairs

### *10.5.1 Accident/Incident Reports*

Accident/incident reports are required, as specified in the Office Pipeline System (OPS) regulations, whenever there is a reportable incident along the pipeline. An incident is defined as an unexpected situation that causes, or could be expected to cause, an operational problem, harm to personnel, or damage to the environment. Reportable situations are defined in the regulations.

### *10.5.2 Operations Logs*

Operations logs record the daily events of operation. These are usually summarized on reports printed from the control system; they include all activities communicated to and from the control room as associated with the pipeline.

### *10.5.3 Work Orders*

Work orders and/or work permits are issued when specific maintenance or other tasks need to be performed. Each work order or permit has its own number and is recorded on a work order or permit log along with a description of the job and other details of the assignment.

These forms are required to ensure systematic maintenance of the pipeline, to assist with scheduling, and to record work done. Work orders and schedules for tasks done at specific time intervals can be triggered by the computer system. Not all pipeline companies use a work order system.

### *10.5.4 Daily Events and Event Logs*

Daily events are recorded in the event log as specific tasks or activities that are performed during the day's operations. These might include repairs, inspections, AOCs, investigations, maintenance, or many other events.

### *10.5.5 Personnel Qualifications*

The OPS regulation provides for formal documentation of the qualifications of personnel responsible for pipeline operations and maintenance in each company. The regulation was designed to reduce the risk of accidents on pipeline facilities caused by human error. Each company must have a written program specific to its own system for the qualification of individuals, including employees, contractors, and subcontractors who work on that system. These qualifications often cover about fifty different safety-related tasks, usually by position, such as pipeline controllers, and corrosion, maintenance,



**Classroom**

**Note the typical components of company policy and procedure manuals.**

**Refer to *49 CFR 195.503*. Review the Operator Qualification Rule and what it means for pipeline operators.**

**Teaching Tip**

**See the Teaching Tip at the end of this module.**

**Classroom**

**Give examples of reportable situations, referring to the OPS regulations if necessary.**

**Ask a trainee why documentation of personnel qualifications is required.**

**Classroom**

**Have trainees identify the information that must be documented for repairs and for corrosion control.**

and operations personnel. The rule requires that a number of specific elements be present in all operator plans, including:

- Personnel must demonstrate the ability to recognize and respond to abnormal operating conditions.
- Companies must keep records that demonstrate compliance to the OPS.

### *10.5.6 Inspection and Repairs*

When repairs are made, the pipe specifications, defect description, stress levels, and action taken must be documented and maintained in case of an audit. Other information, in accordance with the requirements of regulatory authorities, must also be prepared. Some documents are forms that must be filled in by operations and maintenance personnel. Others are forms, reports, and summaries that are printed out periodically through the control system.

Federal guidelines require that all inspections, treatments, and repairs be documented. Some of the documentation required for corrosion control includes:

- Cathodic protection system inspections
- In-line inspection results
- Pipeline coating inspections
- Incident reports
- Maintenance reports

Record-keeping is a huge and important task. Tremendous amounts of inspection and monitoring data are generated over the life of a pipeline, including:

- Development of written pipeline integrity procedures and coordination with operation, maintenance, contingency, and emergency response procedures
- Identification and implementation of repairs and modifications
- Program evaluation

Many software programs are now available to make record-keeping easier and to help companies stay in compliance with regulations. Failure to maintain good documentation can result in significant penalties for your company.

## Summary

The term *pipeline* has many meanings. In the simplest terms, it applies to the pipe and fittings used to construct facilities for transporting commodities. In a broader sense, it applies to all of the facilities in addition to the pipe and fittings that are needed to transport the commodities from the point of origin to the point of delivery. Pipeline facilities include the means of measuring, controlling, and accounting for quantities and qualities of product delivered from the custody of the original transporter to the shipper.

In order to efficiently operate and maintain the pipeline facilities, personnel must understand the basic principles of pipeline hydraulics when working with liquid pipelines, and pipeline pneumatics when working with gas pipelines. Personnel must also familiarize themselves with basic pipeline equipment and its proper use.

Electrical power systems supply power to the pipeline equipment and facilities. Electrical and instrumentation technicians maintain and repair this equipment to ensure that the pipeline remains in service. Another major task of pipeline personnel is to monitor corrosion and practice measures to prevent or minimize the presence of corrosion on the pipeline and equipment.

Each pipeline company has published operations and maintenance and emergency procedures. These procedures are based on federal, state, and local regulations and on industry standards. All pipeline personnel must familiarize themselves with these standards in order to operate the pipeline in a manner that will ensure the safety of the employees and the general public and protect the environment.



Instructor's Notes:

**Review Questions**

1. Oil and gas are formed from _____.
   a. decaying organic matter
   b. buried rocks
   c. salt domes
   d. groundwater

2. Different types of pumps or valves are used to cap the well, depending on the _____ of the fluid being withdrawn.
   a. temperature
   b. destination
   c. properties
   d. amount

3. Lines from the oil fields are called _____ lines.
   a. field
   b. main
   c. distribution
   d. gathering

4. The pipe must be        tested to verify the pipeline integrity.
   a. strength-
   b. X-ray-
   c. hydrostatically
   d. sonar-

5. Different petroleum products can be shipped consecutively in the same pipeline. This is called _____.
   a. bunching
   b. mixing
   c. batching
   d. blending

6. Large volumes of natural gas are typically stored in _____.
   a. underground storage fields
   b. surge tanks
   c. relief tanks
   d. open roof tanks

7. Oil containing a large amount of sulfur compounds is called a _____.
   a. sweet crude
   b. sour crude
   c. paraffin
   d. light crude

8. There is a set of _____ for each hazardous substance, listing its properties and hazards.
   a. chemical descriptions
   b. OSHA labels
   c. product papers
   d. MSDSs

9. The marine environment is particularly sensitive to _____.
   a. high pressure
   b. overproduction
   c. oil spills
   d. vapor accumulation

10. A(n) _____ depicts the functional layout of a piping system and its components as clearly and accurately as possible.
    a. P&ID
    b. alignment sheet
    c. strip map
    d. blueprint

11. _____ is the software used to monitor and control the flow of millions of barrels of oil and products through the pipeline.
    a. RTUS
    b. SCADA
    c. SEACA
    d. PLCS

12. The science of the behavior of fluids at rest or in motion is referred to as _____.
    a. pneumatics
    b. hydraulics
    c. compressibility
    d. viscosity

13. Boyle's law states that when temperature is kept constant and volume is decreased, pressure will _____.
    a. decrease
    b. increase
    c. stay the same
    d. not be affected

14. In a typical positive-displacement pump, a _____ slides back and forth in a cylinder.
    a. rotor
    b. vane
    c. piston
    d. blade



**Classroom**

**Go over the Review Questions. Answer any questions trainees may have on materials covered to this point.**

**Examination**

**Administer the Module Examination. Be sure to record the results of the Exam on Craft Training Report Form 200 and submit the results to the Training Program Sponsor.**

15. _____ are used to control the rate of flow in a line, to open or shut off a line, or to serve as automatic safety devices.
   a. Motors
   b. Valves
   c. Pumps
   d. Manifolds

16. Tanks must be calibrated for _____.
   a. strength
   b. sealing ability
   c. capacity
   d. wall thickness

17. _____ are used to join three pipe ends.
   a. Flanges
   b. Unions
   c. Elbows
   d. Tees

18. The pipeline uses _____ to control the flow path of oil and petroleum products.
   a. manifolds
   b. pumps
   c. tanks
   d. vents

19. Various _____ are used along the pipeline to monitor the volume of fluid.
   a. gauges
   b. tests
   c. meters
   d. provers

20. Communications failure is one example of an _____.
   a. emergency
   b. AOC
   c. accident
   d. interruption



Instructor's Notes:

GLOSSARY

# Trade Terms Introduced in This Module

*Anomaly:* A defect in the pipeline or pipeline coating caused by corrosion, wear, or other external means.

*Barrel:* The common measure of petroleum products: one barrel equals 42 U.S. gallons or 0.15899 cubic meters or 9,702 cubic inches.

*Cathodic protection:* A means of preventing the destructive electrochemical process of corrosion of metal by using the metal as the cathode of a cell with a sacrificial anode.

*Corrosion:* The gradual destruction, dissolving, or wearing away of a substance, usually a metal or an alloy, caused by a chemical or electrochemical reaction with its environment. Corrosion can be a problem in different locations along the pipeline, both externally and internally.

*Crude oil:* Unrefined liquid petroleum. Crude oil ranges in color from yellow to black, and may have a paraffin, asphalt, or mixed base. Crude oils may be referred to as heavy or light according to API gravity, with the lighter oils having the higher gravities.

*Cubicle:* An enclosure for high-voltage equipment.

*Department of Transportation (DOT):* The federal agency that sets standards for, and regulates, transportation.

*Hydrostatic testing:* The most common final quality check on a complete pipeline. Water is pumped into the line and pressured up to a specified level. The pipe is checked for leaks or ruptures.

*Main line:* The primary pipe between distant points; it is also called the trunk line.

*Manifold:* An accessory system of piping to a main piping system that serves to divide the flow into parts, to combine several flows into one, or to reroute a flow to several possible locations.

*Material safety data sheet (MSDS):* A form containing information on a potentially hazardous material, including names, contents, test results, characteristics, potential hazards, and handling procedures.

*Natural gas:* A highly compressible, highly expandable mixture of hydrocarbons with a low specific gravity. It occurs naturally in a gaseous form. Besides hydrocarbon gases, it may contain nitrogen, helium, carbon dioxide, hydrogen sulfide, and water vapor. The gases are variable in form and may be found as gases or liquids under various conditions of pressure and temperature. Natural gas liquids include butane, propane, and natural gasoline.

*Natural gas liquids (NGLs):* The portions of reservoir gas that are liquefied at the surface in lease separators, field facilities, or gas plants.

*Petroleum:* Unrefined liquid hydrocarbons, including *natural gas liquids (NGLs)*.

*Pig:* A scraping tool with blades or brushes that is forced through a pipeline to clean out accumulations of wax, scale, and debris from the pipe walls; it is also called a line scraper or go-devil.

*Product:* Any fluid, refined or not, that is transported through a pipeline.

*Refined products:* Manufactured petroleum products generated from crude oil.

*Refinery:* System of process units used to convert crude petroleum into fuels, lubricants, and other petroleum-derived products.

*Seismic activity:* The phenomena of earth movements, earthquakes, and earth vibrations, including those that are artificially induced.

*Tank:* A metal, plastic, or wooden container used to store a liquid. Three types include mud tanks for drilling, production tanks, and storage tanks.

*Throughput:* The volume of product that goes through a section of equipment during a specified amount of time.



Instructor's Notes:

The following are suggested activities or instructional methods to help you teach the material in this AIG.

## General

When you call on someone to answer a question, the rest of the class relaxes or even tunes out because they expect that the question and answer will take place only between you and the trainee you called on. Instead, use this technique to involve more trainees in answering questions and to keep them on their toes.

1. Ask trainees to define a term or explain a concept.
2. After one trainee has answered, ask a trainee seated nearby if the answer is right. Then ask whether a trainee in the back of the room agrees.
3. Ask trainees to explain why they think an answer is right or wrong.
4. Use the session to clear up incorrect ideas and encourage trainees to learn from their mistakes.

## Section 2.2.0 *Refined Products*

This Quick Quiz will familiarize trainees with the types of refined products. You will need photocopies of the quiz provided on the following page. Trainees will need pencils. If you allow trainees to use the Trainee Guide, decrease the amount of time you give them to complete the quiz.

1. Make as many photocopies as you will need of the quiz.
2. Give trainees between 5 and 10 minutes to complete the quiz.
3. Go over the answers to the quiz.
4. Ask trainees if they have questions about the types of refined products.

## Section 2.2.0 *Answers to Quick Quiz*

1. d. Consumer-ready natural gas
2. e. Fuel oil
3. c. Gasoline
4. b. Diesel
5. a. Jet fuel

## Quick Quiz *Refined Products*

For each description listed, identify the refined product the text best describes. Write the corresponding letter in the blank provided.

_____ **1.** Made up primarily of methane, this refined product's heat-value range is 1,020 to 1,045 Btus per cubic foot.

_____ **2.** This refined product comes in two grades: No. 1 and No. 2.

_____ **3.** This refined product is a volatile, flammable liquid hydrocarbon refined from crude oil.

_____ **4.** This refined product is a light hydrocarbon mixture, with a boiling range just above that of kerosene.

_____ **5.** This refined product is a highly refined kerosene petroleum distillate.

a. Jet fuel
b. Diesel fuel
c. Gasoline
d. Consumer-ready natural gas
e. Fuel oil

## Section 2.3.0 *Special Products*

This exercise will familiarize trainees with special products. For this exercise, trainees will need their Trainee Guide, pencil, and paper. Allow 45 minutes for this exercise.

1. Divide trainees into small groups. Assign each group one or two special products (ensure that each special product is assigned).
2. Have each group write each of their special products at the top of a blank sheet of paper, one product per page.
3. Have each group fill in as many of the following categories as possible for each product:
   - Physical characteristics (color, odor, gas, or liquid)
   - Purpose (what it is used for)
   - Flammability
   - Safety concerns
   - Results of human exposure

   You may add more categories if you wish.
4. Post all the sheets on the board or wall. Have each group come to the front of the classroom to give a brief presentation on their products. Ask trainees if they have any information to add to each presentation. Answer any questions trainees may have.

## Section 6.1.0 *The Ideal Gas Laws*

This exercise will help trainees understand the ideal gas laws. For this exercise, trainees will need their Trainee Guide, sheets of paper, posterboard (or other large sheet of paper), pencil, and colored markers. Allow 45 minutes for this exercise.

1. Divide trainees into small groups. Assign a gas law to each group (each law may be assigned to more than one group).
2. Ask each group to create a poster illustrating the assigned gas law. Emphasize the importance of a clear and simple design with a limited amount of text. Suggest using T, P, and V as symbols. Remind each group to include a legend explaining any symbols used. Suggest that trainees sketch their design on scrap paper first.
3. Invite each group to present their poster to the class. Ask the class to evaluate the poster: does it clearly and accurately explain the law? You may wish to hang the posters up for the remainder of the module.

## Section 7.8.0 *Fittings*

This exercise will help trainees identify various types of fittings. For this exercise, you will need at least one sample of each type of fitting listed in *Figure 33* (a total of nine). Alternatively, you may use Transparency 16, *Figure 33*, with the callouts covered. Trainees may work alone or in groups. Trainees will need their Trainee Guide, pencil, and paper to take notes for discussion. Allow 45 minutes for this exercise.

1. Lay out an assortment of numbered fittings (at least one of each type) on a table. If possible, lay out several identical sets of fittings on several tables. Alternatively, you could show Transparency 16.
2. Have trainees identify the type of each fitting. Ask trainees to write one fact about each fitting.
3. Review the answers. If necessary, review any problem areas. Answer any questions trainees may have about fittings.

## Section 10.5.0 *Documentation*

This exercise will familiarize trainees with various types of documentation and the information contained in each. If possible, obtain samples of each type of documentation listed in Section 10.5.0. For longer documentation types, use representative pages. Make one copy of each for every two trainees. Trainees will need their Trainee Guide, pencil, and paper to take notes for discussion. Allow between 30 and 45 minutes for this exercise.

1. Divide trainees into pairs. Distribute a copy of each type of documentation to each pair.
2. For each documentation sample, have trainees identify what type it is and take notes on the main information it requires.
3. Review trainees' findings. For each type of documentation, ask trainees why the documentation is important. What might happen if actions were not documented? Also ask why each piece of information on the forms is necessary or useful.
4. Answer any questions trainees may have about documentation.

# MODULE 66101-02 — ANSWERS TO REVIEW QUESTIONS

| Answer | Section |
|---|---|
| **1.** a | 1.1.1 |
| **2.** c | 1.1.3 |
| **3.** d | 1.2.0 |
| **4.** c | 1.2.2 |
| **5.** c | 1.2.4 |
| **6.** a | 1.3.0 |
| **7.** b | 2.1.1 |
| **8.** d | 2.4.0 |
| **9.** c | 2.4.0 |
| **10.** a | 3.1.0 |
| **11.** b | 4.3.0 |
| **12.** b | 5.0.0 |
| **13.** b | 6.1.1 |
| **14.** c | 7.1.2 |
| **15.** b | 7.3.0 |
| **16.** c | 7.5.0 |
| **17.** d | 7.8.0 |
| **18.** a | 7.9.0 |
| **19.** c | 7.11.1 |
| **20.** b | 10.2.0 |

## CONTREN™ LEARNING SERIES — USER UPDATES

The NCCER makes every effort to keep these textbooks up-to-date and free of technical errors. We appreciate your help in this process. If you have an idea for improving this textbook, or if you find an error, a typographical mistake, or an inaccuracy in NCCER's Contren™ textbooks, please write us, using this form or a photocopy. Be sure to include the exact module number, page number, a detailed description, and the correction, if applicable. Your input will be brought to the attention of the Technical Review Committee. Thank you for your assistance.

*Instructors* – If you found that additional materials were necessary in order to teach this module effectively, please let us know so that we may include them in the Equipment/Materials list in the Instructor's Guide.


**Write:** Curriculum Revision and Development Department
National Center for Construction Education and Research
P.O. Box 141104, Gainesville, FL 32614-1104

**Fax:** 352-334-0932

**E-mail:** curriculum@nccer.org


Craft ____________ Module Name ____________

Copyright Date ____________ Module Number ____________ Page Number(s) ____________

Description ____________

____________

____________

____________

(Optional) Correction ____________

____________

____________

(Optional) Your Name and Address ____________

____________

____________

Module 66102-02

# *Liquid Pipeline General Abnormal Operating Conditions*

# Liquid Pipeline General Abnormal Operating Conditions
# Annotated Instructor's Guide

**Module 66102-02**

## MODULE OVERVIEW

This module reviews identification and response to abnormal operating conditions. Trainees will learn how to recognize and react to a variety of potentially dangerous conditions.

## PREREQUISITES

Please refer to the Course Map in the Trainee Module. Prior to training with this module, it is recommended that the trainee shall have successfully completed the following:

Core Curriculum; Pipeline Core, Module 66101-02

## OBJECTIVES

Upon completion of this module, the trainee will be able to:

1. Recognize and react to abnormal facility conditions.
2. Recognize and react to activation of a safety device.
3. Recognize and react to communications failures.
4. Recognize and react to power interruptions.
5. Respond appropriately to fire, explosions, and natural disasters.
6. Recognize and react to pipeline system damage.
7. Recognize and react to unexpected hazardous liquid or carbon dioxide ($CO_2$) leaks.
8. Recognize and react to unexplained pressure deviations.

## PERFORMANCE TASKS

This is a knowledge-based module — there is no performance testing.

## NCCER STANDARDIZED CRAFT TRAINING PROGRAM

The National Center for Construction Education and Research (NCCER) provides a standardized national program of accredited craft training. Key features of the program include instructor certification, competency-based training, and performance testing. The program provides trainees, instructors, and companies with a standard form of recognition through a National Craft Training Registry. The program is described in full in the *Guidelines for Accreditation*, published by the NCCER. For more information on standardized craft training, contact the NCCER by writing us at P.O. Box 141104, Gainesville, FL 32614-1104; calling 352-334-0911; or e-mailing info@nccer.org. More information may be found at our Web site, www.nccer.org.

## HOW TO USE THIS ANNOTATED INSTRUCTOR'S GUIDE

Each page presents two sections of information. The larger section displays each page exactly as it appears in the Trainee Module. The narrow column ties suggested trainee and instructor actions to each page and provides icons to call your attention to material, safety, audiovisual, or testing requirements. The bottom of each page includes space for your notes.

If you see the Teaching Tip icon, that means there is a teaching tip associated with this section. Also refer to any suggested teaching tips at the end of the module.

## SAFETY CONSIDERATIONS

Ensure that the trainees are equipped with appropriate personal protective equipment. Emphasize the importance of following all safety precautions and procedures when working with power tools.

## PREPARATION

Before teaching this module, you should review the Module Outline, Objectives, and the Materials and Equipment List. Be sure to allow ample time to prepare your own training or lesson plan and gather all required equipment and materials.

## MATERIALS AND EQUIPMENT LIST

Transparencies
Markers/chalk
Blank acetate sheets
Transparency pens
Pencils and scratch paper
Module Examinations*
Overhead projector and screen
Whiteboard/chalkboard

Appropriate personal protective equipment
Copies of your local code
Photos of damaged pipeline supports or other pipeline components
Copies of a sample Safety-Related Condition Report Form
Copies of the *Code of Federal Regulations, Title 49, Part 195*

* Located in the Test Booklet packaged with this Annotated Instructor's Guide.

## TEACHING TIME FOR THIS MODULE

An outline for use in developing your lesson plan is presented below. Note that each Roman numeral in the outline equates to one session of instruction. Each session has a suggested time period of 2½ hours. This includes 10 minutes at the beginning of each session for administrative tasks and one 10-minute break during the session. Approximately 5 hours are suggested to cover *Liquid Pipeline General Abnormal Operating Conditions.* You will need to adjust the time required for hands-on activity and testing based on your class size and resources.

| Topic | Planned Time |
|---|---|
| **Session I. Responding to Abnormal Operating Conditions, Part One** | |
| A. Introduction | ___________ |
| B. Responding and Reacting to Abnormal Facility Conditions | ___________ |
| 1. Laboratory | ___________ |
| C. Responding and Reacting to Activation of a Safety Device | ___________ |
| D. Recognizing and Reacting to Communications Failures | ___________ |
| E. Recognizing and Reacting to Power Interruptions | ___________ |
| 1. Laboratory | ___________ |
| **Session II. Responding to Abnormal Operating Conditions, Part Two** | |
| A. Responding Appropriately to Fire/Explosion | ___________ |
| B. Recognizing and Reacting to Pipeline Damage | ___________ |
| 1. Laboratory | ___________ |
| C. Recognizing and Reacting to Unexpected/Unintended Release or Encounter of Hazardous Liquid | ___________ |
| D. Recognizing and Reacting to Unexpected Carbon Dioxide Encounter | ___________ |
| E. Injuries As a Result of Unexpected/Unintended Release or Encounter | ___________ |
| 1. Laboratory | ___________ |
| F. Recognizing and Reacting to Unexplained Pressure Deviations | ___________ |
| G. Summary | ___________ |
| 1. Summarize module | ___________ |
| 2. Answer questions | ___________ |
| H. Module Examination | ___________ |
| 1. Trainees must score 70% or higher to receive recognition from the NCCER. | |
| 2. Record the testing results on Craft Training Report Form 200 and submit the results to the Training Program Sponsor. | |

Module 66102-02

# *Liquid Pipeline General Abnormal Operating Conditions*

Instructor's Notes:

Instructor's Notes:

# Introduction to the Pipeline Operator Qualification Rule, API, and NCCER

*It could be said that petroleum products keep America running. Petroleum products like oil and natural gas are the principal sources of energy in the United States and are used in many manufacturing and chemical processes. Petroleum products are primarily transported by pipeline. There are more than a million miles of liquid and natural gas pipeline in this country, ranging in diameter from two inches to four feet. These pipelines bring crude oil and natural gas from production fields to processing plants and refineries—and ultimately to the consumer.*

## Pipeline Operator Qualification

Petroleum products are considered hazardous, so pipeline personnel must take extra precautions and follow specific procedures when working with them. To help ensure the safe handling and transportation of petroleum products, the United States Department of Transportation (DOT) mandates safe practices as federal regulations.

One such regulation, the Operator Qualification (OQ) rule, requires pipeline operators to develop a written qualification plan to evaluate their personnel and their contract personnel (see *49 CFR Part 192 Subpart N—Gas,* and *49 CFR Part 195 Subpart G—Liquid*). The evaluation hinges on the operator's ability to perform certain covered tasks and to be able to recognize and respond to abnormal operating conditions that may be encountered while performing these covered tasks.

### Covered Tasks

A pipeline performance task is deemed *covered* if the activity passes the following four-part test:

1. Is it performed on a pipeline facility?
2. Is it an operations or maintenance task?
3. Is it performed as a requirement of *49 CFR Part 195 (Liquid)* or *49 CFR Part 192 (Gas)*?
4. Does it affect the operation or integrity of the pipeline facility?

If the answer to all four questions is *yes*, then the activity is considered *covered* and must figure into the evaluation process of pipeline personnel.

## American Petroleum Institute

The OQ rule and its related covered tasks provided the catalyst to develop a unified pipeline training and assessment program—so that pipeline operators and owners have ready access to the assessment and training tools they need. This effort was the direct result of the vision and leadership provided by The American Petroleum Institute (API) and API members.

Efforts to unify the industry are nothing new to API. Indeed, the impetus for forming the Institute in 1919 was the need to standardize engineering specifications for drilling and production equipment. Today, the most pressing issues revolve around public perceptions and government policies toward the industry. Speaking with one voice on these issues has become as essential as having interchangeable parts in the field. API's members determine the petroleum industry's positions on public policy as well as the standards that govern its day-to-day operations worldwide.

### Consortium on Operator Qualification

API assembled a focus group whose primary purpose was to identify covered tasks and provide compliance recommendations to pipeline operators.

The Consortium on Operator Qualification (COOQ) became the guiding force for the identification and development of covered tasks for the liquid pipeline industry. Their work product, API publication 1161, *Guidance Document for the Qualification of Liquid Pipeline Personnel,* formed the basis for the resulting Pipeliner Training and Assessment Program.

The covered task list created in API publication 1161 has now evolved to include gas covered tasks and is maintained by NCCER. The most current covered task list can be downloaded at www.nccer.org.

## National Center for Construction Education and Research

With the guidance and support of API, its members, and the Consortium on Operator Qualification, the National Center for Construction Education and Research (NCCER) was able to design and implement the Pipeliner Training and Assessment Program.

NCCER is a not-for-profit 501(c)(3) education foundation established in 1995 by the world's largest and most progressive construction and maintenance contractors and national contractor associations. NCCER was founded to address the

severe workforce shortage facing the construction and maintenance industries and to develop a standardized training process and curricula. Today, NCCER is supported by hundreds of leading construction and maintenance companies, manufacturers, and national associations and developed more than 35 construction and maintenance titles.

NCCER also maintains a National Registry, which contains the records of trainees and instructors who have received NCCER training. All completion information for every trainee is sent to the NCCER and filed within the Registry. The Registry can then confirm training and skills for workers as they move from company to company, state to state, or even within their own company.

### *Standardized Curriculum (NCCER's Contren™ Learning Series)*

NCCER curricula are developed using the expertise of industry subject matter experts (SMEs). The use of SMEs as an integral part of every new development or revision ensures that the curriculum is current, industry-driven, and indicative of industry practices. The development and/or revision process is facilitated from start to finish by NCCER project managers, who work closely with the SMEs and a select team of technical writer(s) on the development and presentation of the curriculum.

### *Pipeline: Five Training Paths*

The pipeline curriculum focuses on essential skills and competencies within five skill areas:

- Corrosion Control
- Electrical and Instrumentation
- Maintenance
- Mechanical
- Operations (Field and Control Center)

SMEs from the pipeline industry have determined the skills and competencies needed based on API's covered tasks and the consensus of the group/industry. Specific company information is generally not incorporated into the curriculum unless the SMEs determine that the information is the industry standard.

### *Training Modules and Covered Tasks*

The smallest unit of instruction in the curriculum is a *module*. Each curriculum is composed of modules (the equivalent of stand-alone chapters) that are logically sequenced into complete *levels* of instruction. Every module is task-oriented and requires the successful completion of a written (knowledge-based) examination and a performance (hands-on) test. Modules may be individually selected and/or combined for efficient task training. Modules in the pipeline curricula, to the extent possible, will align with API-developed covered tasks. While not all modules will relate to covered tasks, all covered tasks will relate to at least one module.

### *Benefits of Training to the Pipeline Marketplace*

Applying NCCER's standardized training process allows pipeline owners and contractors in the United States to establish the benchmark for acceptable skill and competency levels within the pipeline industry. This process will assist training organizations in developing and/or improving standardized pipeline training programs that effectively match their resources and operating conditions while responding to regulatory demands and to the needs of the industry. The process will also recognize training organizations that meet the minimum standards of competence, warranting regulatory, public, and professional confidence. This in turn will foster national unity within the pipeline industry for the mutual benefit of individual technicians, owners, contractors, and ultimately pipeline-users and society.

## Additional Resources

To review the Pipeliner Training and Assessment Program Implementation and Resource Guide and the latest version of pipeline covered tasks, visit NCCER's Web site at www.nccer.org, or contact:

**National Center for Construction Education and Research**
PO Box 141104
Gainesville, FL 32614-1104
(888) 622-3720
Web site: www.nccer.org

For more information on API, please contact:

**American Petroleum Institute**
1220 L Street, Northwest
Washington, D.C. 20005
(202) 682-8000
Web site: www.api.org

To order *API Pub 1161*, contact Global Engineering Documents by calling 1-800-854-7149 or visiting their Web site at www.global.his.com.

Additional sources of information on the pipeline operator qualification regulation include:

**DOT – Office of Pipeline Safety**
www.ops.dot.gov

**Transportation Safety Institute**
www.tsi.dot.gov



Instructor's Notes:

## COURSE MAP

This course map shows each curriculum path in the Pipeliner Training program. The suggested training order begins on the left with the Pipeline Core modules and proceeds to the right. Which curriculum path you follow depends on your areas of specialty and interest. On a given curriculum path, skills levels increase as you advance to the right on the course map, although some are single-level curricula. The local Training Program Sponsor may adjust the training order.

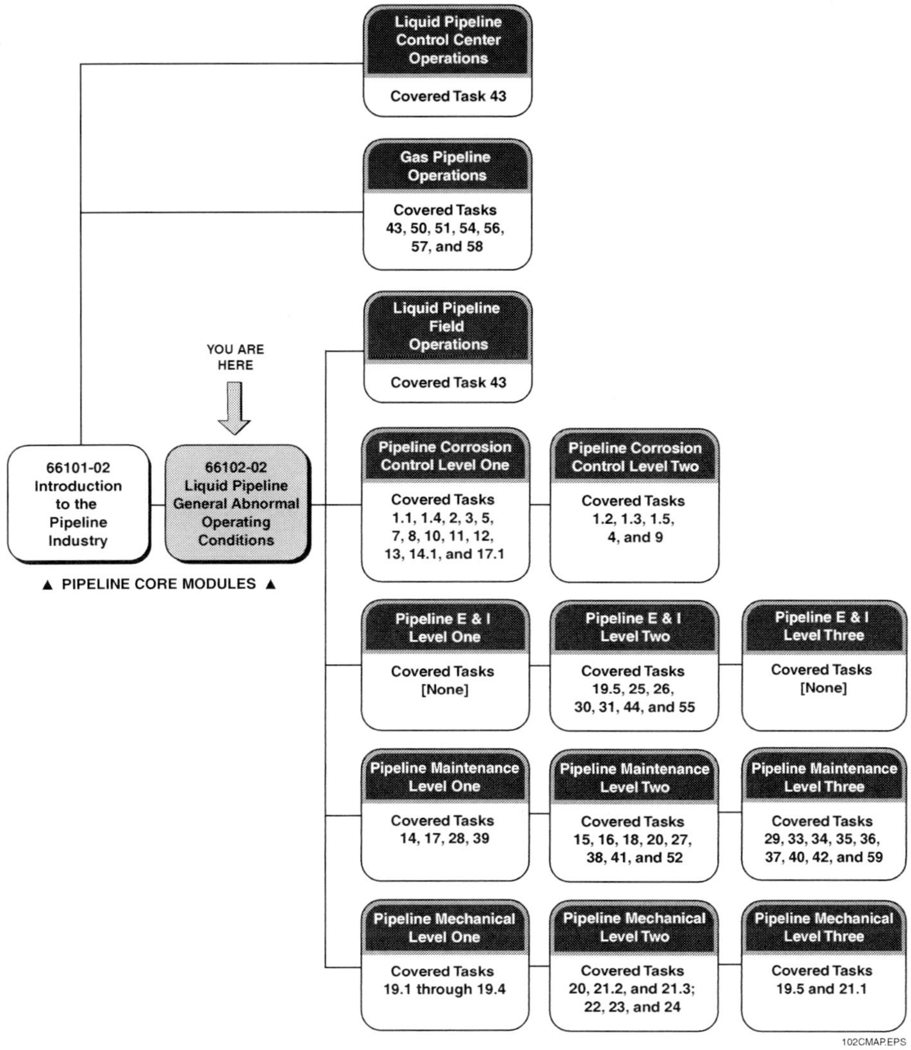






**Homework**

**Assign reading of Module 66102-02.**

Instructor's Notes:

## MODULE 66102-02 CONTENTS

## MODULE 66102-02 CONTENTS (Continued)



### Figures



### Table





Instructor's Notes:

MODULE 66102-02

# Liquid Pipeline General Abnormal Operating Conditions

## Objectives

When you have completed this module, you will be able to do the following:

1. Recognize and react to abnormal facility conditions.
2. Recognize and react to activation of a safety device.
3. Recognize and react to communications failures.
4. Recognize and react to power interruptions.
5. Respond appropriately to fire, explosions, and natural disasters.
6. Recognize and react to pipeline system damage.
7. Recognize and react to unexpected hazardous liquid or carbon dioxide ($CO_2$) leaks.
8. Recognize and react to unexplained pressure deviations.

## Prerequisites

Before you begin this module, it is recommended that you successfully complete the following: Core Curriculum; Pipeline Core, Module 66101-02.

## Required Trainee Materials

1. Pencil and paper
2. *Code of Federal Regulations, Title 49, Part 195*

## 1.0.0 ◆ INTRODUCTION

**Abnormal operating conditions (AOCs)** do not happen frequently on the pipeline. Companies have programs, procedures, training, safety equipment, warning devices, and fail-safe devices on practically every facet of operations. Computers and electronic controls are used to operate much of the pipeline and eliminate many problems and errors before they occur. However, the operator has the responsibility for making certain that when equipment does fail, the pipeline is kept in a safe condition. The operator must make use of training, skills, and good judgment until solutions are found to maintain safe operation.

Time is of the essence in such conditions. Personnel are expected to take decisive action to overcome the abnormal condition, maintaining the safety of pipeline operations. Personnel must follow company procedures. As regulation *49 CFR 195.402* states, each company must ensure its employees maintain a thorough knowledge of its operations, maintenance, and emergency manuals.

If it is necessary to deviate from company policy for the safety and protection of personnel, the public, the environment, or equipment, then the operator is expected to handle the situation. The operator goes through several logical procedures in making decisions and performing actions. Although not every procedure is included in every action to secure the safety and operation of the pipeline, the list of actions might be divided as follows:

- Recognize the AOC.
- Investigate the cause of the alert.
- Perform the actions planned.
- Make the appropriate notifications.

### 1.1.0 Definition

An abnormal operating condition is defined by federal pipeline safety law *49 U.S.C. 60102* and pipeline safety regulations *49 CFR 192.605* and *49 CFR 195.402*. AOCs are also specified by federal law in *49 CFR 195.503*. This module will concentrate on abnormal operating conditions (AOCs)



**Materials**

**Ensure that you have everything required to teach the course. Check the Materials and Equipment List at the front of this module.**

**Classroom**

**Explain that terms shown in bold are defined in the Glossary at the back of this module.**

**Audiovisual**

**Show Transparency 1, Course Objectives.**

**Teaching Tip**

**See the general Teaching Tip at the end of this module.**

**Classroom**

**Explain the importance of proper reaction to abnormal operating conditions.**

**Define abnormal operating conditions. Refer to the appropriate sections of the CFR.**

**Classroom**

**Refer trainees to Table 1. Discuss the examples of and reactions to various abnormal operating conditions.**

related to *49 CFR 195 – Transportation of Hazardous Liquids by Pipeline*. An AOC is any condition identified by the individual that may indicate failure of a component or deviation from normal operations. It may also be a condition exceeding design limits or resulting in hazards to persons, property, or the environment. The Operator Qualification rule requires that an individual be able to recognize and respond to any AOC that could happen while performing a covered task. *Table 1* shows a layout of abnormal operating conditions.

**Table 1** Layout of Abnormal Operating Conditions (1 of 2)

| Abnormal operating condition (Recognize and react to...) | Example | Reaction |
|---|---|---|
| Abnormal facility condition | During the course of performing a covered task, the individual could encounter one of the following:<br>• Exposed pipe<br>• Missing or damaged pipeline supports<br>• Incorrect ANSI pressure rating of a component or material type<br>• Disbonded coatings<br>• Unauthorized equipment on the right-of-way<br>• Sunken floating tank roof | Determine the extent of damage, perform repairs if properly trained to do so, make the appropriate notifications as required by company policy, and complete any required documentation. |
| Activation of a safety device | During the course of a maintenance or operating activity, the following safety devices could activate:<br>• Pressure relief valve relieves to tank/sump/atmosphere<br>• Emergency or abnormal shutdown<br>• High-pressure shutdown<br>• Hazardous atmosphere alarm indicating the presence of an LEL(20%) or HEL(40%) explosive atmosphere<br>• High sump level or shutdown alarm<br>• High tank shutdown alarm<br>These devices are typically designed to operate and reduce/eliminate a hazardous situation. | Contact the operations or control center and make the appropriate company notifications, minimize the situation if possible, and complete any required documentation. |
| Communications failure | During the course of operations, the operator could have a loss of the following communication devices:<br>• Radio communications<br>• Telephone<br>• SCADA communication<br>• Satellite | Immediately contact the control center, make the appropriate notifications as required by company policy, and complete any required documentation. |
| Power interruption | During the course of operating a pipeline component, the individual could experience a loss of commercial or local AC power due to the following:<br>• Utility power failure due to equipment problems or storms<br>• Station or terminal power problems (for example, motor overload, breaker trips) | Inform the power company of the outage if applicable, make the appropriate notifications as required by company policy, and complete any required documentation. |

102T01A.EPS



Instructor's Notes:

**Table 1** Layout of Abnormal Operating Conditions (2 of 2)

| Abnormal operating condition (Recognize and react to...) | Example | Reaction |
|---|---|---|
| Fire, explosion, or natural disaster | During the course of a maintenance or operations activity, the individual could discover a fire or explosion. This fire or explosion could be caused by the individual during the performance of a task due to things such as:<br>• Ignition of a hazardous vapor<br>• Unintentional detonation<br><br>Fire and explosion may also result from natural disasters, such as:<br>• Lightning strike<br>• Flooding<br>• Landslide<br>• Tornado | Notify local 911 emergency personnel; limit access to area as much as possible; make the appropriate notifications as required by company policy, and complete any required documentation. |
| Pipeline system damage | During the course of a maintenance or operations activity, the individual could encounter or cause damage to the pipeline such as:<br>• Damaged or inoperable pipeline components (pipe, valves, pumps, tank gauges) from external force<br>• Gouges/dents/pitting<br>• Damaged pipe support<br>• Wrinkles/buckle/cracks<br>• Tank shell damage<br>• Sunken roof, leaking pontoon<br>• Cracked or significantly corroded flange or bolts<br>• Damage due to flooding or land movement | Anticipate the proper response for unexpected hazardous liquid or carbon dioxide encounter and for fire/explosion if appropriate; determine the extent of the damage; make the appropriate notifications as required by company policy, and complete any required documentation. |
| Unexpected hazardous liquid or carbon dioxide encountered or unintended release | During the course of maintaining or operating the pipeline facility, the individual could encounter an unexpected presence of hazardous liquid or carbon dioxide, or could cause an unintended release during the performance of a task. This does not include the controlled transfer, sampling, venting, draining, or handling of hazardous liquid or carbon dioxide. The individual could be expected to respond to:<br>• Hydrocarbon release<br>• Vapors | Notify local 911 emergency personnel and/or appropriate persons/authorities according to company procedure, and complete any required documentation. |
| Unexplained pressure deviations | During the course of maintaining or operating the pipeline facility, the individual could observe the following pressure deviations:<br>• Increased pressure<br>• Decreased pressure<br>• High-pressure limit<br>• Low-pressure limit<br>• Loss of line pressure | Follow troubleshooting procedures, perform shutdown of the facility if necessary, make the appropriate notifications as required by company policy, and complete any required documentation. |

102T01B.EPS

Classroom

**Ask a trainee to give several examples of abnormal facility conditions.**

Audiovisual

**Show Transparency 2 (Figure 1). Explain how RTDs work. Ask trainees to explain how the two other types of alarms work.**

Classroom

**Ask a trainee to explain the proper response to damaged pipeline supports.**

Demonstration

**Show photos of damaged pipeline supports or other pipeline components. Discuss the possible consequences of this damage.**

## 2.0.0 ◆ RESPOND AND REACT TO ABNORMAL FACILITY CONDITIONS

Abnormal facility conditions relate to problems involving the physical structure of the pipeline. Examples of abnormal facility conditions include exposed pipe, missing or damaged pipeline supports, evidence of corrosion, missing signs/markers, equipment stalled over the line, hazardous atmosphere alarms, or resistance temperature device (RTD) alarms. Incorrect data caused by equipment malfunction, old or incorrect maps or survey records, as well as the incorrect interpretation of the data also create AOCs.

In almost all incidences of noted abnormal facility conditions, personnel should determine the extent of damage if applicable, perform repairs within a reasonable time if properly trained to do so, make the appropriate notifications as required by company policy, and complete the required documentation.

### 2.1.0 Respond to High Pump Case Temperature/Pressure Alarm

Main-line pumps are fitted with RTDs that give temperature readings for the pump case and product temperature. The devices are preset to signal an alarm in the control center when the temperature reaches a critical level and will shut the pump down automatically if some action cannot be taken to lower the temperature. When a pump case temperature reaches such a critical level and automatically shuts down, it is considered an AOC. *Figure 1* shows pump case temperature RTDs.

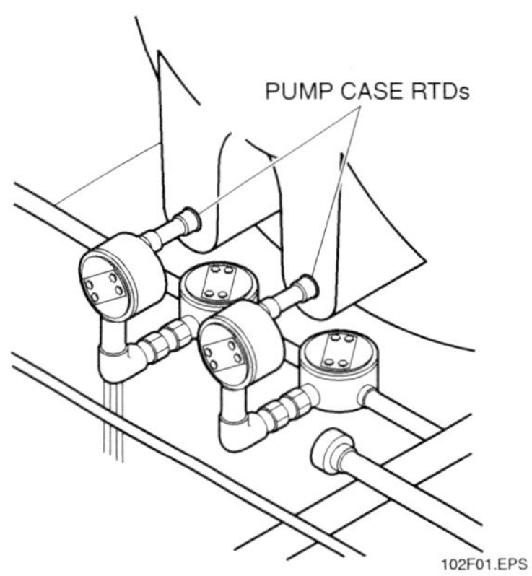


*Figure 1* ◆ Pump case temperature RTDs.

Most main-line pumps are set to shut down on low pressure. Companies fit pumps with differential pressure switches and temperature switches to protect the equipment from damage caused by low pressure, no pressure, or no flow and to avoid exceeding **maximum operating pressure (MOP)**.

Pumps also have high discharge pressure alarms that will shut down the pump when the discharge pressure exceeds a preset limit. Many times, this indicates a downstream line obstruction that is not allowing the product to flow properly from the pump.

Another type of alarm that can be received from a pumping unit is a seal failure alarm. If responding to a seal failure alarm, operations and/or maintenance personnel should investigate the validity of the alarm, check the station sump level, and notify the appropriate personnel. This may not necessarily result in an AOC, but it would require control center personnel to document the alarm and notify the appropriate personnel to check and possibly replace the seal. The control center operator should never attempt to restart the pump before investigating the cause of the alarm and responding to the alarm per company policy.

### 2.2.0 Respond to Damaged Pipeline Supports

Damaged pipeline supports (see *Figure 2*) can be serious problems on the pipeline but may not require immediately shutting down the system. A report detailing the damage is normally submitted by maintenance or inspection personnel who are traversing the pipeline. Action is determined by the extent of the damage.

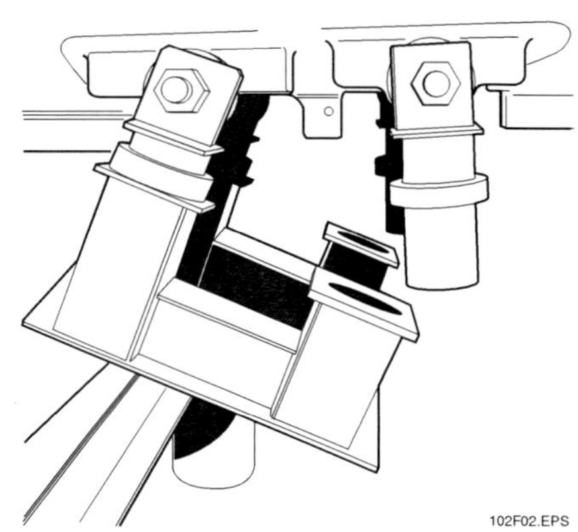


*Figure 2* ◆ Damaged pipeline support.



Instructor's Notes:

If the support is damaged, the pipeline may have to be shut down. If the damage is not critical, it may be possible to continue operation under heightened awareness. Such a decision is a judgement call on the part of the operator in collaboration with qualified maintenance or inspection personnel. The operator must submit a report to appropriate personnel so that repairs can be scheduled.

#### *2.2.1 Reaction*

If the damage to the support is critical or indicative of a possible release, the individual making the discovery must contact the control center or the appropriate company personnel immediately if the pipe is in danger of excess movement. The operator and/or call taker should then follow company policy precisely to investigate the condition and dispatch qualified personnel to the scene. Until the arrival of qualified personnel, the individual at the scene should stay a safe distance away from the site, evacuate the area if necessary, and make the appropriate notifications as required by company policy.

### 2.3.0 Documentation

The documentation and records required following an AOC depend upon the size of the incident and whether a release occurred. Any material defect or physical damage that impairs the serviceability of a pipeline requires the operator to complete a safety-related condition report and the designated company official to submit the report to government agencies (*CFR 195.55*). *Figure 3* shows a typical form for a safety-related condition report.

> **NOTE**
> A safety-related condition report is only required if the condition is not corrected within a given amount of time, or occurs within a certain distance of residences, streams, or other public areas, or if the condition is indicative of general corrosion.

## 3.0.0 ◆ RESPOND AND REACT TO ACTIVATION OF A SAFETY DEVICE

Safety devices are installed to protect the safety of pipeline operations and equipment. Any time a safety device activates, an alarm is sent to the control center. Safety devices include pressure relief systems, pressure limiting devices, high pump case pressure alarms, overfill protection devices, and main-line pressure shut-down alarms.

Pumps also have high discharge pressure alarms that will shut down the pump when the discharge pressure exceeds a preset limit for the pipeline.

> **CAUTION**
> An individual's actions during the performance of a covered task could cause a safety device to activate. The activation represents a deviation from normal operations, resulting in an AOC.

### 3.1.0 Respond to the Activation of a Pressure Relief System

Activation of **relief devices** along the pipeline system is an abnormal occurrence. Relief valves can discharge into tanks or vessels that are designed and constructed in compliance with *API Standard 650* or *API 2510* for pressure vessels. Relief valves (see *Figure 4*) can be installed on pipeline or pressure **breakout tanks**.The purpose of a relief valve is to protect the pipeline or pressure tank from overpressure. These tanks can also be used to temporarily absorb **surge pressure** from the main-line and manifold pressure relief valves to lower pressure on the line. An abnormal operating condition exists when a pressure relief occurs.

#### *3.1.1 Reaction*

Each relief event should be recognized and documented by operations personnel. If relief becomes excessive, operations personnel must immediately begin to investigate and make adjustments to ensure that the pressure of the breakout tank or pipeline does not exceed 110% of the MOP.

If the alarm is signaled and the pressure remains high, control center operations personnel must take immediate action to reduce pressure. A pressure reduction is usually achieved by shutting down the upstream pumping station, closing a block valve, or shutting down the pipeline system.

#### *3.1.2 Documentation*

Any malfunction or operating error that causes the pressure of a pipeline to rise above 110% of its maximum operating pressure may require the operator to complete a safety-related condition report and the designated company official to submit the report to government agencies (*CFR 195.55*).



**Classroom**

**Ask a trainee to explain when a safety-related condition report form is needed.**

**Have trainees name various types of safety devices.**

**Safety**

**Explain some of the actions that could cause a safety device to activate.**

**Classroom**

**Ask a trainee to explain what a relief valve is and what it is used for.**

**Discuss the appropriate reaction to overpressure. Ask trainees if they have any questions.**

**Refer trainees to 49 CFR 195. Discuss when reports of overpressure are required.**

**Safety-Related Condition Report**

Date: ________________ Government Letter #: ________

APSC FILE #: ______________

Information Resources Manager
Office of Pipeline Safety
Research and Special Programs Administration
U.S. Department of Transportation
Room 2335
400 Seventh Street SW
Washington, D.C. 20590

Safety-Related Condition Report

*49 CFR 195.55*
Subject:

**Operator:**

**Date of Report:**

**Person Submitting Report:** Senior Vice President
Business Unit

**Person Determining Condition:** Name:

Title:

Telephone:

**Date Condition Discovered:**

**Date Condition Determined:**

**Location of Condition:**

**Description of Condition:**

**Corrective Action:**

Sincerely,

Sr. Vice President, Business Unit



102F03.EPS

*Figure 3* ◈ Safety-related condition report form.



Instructor's Notes:

102F04.EPS

*Figure 4* ◆ Relief valve.

When filing a safety-related condition report, management must include a description of the condition, including the circumstances leading to its discovery, any significant effects of the condition on safety, and the name of the commodity being transported or stored. Management must also describe the corrective action taken (including reduction of pressure or shutdown) and the planned follow-up or future corrective action, including the anticipated schedule for starting and concluding such action.

### 3.2.0 Respond to Unintended Valve Closure or Shutdown

The unintended closure of a main-line block valve, remote block valve, manual block valve, pump station suction or discharge valve, and certain terminal valves (manifold/tank) will result in an automatic shutdown of the pipeline. In these cases, the operator should inspect the equipment to determine why closure occurred and reset/open/close the valve when it is safe to do so. Such occurrences may not cause an AOC, but if the event causes the pipeline to exceed design limits in any way, its classification changes to abnormal.

### 3.3.0 Testing Safety Devices

Safety devices provided at each pump station must be tested under conditions approximating actual operations (*CFR 195.262*). If, for example, the control room calls to check the **lower explosive limit (LEL)** alarm at a pumping station, it must first shut down the station and then check the LEL alarm at the station.

If, while testing a safety device, an alarm indicates a malfunction not related to the device being tested, operations and/or maintenance personnel should contact control center operations, make appropriate company notifications, and complete the required documentation.

> **WARNING!**
> If the hazardous atmosphere alarm indicates the presence of a LEL of 20%, evacuate the area.

## 4.0.0 ◆ RECOGNIZE AND REACT TO COMMUNICATIONS FAILURES

Failure of communications usually refers to a loss of contact between the control center **supervisory control and data acquisition (SCADA)** system and remotely controlled operating components on the pipeline. Essentially, operations personnel lose the ability to remotely monitor and control the pipeline facility. Communication failures may include:

- Intermittent facility outages
- Loss of the SCADA system

Sometimes the communications with various components can be verified, but the affected control system will not work. In these cases, the problem involves other components of the control system that may be mechanical, electrical, or hydraulic.

The failure to communicate with a remote control valve operator is considered a communications AOC because the valve cannot be opened or closed by the operator. The valve could also have a mechanical failure that prevents it from being operated. Sometimes electrical components or connections burn out, and if this happens on some pump drivers, they would not be able to respond in order to start up or shut down. Control of equipment is lost in this situation and could happen in many others.

### 4.1.0 Reaction

If it is discovered that there is a loss of a communications device, personnel should immediately contact the control center to verify the condition. Operations and/or other appropriate personnel should then determine the exact cause and take decisive action to control the condition, while always following company policies and procedures. In some instances, operations personnel might be able to bypass the offending component until appropriate maintenance can be performed. In other cases, the situation may warrant a shutdown.



**Laboratory**

**Refer to Figure 3. Distribute copies of a sample Safety-Related Condition Report form to pairs of trainees. Create a scenario and have them fill out the form. Discuss the results. See the teaching tip at the end of this module.**

**Classroom**

**Ask a trainee to explain the circumstances in which an unintended valve closure is considered to be an abnormal event.**

**Safety**

**Emphasize the importance of evacuation when the LEL reaches 20 percent.**

**Classroom**

**Have a trainee explain what a communications failure is. Ask for examples.**

**Discuss the appropriate reaction to the loss of a communications device.**

**Classroom**

**Ask a trainee when a pipeline must be shut down as a result of a communications device failure.**

**Discuss the effects of a power interruption on a pipeline.**

**Laboratory**

**Divide trainees into small groups. Present several scenarios of power interruptions. Have the groups determine the appropriate responses to each.**

**Homework**

**Assign the reading of Sections 6.0.0–11.0.0. If desired, have trainees complete the Review Questions for discussion during the next session.**

**Materials**

**Ensure that you have everything required for demonstration and testing during this session.**

**Classroom**

**Ask a trainee to explain how flame detectors work. Have others explain how fire alarms and fire detectors work.**

### 4.2.0 Notifications

If the component can be operated manually, the problem can be alleviated until repairs can be made. Operations must monitor pressure, however, and if 110% of MOP is exceeded, the pipeline must be shut down without discussion. If the pressure remains within limits or the component is not one that will affect the operation directly, the condition must still be monitored closely if the pipeline is to stay in operation. Even after normal conditions have been restored, operations personnel must continue to monitor the condition to make certain that repairs have been successful.

## 5.0.0 ◆ RECOGNIZE AND REACT TO POWER INTERRUPTIONS

A loss of power impacts the abilities of the control center or facility operations to safely control and monitor the pipeline facility. Virtually every component of the pipeline is in some way dependent upon electricity. If power goes off, electronic indicators will be lost. The SCADA system may be off. Pumps will shut down. Remote control valves will not operate, making control of the system difficult.

The pipeline may have a problem if the power failure is of a local nature because only a part of the system will shut down without electricity. Company policy and procedures must be followed. Area-wide power failure may also cut telephone communications and isolate the affected station. At the receiving terminal, a shipment that is in process may not have storage and cannot be shipped further without pumps. A tank being filled that cannot be switched may reach a high level or possibly overflow.

When power failure occurs, some systems can be protected by direct current (DC) backup power supplies, but most stations, terminals, and equipment are operated on alternating current (AC) or are equipped with uninterruptible power systems (UPS), a battery backup that converts DC to AC. The UPS or back-up generators supply emergency AC power to allow minimum operating capabilities or at least the ability to safely shut down the facility until the power is restored.

### 5.1.0 Reaction

Reaction to power failure is determined by what part and how much of the pipeline is affected. If only a single component or station is involved, it may be possible to adjust line pressure and flow to keep the pipeline in operation.

Suppose power interruption occurs while operating a valve. In this scenario, the crew/individual should be prepared to operate the valve manually, either by hand or by another type of operator.

Suppose a pipeline system with multiple stations is flowing with all units running. Power is lost at a station that is needed to keep pressure below MOP. In this situation, the control center operator would have to shut down the pipeline system to avoid overpressurizing the system. The terminal operator must be sure to switch off any pumps that were running so that when power returns, they will not start up. Yard personnel must be stationed outside so valves can be operated manually. If a shutdown is not required, the shipment will continue to arrive, and the schedule must be followed and the tasks completed manually.

## 6.0.0 ◆ RESPOND APPROPRIATELY TO FIRE/EXPLOSION

Some pipelines and facilities are fitted with detection devices that send an alarm when certain conditions occur. Such equipment includes flame detectors, fire detectors, and seismic detectors. Reaction to these types of alarms may require evacuation of facilities.

Flame detectors (see *Figure 5*) are thermal detectors that have normally open contacts. They work by closing their contacts when exposed to rapidly rising temperature, as would be experienced in the presence of open flames. When the contacts close, an alarm is signaled on the console in the control center and on the fire control panel in the station.

Fire alarms are activated by ultraviolet (UV) detectors or by manual pull-switch. Fire detectors are UV detectors that are set off by the light given off by open flame. Their response is almost instantaneous, and they signal fires in the very beginning stages. UV detectors are connected to a firefighting system and immediately trigger a dump of a fire suppression material into the area.

> **NOTE**
> Verify that a hazardous alarm test has been performed before performing any work, such as welding or cutting, opening electrical boxes, or operating non-explosion-proof tools in hazardous classified areas, that may produce an ignition source.



Instructor's Notes:

102F05.EPS

*Figure 5* ◆ Flame detector.

### 6.1.0 Reaction

Reaction to a flame detector or fire detector alarm must be immediate and decisive. Personnel must first verify that it is not a false alarm. A false alarm or an expected alarm due to maintenance activities requires no further action unless, in the judgment of operations, the system needs to be investigated. Alarms often appear on the control center control panel or SCADA screen.

Confirmation of the alarm gives the exact location of an active fire. Company policy dictates procedures for such emergency action and it must be followed. Procedures are site-specific and may allow for some flexibility in responding. However, maintenance and operations personnel should not need to consult a manual in case of an alarm. All personnel must be adequately trained and familiar with emergency procedures in advance. It is important to take necessary action, including a shutdown of the pipeline if required to control the emergency.

> **NOTE**
> According to *CFR 195.403*, operators shall establish and conduct a continuing training program to instruct emergency personnel in the proper use of firefighting procedures and equipment, in the identification of hazardous conditions likely to cause emergencies, in the steps necessary to control accidental releases, and in the characteristics of hazardous liquids or carbon dioxide transported.

The safety of personnel is always top priority. Protection of property is always second. Attention must be totally focused on the emergency. Response must be implemented relative to the type of fire in progress.

If the fire is a small or incipient (beginning) fire, personnel should use the available fire extinguishing equipment to control the situation. Personnel should respond in the way they have been trained. Such fires are usually the result of faulty electrical equipment or a smaller item such as a trash can fire. Responders should make certain to use the proper type of fire extinguisher for the type of fire.

If the fire is larger than the immediate workforce can contain, operations personnel should minimize the situation, notify appropriate personnel and agencies, and complete the required documentation. For example, during an electrical storm, if lightning strikes a full tank at the facility and the roof seal catches on fire, operations should follow company policy and contact the fire department immediately to contain the fire.

### 6.2.0 Respond Appropriately to Natural Disasters

Natural disasters include landslides, earthquakes, flooding, or other dangerous weather patterns that could be precursors to soil and pipeline movement. The impact of moving soil may result in forces sufficient enough to cause pipeline failure or product release.

Additional stresses are imposed on piping by soil movement or subsidence. Rains resulting in saturated soil conditions may also result in soil movement in some areas. Pipeline personnel should be alert to weather patterns and conditions that could cause soil movement and have personnel available for patrolling, preventive maintenance activities, and emergency response actions such as shutdown, isolation, and containment in case of a pipeline failure.

Patrols should be designed to effectively evaluate right-of-way (ROW) conditions in areas susceptible to soil movement that might jeopardize the operation of the pipeline. Personnel should evaluate areas of potential soil movement and, as appropriate, monitor soil subsidence and take preventive actions to ensure the safe operation of the pipeline.

Finally, pipeline personnel should evaluate the accessibility of pipeline facilities, such as valves, needed to isolate pipeline segments or sections of pipelines that might be jeopardized by soil subsidence.

**Classroom**

**Discuss the appropriate reaction to a fire alarm. Ask a trainee to identify the top priority when responding to an alarm.**

**Ask trainees to identify some of the natural disasters that can affect pipelines. Discuss the preventive action to be taken for each.**

**Classroom**

**Discuss the various conditions that might require an emergency shutdown.**

**Have a trainee describe situations that require evacuation. Ask another trainee to describe the required steps for every company evacuation procedure.**

**Laboratory**

**Identify and discuss procedures for public evacuation and notification in the event of a major release.**

**Classroom**

**Ask trainees how to identify pipeline damage.**

### 6.3.0 Perform Emergency Shutdown

Events occurring on the pipeline that present immediate hazard to persons or property require an emergency shutdown. An event might be caused by human error, mechanical failure, electrical failure, or an accident on the pipeline. It might also be brought about by threats from weather or civil disobedience. Regardless of the reason, the line must be shut down quickly and safely. Each company has a standard operating procedure for such an action. (In case of a pump shutdown from low pressure, there may be some time to discuss the move with appropriate management if the line has enough flexibility to withstand the stress.)

However, once the need is confirmed, decisive action must be taken immediately.

> **NOTE**
>
> In emergency conditions such as a possible release, the pipeline operator has full responsibility and authority to make shutdown decisions and take appropriate action. These actions are governed by the company release policy above all other considerations.

In the case of a fire, explosion, or release, discussion time is obviously not an option. The pipeline must be isolated by operations personnel by safely shutting down the originating station and continuing downstream in a safe manner until all pumps are offline. Operations personnel will then close all block valves to complete the shutdown. The responsible operations and maintenance personnel should be notified.

Threats from natural disaster or civil disobedience may come without warning. If an individual witnesses these kinds of threats to the pipeline, the supervisor or control center should be notified immediately. Such affected facilities should be isolated. Once facilities are isolated, pipeline shutdown may follow.

### 6.4.0 Evacuate the Facility

Safety of personnel and the public is the first priority in any event. Regardless of the emergency, if there is a threat of injury or death, all affected persons must be evacuated. The operator must evaluate every emergency to determine possible danger to the people involved. Events that could result in the need for an evacuation are fire, explosion, release of hydrocarbon or other hazardous vapors, seismic activity, and other natural phenomena.

The responsibility to maintain the safety of personnel and the public cannot be taken lightly. If the determination is made that a hazardous atmosphere or condition exists, the command to evacuate must be given. In the case of a release of product, the concentration of vapors downwind should be tested repeatedly to determine if there is a threat of explosion. If a hazardous atmosphere is traveling toward a populated area, the area must be warned and possibly evacuated.

#### *6.4.1 Company Procedures*

Companies have standard procedures in place to guide evacuations. By law, Emergency Response Plans must be in place and evaluated annually. Facility layouts with evacuation maps are required to be posted in every area or department. Safe areas are designated for all personnel. Once the command to evacuate is given, designated persons are required to call a roll of each employee name and check off who is present in the safe area. Every person must be accounted for.

#### *6.4.2 General Public*

In case of a release of product, the atmosphere downwind must be tested repeatedly. A vapor cloud traveling toward a populated area requires that the people be warned and possibly evacuated to a safe area. That area must be determined by the direction in which the hazardous vapors are traveling. Law enforcement and fire and rescue personnel must be solicited to aid in directing people away in a route that will give them the most protection.

In a major release, company records must be consulted so that potentially affected isolated people can be contacted in person. The path of the vapors or cloud should be plotted continuously so that if there are people, animals, or roads that the vapors will cross, the evacuation can be expanded. All roads leading to the area must be blocked until the danger has passed. All means should be employed to keep the product from reaching sewers, underground caverns, or waterways.

## 7.0.0 ◆ RECOGNIZE AND REACT TO PIPELINE DAMAGE

Pipeline damage occasionally occurs. Damaged and inoperable components, such as valves, pipe, or pumps, should be reported. One case may be a pipeline being hit or struck by excavation equipment. In this case, personnel may see gouges, scrapes, dents, or even buckling, but some signs are less obvious. A sheen on the water that covers submerged pipelines could be an indication of



*Instructor's Notes:*

valve damage. A frost ball or vapor cloud that is discovered along the right-of-way may indicate a pipeline release of **highly volatile liquid (HVL)**. Pipelines in some areas have been shot by hunters accidentally or by other people intentionally. Vandalism is also a potential cause of damage.

### 7.1.0 Reaction

The first reaction should be to determine the extent of the damage. Some of these incidents are serious enough to require an emergency shutdown. A shutdown would more likely be involved if the pipeline were struck or damaged in such a way as to involve a major release and/or explosion. If the pipeline were struck and there were no release, then the abnormal operating condition would indicate a pipeline anomaly.

The least desirable way to detect a release is by receiving notification from a third party, yet this is the most important call an operator can receive. Conditions of damaged pipe are frequently brought to the company's attention through a telephone contact from outside the company. Frequently, the call comes from an aerial patrol of the pipeline ROW or an individual landowner who contacts the One-Call notification system to report a release, damage, or other such problem.

Personnel should always anticipate the proper response to potential emergency situations, make the appropriate notifications per company policy, and complete the required documentation.

Examples of times that a pipeline would have to be shut down due to damage are if an excavator operating a bulldozer gouges the pipeline causing a release, or if a field operator loses control of a vehicle and dents a large section of the pipeline or runs into a valve. In these situations, control center operations should be notified immediately. In some cases when a known release is involved, the control center operator may be able to drain a portion of the pipeline into a delivery system to lessen the amount of product that could be released from the pipeline.

If qualified and trained to do so, personnel should perform an initial site assessment, determine the location of the release, isolate it, and contact 911 emergency personnel if necessary. Regardless, control center operations should dispatch qualified personnel to assess the conditions. If the investigation reveals an immediate hazard, the line must be shut down in a safe manner as rapidly as possible. On multiple pump station systems, the pump units must be shut down in the order of upstream units first and then downstream units.

### 7.2.0 Abnormal Movement or Loading of an In-Service Pipeline

Great care must be taken during the process of moving in-service pipelines. Additional stress could be placed on the pipeline and could lead to pipe failure. When in-service pipe is lowered, the ambient temperature is also a critical factor. As the weather changes, the pipe may expand or contract and expand.

## 8.0.0 ◆ RECOGNIZE AND REACT TO UNEXPECTED/UNINTENDED RELEASE OR ENCOUNTER OF HAZARDOUS LIQUID

An operator may suddenly observe product being released or sprayed from a manifold, valve, or ruptured piece of equipment. A pump could start accidentally during a repair process, resulting in an increase in pressure in the pipeline. An isolation device might fail. The pipe may be of insufficient wall thickness (corrosion). There could also be an occasion when product vapor is discovered to be exceedingly strong in an unexpected area or a pool of product is observed during ROW inspections.

Such discoveries may signal the presence of a release. A major release in the field, terminals, or tank farms demonstrates many of the same characteristics. Decisive action must be taken to control the event as quickly as possible. If it is not controlled quickly, explosion, fire, and injury is imminent, as well as pollution of the environment and loss of product.

### 8.1.0 Reaction

Reactions to product release are determined by the type of facility or location involved. If the hazard is discovered in a building or enclosed space, it must be handled somewhat differently than if it is discovered outside.

#### *8.1.1 Enclosed Areas*

If the hazard is in a building or enclosure, personnel must get to a safe location immediately. Manually operated doors and windows should be opened to allow vapors to dissipate. The source of the product release must be determined quickly.

The nearest cutoff valve should be closed. If a major line or equipment is involved and can be operated from the control center, it is better for personnel to move to a safe location and call the control center operator to have it closed remotely.



**Classroom**

**Discuss proper reactions to specific types of pipeline damage. Ask trainees if they have any questions.**

**Ask trainees to name possible causes of unintended/unexpected release.**

**Teaching Tip**

**See the Teaching Tip at the end of this module.**

Classroom

**Refer trainees to *49 CFR 195.50*. Discuss circumstances in which accident reports are required.**

No electrical equipment should be operated in the area. If power cannot safely be cut off at the switchgear, operations personnel must call the electrical supplier and have power cut off from outside the facility. Personnel should only be allowed to enter the area using appropriate personal protective equipment (PPE).

> **NOTE**
> To identify the hazards associated with the chemical being released, consult the material safety data sheets (MSDS).

### *8.1.2 Outdoor Areas*

After first assuring the protection of the public, report the release to control center operations and/or appropriate personnel. The source of product being released must be discovered and cut off. Sources of ignition must be avoided and the power cut off. Personnel must remove themselves to a safe location. Vapor associated with the release must be allowed to dissipate before personnel attempt to enter the area. If an odor of gasoline has been detected even a few hundred feet from the source, control center operations must be contacted to shut down the line. If the leak is from equipment at a remote station, drivers should approach the equipment from the upwind side of the station.

> **WARNING!**
> If you are witnessing a release and there is a downed person in the area of the release, notify the dispatch control center to call 911 and clear the area.

> **NOTE**
> Process Safety Management (PSM) stipulates that employers must ensure that contract employees are trained in the work practices necessary to perform their job safely.

Personnel should only be allowed to enter the area using appropriate PPE. Repairs must be carried out with the permits and documentation as required. Product on the ground will require vacuum equipment and/or other means of cleanup.

### *8.1.3 Response*

Responses to both enclosed and outdoor releases would include some or all of the following:

- Open all windows and doors.
- Find the source of the release.
- Use appropriate PPE.
- Avoid opening switchgear doors.
- Operate no electrical switches or equipment including lights.
- Exit any area where meters or smells indicate vapors are increasing.
- Have the power company cut power to shut down any pumps running if required before the problem is solved. (Go to an outside telephone to call the power company.)

## 8.2.0 Notifications

Appropriate personnel must be notified if conditions advance to dangerous levels. In case of a pipeline shutdown or other major action, appropriate personnel must be notified as soon as possible. If it is determined that product has escaped or if there has been an injury or major property damage, government agencies must receive appropriate reports.

An accident report is required for each failure in a pipeline system in which there is a release of hazardous liquid or carbon dioxide resulting in any of the following (see *CFR 195.50*):

- Any emergency that includes fire, explosion, or injury
- Releases of five gallons or more of HVLs
- All releases on water
- Releases that exceed five or more barrels of product
- Death of any person
- Bodily harm
- Estimated property damage exceeding $50,000

The accident report should be filed with the National Response Center (NRC) no later than 30 days after the discovery of the incident.



Instructor's Notes:

## 9.0.0 ◈ RECOGNIZE AND REACT TO UNEXPECTED CARBON DIOXIDE ENCOUNTER

Carbon dioxide is a colorless, odorless, gaseous compound used as a propellant in many domestic applications, such as fire extinguishing equipment, and is a by-product of combustion. Being heavier than air, it can collect in low-lying areas where it may displace oxygen; exposure can result in suffocation and death. A leak detector can detect a hazardous atmosphere, but without it, personnel can be overcome without realizing the problem. Carbon dioxide is often a by-product of incomplete burning and can be a deadly hazard if personnel are not aware of its possible existence. Some pipeline operations pump carbon dioxide commercially, and in such operations, the hazards are increased.

### 9.1.0 Reaction

In case of a fire, the operator must make certain that no one is allowed to enter the building or area after the fire is out unless the proper gas detection meter is carried along with appropriate PPE. The meter must be calibrated for detection of carbon dioxide immediately before entry. Of special concern until the atmosphere has been tested for carbon dioxide would be low-lying areas or lower rooms and compartments of buildings involved.

The operator must make certain that the first personnel to enter such questionable atmospheres are wearing self-contained breathing apparatus (SCBA). Cases have been noted in which co-workers have discovered associates lying in such areas unconscious. Rushing to their aid, the co-workers were also overcome by carbon dioxide.

In such emergencies, personnel who are trained and qualified in the use of SCBAs and are trained in rescue must enter first. Other workers must be trained to understand the danger and to stay a safe distance away until the deadly atmosphere has been cleared. In the case of a building or compartment, signs must be posted at the doors warning of the dangerous atmosphere. The operator must make certain that all such areas are plainly marked so that personnel will not enter.

## 10.0.0 ◈ INJURIES AS A RESULT OF UNEXPECTED/UNINTENDED RELEASE OR ENCOUNTER

It is the operator's duty to call for emergency rescue and medical assistance as soon as the incident is known. The operator must make certain that any injured or unconscious personnel are provided medical attention as quickly as possible.

Company procedure must be strictly followed in such incidents. In some cases, if there are physical injuries, the person must only be moved by trained and qualified personnel. In most companies, properly trained personnel are allowed to apply cardiopulmonary resuscitation (CPR) and first aid until an ambulance arrives.

### 10.1.0 Notifications

The supervisor must make certain that the incident and the injury are properly recorded and documented. In the case of personnel being injured, reports must be made to the Occupational Safety and Health Administration (OSHA) as required by the federal standard. The death of an employee requires notification of the NRC immediately and OSHA within a specified number of hours. A written accident report must be submitted to the NRC as soon as possible but no longer than 30 days from the time of the incident.

When the problem has been solved, the operator must complete the required documentation. Included in the list of documentation that must be completed are:

- Incident Investigation and Analysis Report
- Incident Report
- Event Notification forms
- Control center logs
- Station logs (if involved)
- Records of recommendations for corrective actions and training
- Documentation of such corrections and training after completion



**Classroom**

**Discuss, from your own experience if possible, specific instances of abnormal operating conditions, such as fires or releases.**

**Emphasize the importance of following the proper procedure when dealing with fires.**

**Ask trainees what must be done in case of injuries resulting from an unexpected or unintended release of product from the pipeline.**

**Laboratory**

**Have trainees name the various pieces of documentation required for reporting injuries. Write their answers on the board and discuss.**

**Classroom**

**Ask a trainee to explain the consequences of an unexplained deviation in pressure. Ask another trainee to identify the appropriate response.**

## 11.0.0 ◆ RECOGNIZE AND REACT TO UNEXPLAINED PRESSURE DEVIATIONS

Unexplained pressure deviations can include an absence of pressure, a sudden increase in pressure, or a sudden decrease in pressure. Deviations in pressure indicate a serious condition on the pipeline. Occurrences such as unintended valve closures cause AOCs that will endanger the pipeline. The condition must be recognized, and reaction must be quick and decisive. If not, an emergency condition could result very quickly.

### 11.1.0 Reaction

If a change in pressure occurs that is different from the expected pressure, personnel should immediately investigate the change. If the cause is not determined, the operating lines should be shut down, referring to the abnormal condition shutdown procedure. Personnel should then proceed with emergency response notification.

### Summary

An AOC is a condition that may indicate a malfunction of a component or deviation from normal operations, which may exceed design limits or result in a hazard to persons, property, or the environment. As required by law, companies have standard operating procedures in place to give personnel direction as to how to alleviate these conditions. Time is of the essence in such conditions. Operations personnel should be able to recognize and react to these conditions, using knowledge of policies and procedures as well as their own skill and good judgment.

In the event of a fire, explosion, or other disaster, operations personnel must be prepared to aid in emergency operations. Whether the AOC is the result of a valve closed inadvertently or a major release of product, personnel must react quickly to ensure the protection of the public and to preserve the safety and efficiency of the pipeline.



Instructor's Notes:

## Review Questions

1. An example of an abnormal facility condition would be _____.
   a. a high-pressure shutdown
   b. unauthorized equipment on the ROW
   c. transformer failure
   d. a lightning strike

2. The first reaction once pipeline damage has been discovered is to _____.
   a. shut down the pipeline facility
   b. contact the control center
   c. evacuate the surrounding area
   d. determine the extent of the damage

3. Main-line pumps have high discharge pressure alarms that will shut down the pump when the discharge pressure _____.
   a. exceeds 100% of the MOP
   b. exceeds a preset limit
   c. is too low
   d. is at null

4. If 110% of maximum operating pressure is being exceeded, operations personnel must _____.
   a. shut down the pipeline
   b. continue to monitor the situation closely
   c. file an accident report form
   d. contact OSHA

5. _____ failures are somewhat interlinked with communications failures because the SCADA system may be involved.
   a. Power
   b. Equipment
   c. Control system
   d. Pump

6. In the case of a release, pipeline operations personnel should first _____.
   a. shut down the originating station, continuing until all pumps are offline
   b. engage the emergency response team
   c. close all block valves to shut down the system
   d. debate the situation with their co-workers

7. An uninterruptible power system _____.
   a. is a device that converts alternating current (AC) to direct current (DC)
   b. provides DC backup in case of power failure
   c. breaks down electric power from a high-voltage source to appropriate levels for distribution
   d. provides protection against primary AC power failure

8. An accident report is required for a failure in a pipeline system in which there is a release of product resulting in _____.
   a. releases of carbon dioxide
   b. releases that exceed five barrels of product
   c. estimated property damage exceeding $10,000
   d. releases of three gallons of HVLs or more

9. Carbon dioxide can collect in low-lying areas and displace oxygen because _____.
   a. it is a by-product of combustion
   b. it is heavier than air
   c. it is 1.5 times less dense than air
   d. it is produced by respiration

10. If an odor of _____ has been detected even a few hundred feet from the source, control center operations must be contacted to shut down the line.
    a. carbon dioxide
    b. propane
    c. gasoline
    d. carbon monoxide



**Classroom**

**Go over the Review Questions. Answer any questions trainees may have on materials covered to this point.**

**Examination**

**Administer the Module Examination. Be sure to record the results of the Exam on Craft Training Report Form 200 and submit the results to the Training Program Sponsor.**

GLOSSARY

# *Trade Terms Introduced in this Module*

***Abnormal Operating Condition (AOC):*** A condition indentified by the pipeline operating company that may indicate malfunction of a pipeline component or deviation from normal pipeline operations. An AOC could indicate a condition exceeding design limits or result in a hazard to persons, property, or the environment.

***Breakout tank:*** A DOT term used to identify a tank that relieves surges in a jurisdictional hazardous pipeline system or receives and stores hazardous liquid transported by a pipeline for subsequent reinjection and continued transportation by the jurisdictional pipeline. The breakout tank is a jurisdictional tank.

***Highly volatile liquid (HVL):*** A hazardous liquid that will form a vapor cloud when released to the atmosphere and has a vapor pressure exceeding 40 pounds per square inch absolute at 100°F.

***Lower explosive limit (LEL):*** The minimum concentration of vapor in the air below which propagation of a flame does not occur in the presence of an ignition source.

***Maximum operating pressure (MOP):*** The maximum internal gauge pressure at which a hazardous liquid or carbon dioxide pipeline system or segment of such may be operated. The highest level of pressure allowed for a particular pipeline as set by the Department of Transportation to regulate the range of safe operations for that pipeline. Broadly speaking, it is set at 80 percent of the original design pressure, considering all components, or the highest pressure the pipeline was operated at during the last five years.

***Relief device:*** Device for diverting a stream of product into a tank to take pressure off the pipeline during an overpressure condition.

***Supervisory Control and Data Acquisition (SCADA):*** The computerized system and process used to gather information from and send control instructions to remote pipeline facilities.

***Surge pressure:*** Pressure produced by a change in velocity of the moving stream that results from shutting down a pump station or pumping unit, closure of a valve, or any other change in velocity of the moving stream.



Instructor's Notes:

# MODULE 66102-02 — TEACHING TIPS

The following are suggested activities or instructional methods to help you teach the material in this AIG.

## General

When you call on someone to answer a question, the rest of the class relaxes or even tunes out because they expect that the question and answer will take place only between you and the trainee you called on. Instead, use this technique to involve more trainees in answering questions and to keep them on their toes.

1. Ask trainees to define a term or explain a concept.
2. After one trainee has answered, ask a trainee seated nearby if the answer is right. Then ask whether a trainee in the back of the room agrees.
3. Ask trainees to explain why they think an answer is right or wrong.
4. Use the session to clear up incorrect ideas and encourage trainees to learn from their mistakes.

## Section 3.1.2 *Documentation*

This exercise will allow trainees an opportunity to practice reporting a safety-related condition. Trainees will need paper and pencil. Allow 20–30 minutes for this exercise.

1. Divide the class into pairs. Provide one member of each pair with a scenario involving a safety-related condition.
2. Have the trainees "report" to their partners on the safety-related condition.
3. The trainees who receive the reports should use the information to write a report, including all necessary information. They may refer to *Figure 3* as an example.
4. Have the pairs switch roles. Provide new scenarios to each pair.
5. Review the reports in class, and discuss the processes of collecting information and writing a complete and accurate report.

## Section 8.1.3 *Response to Unexpected or Unintended Releases*

This exercise will familiarize trainees with proper response to unexpected and unintended releases. You will need an overhead projector, blank acetate sheets, and a transparency marker. Trainees will need pencils and paper to take notes for later discussion. Allow 30–45 minutes for this exercise.

1. Divide the class into two teams. Tell trainees that you will be stating several kinds of indoor and outdoor releases. The first team to identify the proper procedure for handling the type of release will be awarded one point. If they get it wrong, the other team will have an opportunity to answer.
2. No individual can answer more than twice.
3. Proceed until a predetermined score has been reached by one of the teams.
4. Discuss the results and review any unclear or inaccurate procedures that were presented during the exercise.

## MODULE 66102-02 — ANSWERS TO REVIEW QUESTIONS

| Answer | Section |
|---|---|
| **1.** b | 1.1.0 |
| **2.** d | 2.0.0 |
| **3.** b | 3.0.0 |
| **4.** a | 4.2.0 |
| **5.** a | 5.0.0 |
| **6.** a | 6.3.0 |
| **7.** d | 5.0.0 |
| **8.** b | 8.2.0 |
| **9.** b | 9.0.0 |
| **10.** c | 8.1.2 |

## CONTREN™ LEARNING SERIES — USER UPDATES

The NCCER makes every effort to keep these textbooks up-to-date and free of technical errors. We appreciate your help in this process. If you have an idea for improving this textbook, or if you find an error, a typographical mistake, or an inaccuracy in NCCER's Contren™ textbooks, please write us, using this form or a photocopy. Be sure to include the exact module number, page number, a detailed description, and the correction, if applicable. Your input will be brought to the attention of the Technical Review Committee. Thank you for your assistance.

*Instructors* – If you found that additional materials were necessary in order to teach this module effectively, please let us know so that we may include them in the Equipment/Materials list in the Instructor's Guide.


**Write:** Curriculum Revision and Development Department
National Center for Construction Education and Research
P.O. Box 141104, Gainesville, FL 32614-1104

**Fax:** 352-334-0932

**E-mail:** curriculum@nccer.org


Craft ______________________ Module Name ______________________

Copyright Date ______________ Module Number ______________ Page Number(s) ______________

Description

______________________________________________

______________________________________________

______________________________________________

______________________________________________

(Optional) Correction

______________________________________________

______________________________________________

______________________________________________

(Optional) Your Name and Address

______________________________________________

______________________________________________

______________________________________________

Module 62103-02

# *Release Identification and Response*

# Release Identification and Response
# Annotated Instructor's Guide

**Module 62103-02**

## MODULE OVERVIEW

This module introduces trainees to the hazards associated with pipeline failure and the release of flammable, explosive, or toxic materials. Topics include company environmental manuals; government regulations; coordination of response to spills; spill prevention, reporting, and containment; soil contamination; and waste handling.

## PREREQUISITES

Please refer to the Course Map in the Trainee Module. Prior to training with this module, it is recommended that the trainee shall have successfully completed the following:

Core Curriculum; Pipeline Core, Modules 66101-02 and 66102-02

## OBJECTIVES

Upon completion of this module, the trainee will be able to:

1. Describe company environmental manuals and regulations.
2. Describe United States Coast Guard responsibilities.
3. Describe spill prevention.
4. Recognize soil contamination.
5. Explain spill reporting and spill containment.
6. Explain proper waste handling.

## PERFORMANCE TASKS

Under the supervision of the instructor, the trainee should be able to do the following using a company's manuals or procedure books:

1. Identify and explain the company's oil spill response plan and emergency procedures.
2. Identify and explain the company's policies on spill prevention.
3. Identify and explain the company's policies on spill reporting and spill containment.
4. Identify and explain the company's policies on proper waste handling.

## NCCER STANDARDIZED CRAFT TRAINING PROGRAM

The National Center for Construction Education and Research (NCCER) provides a standardized national program of accredited craft training. Key features of the program include instructor certification, competency-based training, and performance testing. The program provides trainees, instructors, and companies with a standard form of recognition through a National Craft Training Registry. The program is described in full in the *Guidelines for Accreditation*, published by the NCCER. For more information on standardized craft training, contact the NCCER by writing us at P.O. Box 141104, Gainesville, FL 32614-1104; calling 352-334-0911; or e-mailing info@nccer.org. More information may be found at our Web site, www.nccer.org.

## HOW TO USE THIS ANNOTATED INSTRUCTOR'S GUIDE

Each page presents two sections of information. The larger section displays each page exactly as it appears in the Trainee Module. The narrow column ties suggested trainee and instructor actions to each page and provides icons to call your attention to material, safety, audiovisual, or testing requirements. The bottom of each page includes space for your notes.

**Teaching Tip**

If you see the Teaching Tip icon, that means there is a teaching tip associated with this section. Also refer to any suggested teaching tips at the end of the module.

## SAFETY CONSIDERATIONS

Ensure that the trainees are equipped with appropriate personal protective equipment. Emphasize the importance of following all safety precautions and procedures when working with power tools.

## PREPARATION

Before teaching this module, you should review the Module Outline, Objectives, Performance Tasks, and the Materials and Equipment List. Be sure to allow ample time to prepare your own training or lesson plan and gather all required equipment and materials.

## MATERIALS AND EQUIPMENT LIST

Transparencies and blank transparencies
Markers/chalk
Pencils and scratch paper
Plastic painter's drop cloth
Shoebox lined with heavy-duty plastic wrap or plastic box of similar size
Small bag of sand mixed with garden soil and rocks
Garden trowel
Wide, shallow plastic bowl
Bottle of water
Flexible plastic tubing
Small bottle of automobile oil
Plastic trash bags
Paper towels
Module Examinations*
Performance Profile Sheets*
Overhead projector and screen
Whiteboard/chalkboard
Appropriate personal protective equipment

* Located in the Test Booklet packaged with this Annotated Instructor's Guide.

## TEACHING TIME FOR THIS MODULE

An outline for use in developing your lesson plan is presented below. Note that each Roman numeral in the outline equates to one session of instruction. Each session has a suggested time period of 2½ hours. This includes 10 minutes at the beginning of each session for administrative tasks and one 10-minute break during the session. Approximately 7½ hours are suggested to cover *Release Identification and Response*. You will need to adjust the time required for hands-on activity and testing based on your class size and resources.

| Topic | Planned Time |
|---|---|
| **Session I. Company Policies and Government Regulations** | |
| A. Company Environmental Policies | ________ |
| 1. Oil Spill Response Plans | ________ |
| 2. Emergency Procedures | ________ |
| B. Department of Natural Resources (DNR) and Environmental Protection Agency (EPA) Regulations | ________ |
| 1. DNR | ________ |
| 2. EPA | ________ |
| a. Prevention | ________ |
| b. Preparation | ________ |
| c. Response | ________ |

**Session II. National Response Center (NRC); Coast Guard; Spill Prevention; and Soil Contamination**

A. NRC ____________
B. Coast Guard ____________
C. Spill Prevention ____________
  1. Damage Prevention ____________
    a. Inspections ____________
    b. Observation ____________
    c. Recognition ____________
    d. Response ____________
    e. Maintenance ____________
D. Soil Contamination ____________

**Session III. Spill Reporting and Containment; Waste Handling; Summary; and Examination**

A. Spill Reporting
  1. Written Reports ____________
  2. Telephonic Notice ____________
  3. Safety-Related Conditions Reports ____________
B. Spill Containment
  1. Minimizing the Amount ____________
  2. Controlling the Spill ____________
C. Waste Handling ____________
D. Summary
  1. Summarize material ____________
  2. Answer questions ____________
E. Module Examination ____________
  1. Trainees must score 70% or higher to receive recognition from the NCCER.
  2. Record the testing results on Craft Training Report Form 200 and submit the results to the Training Program Sponsor.
F. Performance Testing ____________
  1. Trainees must perform each task to the satisfaction of the instructor to receive recognition from the NCCER.
  2. Record the testing results on Craft Training Report Form 200 and submit the results to the Training Program Sponsor.

Module 62103-02

# *Release Identification and Response*

Instructor's Notes:

Instructor's Notes:

# Introduction to the Pipeline Operator Qualification Rule, API, and NCCER

*It could be said that petroleum products keep America running. Petroleum products like oil and natural gas are the principal sources of energy in the United States and are used in many manufacturing and chemical processes. Petroleum products are primarily transported by pipeline. There are more than a million miles of liquid and natural gas pipeline in this country, ranging in diameter from two inches to four feet. These pipelines bring crude oil and natural gas from production fields to processing plants and refineries—and ultimately to the consumer.*

## *Pipeline Operator Qualification*

Petroleum products are considered hazardous, so pipeline personnel must take extra precautions and follow specific procedures when working with them. To help ensure the safe handling and transportation of petroleum products, the United States Department of Transportation (DOT) mandates safe practices as federal regulations.

One such regulation, the Operator Qualification (OQ) rule, requires pipeline operators to develop a written qualification plan to evaluate their personnel and their contract personnel (see *49 CFR Part 192 Subpart N—Gas,* and *49 CFR Part 195 Subpart G—Liquid*). The evaluation hinges on the operator's ability to perform certain covered tasks and to be able to recognize and respond to abnormal operating conditions that may be encountered while performing these covered tasks.

## *Covered Tasks*

A pipeline performance task is deemed *covered* if the activity passes the following four-part test:

1. Is it performed on a pipeline facility?
2. Is it an operations or maintenance task?
3. Is it performed as a requirement of *49 CFR Part 195 (Liquid)* or *49 CFR Part 192 (Gas)*?
4. Does it affect the operation or integrity of the pipeline facility?

If the answer to all four questions is *yes,* then the activity is considered *covered* and must figure into the evaluation process of pipeline personnel.

## *American Petroleum Institute*

The OQ rule and its related covered tasks provided the catalyst to develop a unified pipeline training and assessment program—so that pipeline operators and owners have ready access to the assessment and training tools they need. This effort was the direct result of the vision and leadership provided by The American Petroleum Institute (API) and API members.

Efforts to unify the industry are nothing new to API. Indeed, the impetus for forming the Institute in 1919 was the need to standardize engineering specifications for drilling and production equipment. Today, the most pressing issues revolve around public perceptions and government policies toward the industry. Speaking with one voice on these issues has become as essential as having interchangeable parts in the field. API's members determine the petroleum industry's positions on public policy as well as the standards that govern its day-to-day operations worldwide.

## *Consortium on Operator Qualification*

API assembled a focus group whose primary purpose was to identify covered tasks and provide compliance recommendations to pipeline operators.

The Consortium on Operator Qualification (COOQ) became the guiding force for the identification and development of covered tasks for the liquid pipeline industry. Their work product, API publication 1161, *Guidance Document for the Qualification of Liquid Pipeline Personnel,* formed the basis for the resulting Pipeliner Training and Assessment Program.

The covered task list created in API publication 1161 has now evolved to include gas covered tasks and is maintained by NCCER. The most current covered task list can be downloaded at www.nccer.org.

## *National Center for Construction Education and Research*

With the guidance and support of API, its members, and the Consortium on Operator Qualification, the National Center for Construction Education and Research (NCCER) was able to design and implement the Pipeliner Training and Assessment Program.

NCCER is a not-for-profit 501(c)(3) education foundation established in 1995 by the world's largest and most progressive construction and maintenance contractors and national contractor associations. NCCER was founded to address the

severe workforce shortage facing the construction and maintenance industries and to develop a standardized training process and curricula. Today, NCCER is supported by hundreds of leading construction and maintenance companies, manufacturers, and national associations and developed more than 35 construction and maintenance titles.

NCCER also maintains a National Registry, which contains the records of trainees and instructors who have received NCCER training. All completion information for every trainee is sent to the NCCER and filed within the Registry. The Registry can then confirm training and skills for workers as they move from company to company, state to state, or even within their own company.

### *Standardized Curriculum (NCCER's Contren™ Learning Series)*

NCCER curricula are developed using the expertise of industry subject matter experts (SMEs). The use of SMEs as an integral part of every new development or revision ensures that the curriculum is current, industry-driven, and indicative of industry practices. The development and/or revision process is facilitated from start to finish by NCCER project managers, who work closely with the SMEs and a select team of technical writer(s) on the development and presentation of the curriculum.

### *Pipeline: Five Training Paths*

The pipeline curriculum focuses on essential skills and competencies within five skill areas:

- Corrosion Control
- Electrical and Instrumentation
- Maintenance
- Mechanical
- Operations (Field and Control Center)

SMEs from the pipeline industry have determined the skills and competencies needed based on API's covered tasks and the consensus of the group/industry. Specific company information is generally not incorporated into the curriculum unless the SMEs determine that the information is the industry standard.

### *Training Modules and Covered Tasks*

The smallest unit of instruction in the curriculum is a *module*. Each curriculum is composed of modules (the equivalent of stand-alone chapters) that are logically sequenced into complete *levels* of instruction. Every module is task-oriented and requires the successful completion of a written (knowledge-based) examination and a performance (hands-on) test. Modules may be individually selected and/or combined for efficient task training. Modules in the pipeline curricula, to the extent possible, will align with API-developed covered tasks. While not all modules will relate to covered tasks, all covered tasks will relate to at least one module.

### *Benefits of Training to the Pipeline Marketplace*

Applying NCCER's standardized training process allows pipeline owners and contractors in the United States to establish the benchmark for acceptable skill and competency levels within the pipeline industry. This process will assist training organizations in developing and/or improving standardized pipeline training programs that effectively match their resources and operating conditions while responding to regulatory demands and to the needs of the industry. The process will also recognize training organizations that meet the minimum standards of competence, warranting regulatory, public, and professional confidence. This in turn will foster national unity within the pipeline industry for the mutual benefit of individual technicians, owners, contractors, and ultimately pipeline-users and society.

## Additional Resources

To review the Pipeliner Training and Assessment Program Implementation and Resource Guide and the latest version of pipeline covered tasks, visit NCCER's Web site at www.nccer.org, or contact:

**National Center for Construction Education and Research**
PO Box 141104
Gainesville, FL 32614-1104
(888) 622-3720
Web site: www.nccer.org

For more information on API, please contact:

**American Petroleum Institute**
1220 L Street, Northwest
Washington, D.C. 20005
(202) 682-8000
Web site: www.api.org

To order *API Pub 1161*, contact Global Engineering Documents by calling 1-800-854-7149 or visiting their Web site at www.global.his.com.

Additional sources of information on the pipeline operator qualification regulation include:

**DOT – Office of Pipeline Safety**
www.ops.dot.gov

**Transportation Safety Institute**
www.tsi.dot.gov



Instructor's Notes:

## COURSE MAP

This course map shows all of the modules in the first level of the Pipeline Maintenance curriculum. The suggested training order begins at the bottom and proceeds up. Skill levels increase as you advance on the course map. The local Training Program Sponsor may adjust the training order.

### PIPELINE MAINTENANCE LEVEL ONE

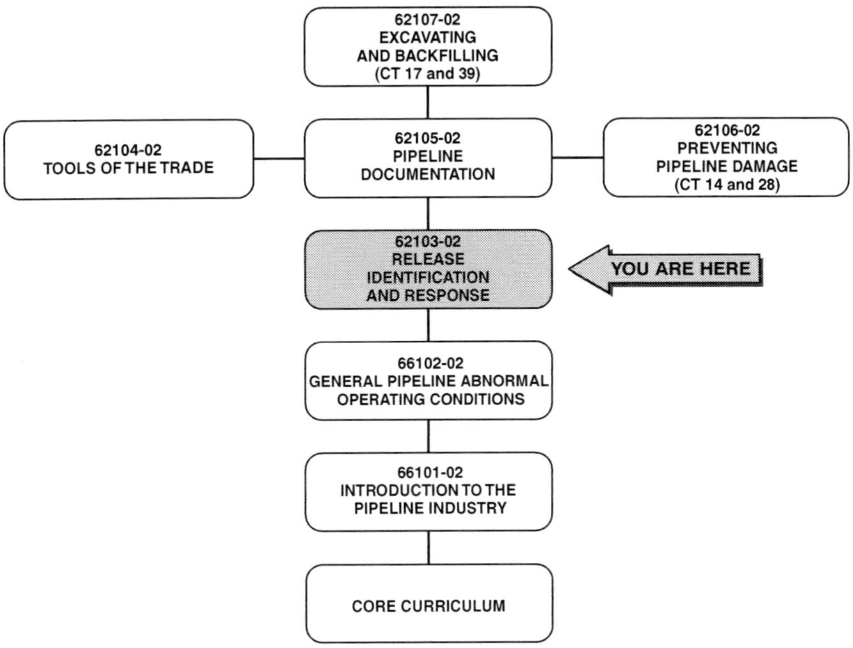


103CMAP.EPS





**Homework**

**Assign reading of Module 62103-02.**

Instructor's Notes:

## MODULE 62103-02 CONTENTS



### Figures

Instructor's Notes:

MODULE 62103-02

# *Release Identification and Response*

### Objectives

When you have completed this module, you will be able to do the following:

1. Describe company environmental manuals, and regulations.
2. Describe United States Coast Guard responsibilities.
3. Describe spill prevention.
4. Recognize soil contamination.
5. Explain spill reporting and spill containment.
6. Explain proper waste handling.

### Prerequisites

Before you begin this module, it is recommended that you successfully complete the following: Core Curriculum; Pipeline Core, Modules 66101 and 66102.

### Required Trainee Materials

1. Pencil and paper
2. Appropriate personal protective equipment

## 1.0.0 ◆ INTRODUCTION

The products that pipelines carry are flammable, explosive, and/or toxic, which means that pipelines could pose a danger to people, animals, and property if these products are released to the environment. The extent of the hazard is a function not only of the characteristics of the products, but also of the pressures at which they are moved and the conditions under which they are released.

Significant quantities of natural gas from a release can ignite and/or explode. However, this gas is lighter than air, and once the release is stopped, the gas may rise and disperse. Crude oil vapors and its products are heavier than air and not so easily removed. Crude oil and petroleum products can contaminate the environment. Petroleum fuels transported in their natural liquid state also pose a fire and pollution hazard. Liquid petroleum gases, such as propane and butane, are transported as liquids under high pressure. If there is a pipeline release, these highly volatile liquids will atomize and vaporize, creating a highly flammable and explosive mixture of gas and mist. This mixture can travel long distances downwind before ignition.

Because of these dangers, federal, state, and local regulations govern nearly all aspects of the pipeline industry. Because of these dangers and because of the regulations, each company in the industry has procedures for preventing and responding to spills.

Every effort is made to prevent pipeline failure and resulting releases. If a release does occur, report it immediately. Immediate action is needed to stop the release and to contain the spill. Since most pipelines are underground, it is important to know how to recognize the signs of releases. Release identification and response are explained in the following sections:

- Company environmental manuals
- **Department of Natural Resources (DNR)** and **Environmental Protection Agency (EPA)** regulations
- National Response Center (NRC) and Coast Guard responsibilities
- Spill prevention
- Soil contamination
- Spill reporting
- Spill containment
- Waste handling



**Materials**

**Ensure that you have everything required to teach the course. Check the Materials and Equipment List at the front of this module.**

**Classroom**

**Explain that the trade terms shown in bold are defined in the Glossary at the end of the module.**

**Teaching Tip**

**See the general Teaching Tip at the back of this module.**

**Audiovisual**

**Show Transparency 1, Course Objectives.**

**Show Transparency 2, Performance Tasks.**

**Classroom**

**Outline the characteristics of pipeline products and their effect when released.**

**Classroom**

**Describe company environmental manuals in general. If appropriate, show your company's manual.**

**Describe how the Oil Spill Response Plan works.**

**Audiovisual**

**Show Transparency 3. Stress the importance of prompt, efficient emergency response.**

**Performance Testing**

**Have each trainee complete Performance Task 1 to your satisfaction. Fill out a Performance Profile Sheet for each trainee.**

**Classroom**

**Discuss how DNRs and the EPA affect the pipeline industry.**

## 2.0.0 ◆ COMPANY ENVIRONMENTAL POLICIES

Company manuals detail the environmental hazards of the company pipeline products and the procedures to be followed for spill prevention and spill response. The company manuals include information similar to that presented in this module, but in more detail and including specific procedures for specific situations that are required to be followed.

Protecting the environment is a critical part of everyday operation. It is important that all employees study the required procedures so that they will know how to do their job in a manner that prevents emergencies and how to react if an emergency occurs.

There are many manuals relating to environmental protection, including the following:

- Oil spill response plans
- Emergency procedures

### 2.1.0 Oil Spill Response Plans

The Oil Spill Response Plan is a plan of action that provides decision-making criteria in the event of an oil spill. This plan satisfies the requirements of *49 CFR Part 194*, also known as the *Oil Pollution Act of 1990 (OPA 90)*. Telephone numbers and addresses of the various response centers and state and federal agencies are listed in the plan.

These plans explain in detail what will be done and by whom in the event of an oil spill. These contain lists of emergency contacts and procedures for all personnel to follow. Required notifications and reports are identified, as are the persons responsible for making the notifications and reports.

### 2.2.0 Emergency Procedures

Emergency procedures will address any emergency situation with the procedures to be followed, the appropriate persons to be contacted, and the proper reports to be made.

An emergency is a situation in which property, the health and safety of human life, and/or the protection of the environment are in jeopardy and prompt corrective action and the summoning of aid are essential. A pipeline emergency is an accident or incident that:

- Endangers life
- Presents a risk to the health and well-being of the public
- Presents a hazard to public and private property
- Pollutes watersheds or bodies of water
- Has the potential to damage the environment
- Endangers wildlife

Company's emergency policies' will typically cover the following topics:

- Receiving report of a potential emergency
- Identifying an emergency
- Classification of an emergency
- Immediate response locations
- Notifying appropriate personnel
- Response to impact of ground surface over the pipeline
- Response to emergencies
- Oil Movement Control Center emergency response
- Incident Command System Organization
- Emergency plan of action
- Post accident review
- Emergency records and documentation
- Analyzing failed components

The foremost consideration and objective of these policies are to take all necessary actions, including evacuation and road/railroad closure, to minimize the public's exposure to injury and potential accidental ignition, to protect the environment, and to establish and maintain the pipeline in a safe condition.

All pipeline personnel must be prepared to act quickly and efficiently when an emergency occurs, making it imperative that they have a thorough understanding of the Emergency Response Procedures Manual and the Oil Spill Response Plan. These manuals cover emergencies and do not preclude using common sense or taking reasonable and prudent actions to mitigate a problem. Employees should always respond in ways that will result in the highest degree of safety. When in doubt about a situation, employees should contact supervision immediately.

## 3.0.0 ◆ DNR AND EPA REGULATIONS

The EPA and appropriate state agencies work together to protect the environment and natural resources of the United States. In some states, the agency is called the Department of Natural Resources (DNR).

To provide the necessary protection, both the EPA and DNRs have established regulations. These regulations affect the pipeline industry.



Instructor's Notes:

### 3.1.0 DNR

Every state has a DNR or similar agency whose goal is to protect, maintain, and encourage wise development of state resources. The DNR in each state manages all state-owned land, water, and natural resources, except for fish and game, on behalf of the people of that state. A state's typical DNR is organized into divisions that specialize in certain areas, such as:

- Agriculture
- Forestry
- Geological and geophysical surveys
- Mining
- Land and water management
- Oil and gas
- Parks and recreation
- Support services

### 3.2.0 EPA

To address the potential threat posed by petroleum and non-petroleum oils, the EPA has established an oil spill program with the goals of preventing, preparing for, and responding to oil spills.

The pipeline industry, using the EPA prevention program, has reduced the number of spills from the total volume handled each year. The EPA has a long history of responding to oil spills, including several major ones, and the lessons learned have helped to improve the country's prevention and response capabilities. The program is administered through the EPA headquarters and the 10 EPA regions.

#### *3.2.1 Prevention*

To prevent oil spills, the EPA requires owners or operators of oil storage facilities to prepare and implement Spill Prevention Control and Countermeasures (SPCC) Plans that detail the facility's spill prevention and control measures. The EPA also enforces the oil spill liability and penalty provisions found in *OPA 90,* which provide incentives to facility owners/operators who take the necessary steps to prevent oil spills. The EPA also conducts on-site facility inspections to ensure that adequate measures are being taken.

#### *3.2.2 Preparation*

Two principles of the EPA's oil spill preparedness program are developing and coordinating contingency plans and conducting oil spill prevention and response training. Well-designed facility, local, area, regional, and national contingency plans assist response personnel in their efforts to contain and clean up any size spill by providing information that the teams will need before, during, and after an oil spill occurs. Training ensures that emergency responders know the what, when, and how of oil spill response.

#### *3.2.3 Response*

Although many spills are contained and cleaned up by the responsible parties, some spills require the assistance of local and state agencies and the federal government. Under the National Contingency Plan, the EPA is the lead federal response agency for spills occurring in inland waters, and the Coast Guard is the lead response agency for spills in coastal waters and deepwater ports.

The EPA tracks all reports of oil spills to United States waters. The EPA learns about the spill through responsible parties, who are required to report the spill to the federal government, or from state and local responders. Once the federal government receives the report, either through the National Response Center (NRC), the EPA, or another agency, it is recorded in the Emergency Response Notification System (ERNS). The ERNS contains historical spill information for the entire country dating from 1986.

## 4.0.0 ◆ NRC AND COAST GUARD RESPONSIBILITIES

The NRC and Coast Guard are closely related. The NRC is staffed by Coast Guard personnel. Both are involved in environmental or public safety emergencies.

### 4.1.0 NRC

The primary function of the NRC is to serve as the sole national point of contact for reporting all oil, chemical, radiological, biological, and etiological discharges into the environment anywhere in the United States and its territories. The reporting call must be made promptly. In addition to gathering and distributing spill data for Federal On-Scene Coordinators (FOSCs) and serving as the communications and operations center for the National Response Team (NRT), the NRC maintains agreements with a variety of federal entities to make additional notifications regarding incidents meeting established trigger criteria.



**Classroom**

**Review areas covered by the DNR. Give trainees contact information for the local DNR.**

**Review the EPA's oil spill program.**

**Describe Spill Prevention Control and Countermeasures (SPCC).**

**Outline the EPA's oil spill preparedness program.**

**Discuss coordination of response by various agencies. Describe the Emergency Response Notification System (ERNS). Ask trainees if they have questions on material covered so far.**

**Homework**

**Assign reading of Sections 4.0.0–6.0.0 for the next class session.**

**Materials**

**Ensure that you have everything required for testing during this session.**

**Classroom**

**Describe the roles of the NRC and Coast Guard in reporting and responding to accidents/incidents.**

**Teaching Tip**

**See the Teaching Tip at the end of this module.**

**Performance Testing**

**Have each trainee perform Performance Task 2 to your satisfaction. Fill out a Performance Profile Sheet for each trainee.**

**Classroom**

**Review the spill prevention measures in maintenance.**

**Review inspection procedures and outline the different approaches (aerial, vehicle, on foot).**

**Stress the importance of being alert to potential problems when working along the pipeline.**

**Audiovisual**

**Show Transparency 4. Describe each safety-related condition and discuss reporting procedures.**

### 4.2.0 Coast Guard

The Coast Guard fields teams to respond to a variety of emergency incidents. Among these are:

- Search and rescue operations
- Law enforcement
- Oil spills
- Hazardous substance releases
- Terrorism
- Marine fires
- Multi-casualty situations

In the event of marine oil spills, the Coast Guard provides law enforcement presence, performs rescue operations as needed, supervises damaged vessel removal, and manages containment and cleanup operations.

## 5.0.0 ◆ SPILL PREVENTION

Spill prevention involves practically every aspect of pipeline construction, operations, and maintenance. Although spill prevention policies are required by regulations, pipeline companies are already very much interested in preventing spills. Spills mean lost production and operations shutdowns. From pipeline design to product delivery, efforts are made to prevent spills.

### 5.1.0 Damage Prevention

Company procedures are documented for every aspect of pipeline operations and maintenance to ensure against damage to the pipeline. Routine inspections and maintenance are targeted for keeping the pipeline operating safely.

Every company will have procedures to prevent damage due to excavation and any subsequent spills. Excavation is the primary outside force causing pipeline damage and spills. These procedures will include contractor listings and information programs, public education, One-Call systems, temporary marking protocols, and special procedures or precautions for boring and crossings by cables or other pipelines.

Specific spill prevention measures in maintenance include the following:

- Routine and thorough inspections
- Continuous observation and awareness
- Recognition of hazards and safety related conditions
- Quick and educated response to problems
- Thorough maintenance

#### *5.1.1 Inspections*

Inspections are routinely carried out along the entire pipeline. These may be done by aerial or vehicle patrol or by foot. The interval between inspections is one to three weeks. These inspections must occur at least 26 times per year. The inspections provide valuable information about what is happening on the pipeline route and may alert the company to impending problems such as encroachment, excavations, or vandalism. Inspectors may also be able to recognize signs of releases and other pipe problems.

The maintenance crew, when in the field for routine or specific maintenance tasks, also does inspections of the site. These inspections cover such things as:

- Condition of warning signs
- Condition of fences
- Soil erosion, settling, and gravel coverage
- Pipeline markers
- Paint conditions
- Chains and locks on valves, gates, and electrical panels
- Signs of releases:
  - Flanged connection
  - Valve gaskets
  - Gear boxes
  - Fittings
  - Plugs
  - Bolts
- Trash and overgrown vegetation
- Signs of vandalism
- Pipeline markers

#### *5.1.2 Observation*

Any time personnel are working along the route in the field or around equipment at facilities, it is good practice to be observant and aware of the surroundings. With experience, workers learn to recognize differences in the sounds and smells of their working environment. These can often give a clue that there is a problem. When looking at the equipment or the ground around the lines, be sure to notice any difference in soil coloring, soil consistency, or vegetation.

#### *5.1.3 Recognition*

Recognizing hazards and safety-related conditions is a way to protect the environment and everyone's safety. To learn to do this, study these



*Instructor's Notes:*

and other modules associated with the hazards of pipeline products, learn the information provided by the company on environmental and health hazards, and practice using this knowledge at work.

All personnel performing operation and maintenance activities should be aware of conditions that potentially are safety related. The following are examples of conditions that, when observed, must be immediately reported to supervision:

- General corrosion
- Localized corrosion pitting
- Corrosion
- Uncontrolled movement or abnormal loading of the pipe by environmental causes
- Material defect or physical damage
- Malfunctions or operating errors
- Releases
- Events impacting the cover over the pipeline
- Conditions that could lead to imminent hazard

#### 5.1.4 Response

Response to problems on the pipeline must be quick and done correctly. If the first person aware of the problem is not trained in how to respond, then his or her job is to inform someone who is. It is usually not a good idea, nor is it approved by the company, to make a stab at solving a problem you don't understand.

#### 5.1.5 Maintenance

Pipeline maintenance is critical to prevent spills. Regulations require regular maintenance of critical parts. Companies, in compliance with these regulations, set schedules and provide specific maintenance procedures. The schedules are often monitored by the company computer control system, which automatically issues work orders and lists the required tasks at the appropriate times. Maintenance workers check these schedules and work orders daily.

## 6.0.0 ◆ SOIL CONTAMINATION

Sometimes the first sign on the surface that there is a problem with the pipeline is soil contamination. Employees working on the pipeline route must be familiar with the signs of soil contamination and be observant of conditions when on site. Some general guidelines for locating contamination from an underground release are:

- Wilted or dead vegetation
- Discolored soil
- Hydrocarbon liquid or frost on the surface
- Discolored or melted snow
- Hydrocarbon smell
- Vapor clouds or shimmering in the air above the surface
- Sounds of blowing, whistling, or trickling

Surface contamination is identified by the presence of free product on the ground, as a pool of petroleum or as a sheen or film on water. Oil and less refined products are black or dirty brown and viscous; they produce a blackish sheen on water and will form a consistent sheen on the surface, even after being disturbed. Gasoline produces a multi-colored sheen on water, representing the refined hydrocarbons and additives. Most petroleum products float on water, but some components of gasoline may mix with the water molecules and be transported with the water.

The most likely indication of ground contamination is stained and odorous soil. Odor may also be present without visible staining. Vegetation around the contamination site may be affected. In some cases of gas releases, insects are drawn to the site.

## 7.0.0 ◆ SPILL REPORTING

The company's environmental and spill response procedures should be reviewed carefully so that appropriate response to spills can be followed right away. Because releases may occur at any time, personnel must be alert to signs indicating a loss of oil or product. Releases may be detected by pipeline monitoring equipment, personnel surveillance, outside parties, pressure changes, or computerized volume balancing. Each emergency situation involving a spill or release may be different; therefore, all employees must be familiar with and trained to respond according to company procedures.

When a spill does occur and is discovered, it must be reported immediately to the control center. Management will initiate response to the spill and be responsible for documenting any accidents and safety-related conditions, as well as reporting to regulatory agencies.

Spills on federal lands must also be reported to the Bureau of Land Management. State and local agencies will also require reports and should be listed in the company manual. Other issues of spill reporting include the following:

- Written reports
- Telephonic notice
- Safety-related conditions reports



**Classroom**

**Stress the importance of rapid and efficient responses to problems.**

**Outline company maintenance procedures. Emphasize the importance of following company procedures and schedules.**

**Ask trainees for examples of guidelines for detecting soil contamination. Write them on the board or on a blank transparency. Ask trainees if they have questions on the material covered so far.**

**Homework**

**Assign reading of Sections 7.0.0–9.0.0 and completion of the Review Questions for the next class session.**

**Materials**

**Ensure that you have everything required for testing during this session.**

**Classroom**

**Outline the various spill reports (written, telephonic, and safety-related conditions). Stress the importance of immediate, accurate reporting.**

**Performance Testing**

**Have each trainee perform Performance Task 3 to your satisfaction. Fill out a Performance Profile Sheet for each trainee.**

**Classroom**

**Refer trainees to Figure 1. Describe how and by whom the accident report is completed.**

**Refer trainees to Figure 2. Give examples of when telephonic notification is required.**

**Discuss the conditions that trigger safety-related conditions reports.**

**Ask a trainee to state the two main areas of concern in spill containment.**

**Describe the controls and methods used to minimize the amount of oil spilled.**

### 7.1.0 Written Reports

All reportable accidents must be reported to the DOT in writing as soon as possible, but not later than 30 days after discovery of the accident. The written report must be filed on DOT form 7000-1 or a facsimile (*Figure 1*). A company representative will collect all relevant information and make this available to the regulatory compliance personnel, who will file the written report. Supplemental reports may be filed if the information changes.

Any failure in the pipeline system in which there is a release of a hazardous liquid resulting in any of the following must also be reported to the appropriate regulatory agency:

- Explosion or fire not intentionally set by the operator
- Loss of five or more barrels of liquid
- Escape to the atmosphere of more than five barrels a day of highly volatile liquids
- Death of any person
- Bodily harm to any person resulting in loss of consciousness, having to be carried away, needing medical treatment, or disability beyond the day of occurrence
- Estimated property damage exceeding $50,000

### 7.2.0 Telephonic Notice

Certain accidents must be reported to the NRC at the earliest possible time – within two hours after discovery unless circumstances prevent it. Because these accidents must be reported quickly, the manager or operations or control room dispatcher will telephone it in to the NRC. To facilitate this, the NRC has a telephonic notice report form that lists information NRC will ask for when an accident is reported (see *Figure 2*).

Telephonic notification must be made for any failure or release that meets any of the following:

- Results in a death or an injury requiring hospitalization
- Results in either an explosion or a fire not intentionally set by the operator
- Results in pollution of any body of water that discolors the surface of the water or shoreline or deposits sludge or emulsion
- Is significant in the judgment of the operator

### 7.3.0 Safety-Related Conditions Reports

Safety-related conditions must be reported if they occur on an active pipeline or pipeline facility that is:

- Within 220 yards of any building of human occupancy or assembly
- Within the right-of-way of an active railroad, paved road, street, or highway
- Offshore
- Where a spill could reasonably be expected to pollute a body of water

The report must be filed within five working days after the condition is determined, but no less than ten days after first discovery, to the DOT and relevant state and local agencies. Reportable safety-related conditions are listed in a previous section, but should be reviewed in more detail in the company manuals.

## 8.0.0 ◆ SPILL CONTAINMENT

When a spill does occur, the two main areas of concern are minimizing the amount and controlling the spill.

### 8.1.0 Minimizing the Amount

Several actions may be taken to minimize the amount of oil spilled through pipeline releases. Often stopping the pipeline pumps stops the spill. Suction pumps can be used to assist gravity. Such pumps are especially effective where elevation differences are small compared with the total lift available. When releases occur at low points on the hydraulic profile, static pressures usually cause continued pressure and spillage.

Controls at automatic pump stations are usually set to shut down the stations and signal an alarm when changes in suction or discharge pressures occur. Block valves on one or both sides of the release are closed to stop flow. Portable pumps and bypass fittings can be used to reconnect the flow around the release to keep the pipeline operating. When internal pressure has been reduced at the site of the release, the outflow is stopped by temporarily closing the hole with clamps or pegs.

The controls and valves designed to shut down flow are required to be kept in operating condition and to be inspected regularly by the DOT. Records of inspections and tests are required to be maintained.



Instructor's Notes:

NOTICE: This report is required by 49 CFR Part 195. Failure to report can result in a civil penalty not to exceed $25,000 for each violation for each day that such violation persists except that the maximum civil penalty shall not exceed $500,000 as provided in 49 USC 60122

Form Approved OMB No. 2137-0047

U.S. Department of Transportation
Research and Special Programs Administration

**ACCIDENT REPORT – HAZARDOUS LIQUID PIPELINE SYSTEMS**

Report Date ______________

No. ______________ (DOT Use Only)

**INSTRUCTIONS**

***Important:*** *Please read the separate instructions for completing this form before you begin. They clarify the information requested and provide specific examples. If you do not have a copy of the instructions, you can obtain one from the Office Of Pipeline Safety Web Page at http://ops.dot.gov.*

**PART A – GENERAL REPORT INFORMATION** Check: ☐ **Original Report** ☐ **Supplemental Report** ☐ **Final Report**

1. a. Operator's OPS 5-digit Identification Number (if known) /__/__/__/__/__/
2. b. If Operator does not own the pipeline, enter Owner's OPS 5-digit Identification Number (if known) /__/__/__/__/__/
   c. Name of Operator ______________
   d. Operator street address ______________
   e. Operator address ______________ City, County, State and Zip Code

**IMPORTANT: IF THE SPILL IS SMALL, THAT IS, THE AMOUNT IS AT LEAST 5 GALLONS BUT IS LESS THAN 5 BARRELS, COMPLETE THIS PAGE ONLY, UNLESS THE SPILL IS TO WATER AS DESCRIBED IN 49 CFR §195.52(A)(4) OR IS OTHERWISE REPORTABLE UNDER §195.50 AS REVISED IN CY 2001.**

2. Time and date of the accident

/__/__/__/__/ hr. /__/__/ month /__/__/ day /__/__/ year

3. Location of accident
*(If offshore, do not complete a through d. See Part C.1)*

a. Latitude: ________ Longitude: ________
*(if not available, see instructions for how to provide specific location)*

b. ______________ City, and County or Parish

c. ______________ State and Zip Code

d. Mile post/valve station O or survey station no. O
(whichever gives more accurate location)
______________

4. Telephone report

/__/__/__/__/__/__/ NRC Report Number /__/__/ month /__/__/ day /__/__/ year

5. Losses *(Estimated)*

**Public/Community Losses reimbursed by operator:**

Public/private property damage $__________
Cost of emergency response phase $__________
Cost of environmental remediation $__________
Other Costs $__________
(describe) ______________

**Operator Losses:**

Value of product lost $__________
Value of operator property damage $__________
Other Costs $__________
(describe) ______________

**Total Costs** $__________

6. Commodity Spilled O Yes O No
(If Yes, complete Parts a through c where applicable)
a. Name of commodity spilled ______________
b. Classification of commodity spilled:
O HVLs /other flammable or toxic fluid which is a gas at ambient conditions
O $CO_2$ or other non-flammable, non-toxic fluid which is a gas at ambient conditions
O Gasoline, diesel, fuel oil or other petroleum product which is a liquid at ambient conditions
O Crude oil

**c. Estimated amount of commodity involved :**
**O Barrels**
**O Gallons (check only if spill is less than one barrel)**

**Amounts:**
**Spilled :** __________
**Recovered:** __________

**CAUSES FOR SMALL SPILLS ONLY (5 gallons to under 5 barrels) :** ***(For large spills [5 barrels or greater] see Part H)***

O **Corrosion** O **Natural Forces** O **Excavation Damage** O **Other Outside Force Damage**
O **Material and/or Weld Failures** O **Equipment** O **Incorrect Operation** O **Other**

**PART B – PREPARER AND AUTHORIZED SIGNATURE**

______________ (type or print) Preparer's Name and Title ______________ Area Code and Telephone Number

______________ Preparer's E-mail Address ______________ Area Code and Facsimile Number

______________ Authorized Signature ______________ (type or print) Name and Title ______ Date ______________ Area Code and Telephone Number

Form RSPA F 7000-1 ( 01-2001 ) *Page 1 of 4*

*Figure 1* ◆ DOT report form (page 1 of 4).

103F01A.PDF

**PART C – ORIGIN OF THE ACCIDENT (Check all that apply)**

1. Additional location information
   a. Line segment name or ID ____________
   b. Accident on Federal land other than Outer Continental Shelf O Yes O No
   c. Is pipeline interstate? O Yes O No

Offshore: O Yes O No *(complete d if offshore)*

d. Area ____________ Block # ____________

State / / / or Outer Continental Shelf ☐

2. Location of system involved (*check all that apply*)
   ☐ Operator's Property
   ☐ Pipeline Right of Way
   ☐ High Consequence Area (HCA)?
   Describe HCA____________

3. Part of system involved in accident
   O Above Ground Storage Tank
   O Cavern or other below ground storage facility
   O Pump/meter station; terminal/tank farm piping and equipment, including sumps
   O Other Specify: ____________
   O Onshore **pipeline**, including valve sites
   O Offshore **pipeline**, including platforms

*If failure occurred on **Pipeline**, complete items a - g:*

a. Type of leak or rupture
   OLeak: O Pinhole O Connection Failure *(complete sec. H5)*
   O Puncture, diameter *(inches)* ____________
   ORupture: O Circumferential – Separation
   O Longitudinal – Tear/Crack, length *(inches)* ____________
   Propagation Length, total, both sides *(feet)* ____________
   ON/A
   OOther ____________

b. Type of block valve used for isolation of immediate section:
   Upstream: ☐ Manual ☐ Automatic ☐ Remote Control ☐ Check Valve
   Downstream: ☐ Manual ☐ Automatic ☐ Remote Control ☐ Check Valve

c. Length of segment isolated ______ ft

d. Distance between valves ______ ft

e. Is segment configured for internal inspection tools? OYes O No

f. Had there been an in-line inspection device run at the point of failure? O Yes O No O Don't Know
   O Not Possible due to physical constraints in the system

g. If Yes, type of device run (*check all that apply*)

| | |
|---|---|
| ☐ High Resolution Magnetic Flux tool | Year run: ______ |
| ☐ Low Resolution Magnetic Flux tool | Year run: ______ |
| ☐ UT tool | Year run: ______ |
| ☐ Geometry tool | Year run: ______ |
| ☐ Caliper tool | Year run: ______ |
| ☐ Crack tool | Year run: ______ |
| ☐ Hard Spot tool | Year run: ______ |
| ☐ Other tool | Year run: ______ |

4. Failure occurred on

| | | |
|---|---|---|
| O Body of Pipe | O Pipe Seam | O Scraper Trap |
| O Pump | O Sump | O Joint |
| O Component | O Valve | O Metering Facility |
| O Repair Sleeve | O Welded Fitting | O Bolted Fitting |
| O Girth Weld | | |

Other *(specify)* ____________

Year the component that failed was installed: / / / / /

5. Maximum operating pressure *(MOP)*
   a. Estimated pressure at point and time of accident: ______ *PSIG*
   b. MOP at time of accident: ______ *PSIG*
   c. Did an overpressurization occur relating to the accident? OYes O No

**PART D – MATERIAL SPECIFICATION**

1. Nominal pipe size *(NPS)* / / / / / in.
2. Wall thickness / / / / / in.
3. Specification ____________ SMYS / / / / / / /
4. Seam type ____________
5. Valve type ____________
6. Manufactured by ____________ in year / / / / /

**PART E – ENVIRONMENT**

1. Area of accident
   O Under pavement O In open ditch
   O Underground O Above ground
   O Inside/under building O Under water
   O Other ____________
2. Depth of cover: ____________ inches

**PART F – CONSEQUENCES**

**1. Consequences** ***(check and complete all that apply)***

a.

| | Fatalities | Injuries |
|---|---|---|
| Number of operator employees: | | ______ |
| Contractor employees working for operator: | | ______ |
| General public: | | ______ |
| **Totals:** | | ______ |

b. Was pipeline/segment shutdown due to leak? OYes O No
   If Yes, how long? ______ days ______ hours ______ minutes

c. Product ignited OYes O No

d. Explosion OYes O No

e. ☐ Evacuation *(general public only)* / / / / / people
   Reason for Evacuation:
   O Precautionary by company
   O Evacuation required or initiated by public official

f. Elapsed time until area was made safe: / / / hr. / / / min.

**2. Environmental Impact**

a. Wildlife Impact: Fish/aquatic O Yes O No
   Birds O Yes O No
   Terrestrial O Yes O No

b. Soil Contamination O Yes O No
   If Yes, estimated number of cubic yards: ____________

c. Long term impact assessment performed: O Yes O No

d. Anticipated remediation O Yes O No
   If Yes, check all that apply: ☐ Surface water ☐ Groundwater ☐ Soil ☐ Vegetation ☐ Wildlife

e. Water Contamination: O Yes O No *(If Yes, provide the following)*
   Amount in water ______ barrels
   Ocean/Seawater O No O Yes
   Surface O No O Yes
   Groundwater O No O Yes
   Drinking water O No O Yes *(If Yes, check below.)*
   O Private well O Public water intake

Form RSPA F 7000-1 ( 01-2001 ) *Page 2 of 4*

103F01B.PDF

*Figure 1* ◆ DOT report form (page 2 of 4).



Instructor's Notes:

**PART G – LEAK DETECTION INFORMATION**

1. Computer based leak detection capability in place? O Yes O No
2. Was the release initially detected by? (check one): O CPM/SCADA-based system with leak detection O Static shut-in test or other pressure or leak test O Local operating personnel, procedures or equipment O Remote operating personnel, including controllers O Air patrol or ground surveillance O A third party O Other (specify) ________
3. Estimated leak duration days ____ hours ____

**PART H – APPARENT CAUSE**

***Important:*** *There are 25 numbered causes in this Part H. Check the box corresponding to the primary cause of the accident. Check one circle in each of the supplemental categories corresponding to the cause you indicate. See the instructions for guidance.*

**H1 – CORROSION**

1. ☐ External Corrosion
2. ☐ Internal Corrosion

(Complete items a – e where applicable.)

a. Pipe Coating: O Bare O Coated

b. Visual Examination: O Localized Pitting O General Corrosion O Other ________

c. Cause of Corrosion: O Galvanic O Atmospheric O Stray Current O Microbiological O Cathodic Protection Disrupted O Stress Corrosion Cracking O Selective Seam Corrosion O Other ________

d. Was corroded part of pipeline considered to be under cathodic protection prior to discovering accident? O No O Yes, Year Protection Started: _/_/_/_/

e. Was pipe previously damaged in the area of corrosion? O No O Yes ⇒ Estimated time prior to accident: _/_/ *years* _/_/ *months* *Unknown* ☐

**H2 – NATURAL FORCES**

3. ☐ Earth Movement ⇒ O Earthquake O Subsidence O Landslide O Other ________
4. ☐ Lightning
5. ☐ Heavy Rains/Floods ⇒ O Washouts O Flotation O Mudslide O Scouring O Other ________
6. ☐ Temperature ⇒ O Thermal stress O Frost heave O Frozen components O Other ________
7. ☐ High Winds

**H3 – EXCAVATION DAMAGE**

8. ☐ Operator Excavation Damage (including their contractors/Not Third Party)
9. ☐ Third Party *(complete a-f)*
   - a. Excavator group: O General Public O Government O Excavator other than Operator/subcontractor
   - b. Type: O Road Work O Pipeline O Water O Electric O Sewer O Phone/Cable O Landowner-not farming related O Farming O Railroad O Other liquid or gas transmission pipeline operator or their contractor O Nautical Operations O Other ________
   - c. Excavation was: O Open Trench O Sub-strata (boring, directional drilling, etc...)
   - d. Excavation was an ongoing activity (Month or longer) O Yes O No If Yes, Date of last contact _/_/_/
   - e. Did operator get prior notification of excavation activity? O Yes; Date received: _/_/ mo. _/_/ day _/_/_/_/ yr. O No — Notification received from: O One Call System O Excavator O Contractor O Landowner
   - f. Was pipeline marked as result of location request for excavation? O No O Yes *(If Yes, check applicable items i - iv)*
     - i. Temporary markings: O Flags O Stakes O Paint
     - ii. Permanent markings: O
     - iii. Marks were (check one): O Accurate O Not Accurate
     - iv. Were marks made within required time? O Yes O No

**H4 – OTHER OUTSIDE FORCE DAMAGE**

10. ☐ Fire/Explosion as primary cause of failure ⇒ Fire/Explosion cause: O Man made O Natural
11. ☐ Car, truck or other vehicle not relating to excavation activity damaging pipe
12. ☐ Rupture of Previously Damaged Pipe
13. ☐ Vandalism

Form RSPA F 7000-1 ( 01-2001 ) *Page 3 of 4*

103F01C.PDF

*Figure 1* ◆ DOT report form (page 3 of 4).

**H5 – MATERIAL AND/OR WELD FAILURES**

**Material**

14. ☐ Body of Pipe ⇒ ○ Dent ○ Gouge ○ Bend ○ Arc Burn ○ Other ________
15. ☐ Component ⇒ ○ Valve ○ Fitting ○ Vessel ○ Extruded Outlet ○ Other ________
16. ☐ Joint ⇒ ○ Gasket ○ O-Ring ○ Threads ○ Other ________

**Weld**

17. ☐ Butt ⇒ ○ Pipe ○ Fabrication ○ Other ________
18. ☐ Fillet ⇒ ○ Branch ○ Hot Tap ○ Fitting ○ Repair Sleeve ○ Other ________
19. ☐ Pipe Seam ⇒ ○ LF ERW ○ DSAW ○ Seamless ○ Flash Weld
○ HF ERW ○ SAW ○ Spiral ○ Other ________

*Complete a-g if you indicate **any** cause in part H5.*

a. Type of failure:
○ Construction Defect ⇒ ○ Poor Workmanship ○ Procedure not followed ○ Poor Construction Procedures
○ Material Defect

b. Was failure due to pipe damage sustained in transportation to the construction or fabrication site? ○ Yes ○ No

c. Was part which leaked pressure tested before accident occurred? ○ Yes, *complete d-g* ○ No

d. Date of test: /__/__/__/__/ yr. /__/__/ mo. /__/__/ day

e. Test medium: ○ Water ○ Inert Gas ○ Other ________

f. Time held at test pressure: /__/__/ hr.

g. Estimated test pressure at point of accident: ________ *PSIG*

**H6 – EQUIPMENT**

20. ☐ Malfunction of Control/Relief Equipment ⇒ ○ Control valve ○ Instrumentation ○ SCADA ○ Communications
○ Block valve ○ Relief valve ○ Power failure ○ Other ________
21. ☐ Threads Stripped, Broken Pipe Coupling ⇒ ○ Nipples ○ Valve Threads ○ Dresser Couplings ○ Other ________
22. ☐ Seal Failure ⇒ ○ Gasket ○ O-Ring ○ Seal/Pump Packing ○ Other ________

**H7 – INCORRECT OPERATION**

23. ☐ Incorrect Operation

a. Type: ○ Inadequate Procedures ○ Inadequate Safety Practices ○ Failure to Follow Procedures
○ Other ________

b. Number of employees involved who failed a post-accident test: drug test: /__/__/__/ alcohol test /__/__/__/

**H8 – OTHER**

24. ☐ Miscellaneous, *describe:* ________
25. ☐ Unknown
○ Investigation Complete ○ Still Under Investigation *(submit a supplemental report when investigation is complete)*

**PART I – NARRATIVE DESCRIPTION OF FACTORS CONTRIBUTING TO THE EVENT** *(Attach additional sheets as necessary)*

Form RSPA F 7000-1 (01-2001) *Page 4 of 4*

103F01D.PDF

*Figure 1* ◆ DOT report form (page 4 of 4).



Instructor's Notes:

**NATIONAL RESPONSE CENTER**
**TELEPHONIC NOTIFICATION**
Response Information for Reporting

**NRC Call/Case #** ____________

The following format was developed from the computer screens the National Response Center (NRC) watchstanders will be using when they question you regarding your telephone notification of a reportable incident. You should have all the information below as complete as possible before making your telephone report to the NRC. You will be asked these questions. You have two (2) hours from the time the incident was discovered to obtain the information and make telephone notification to the NRC. All information reported to the NRC shall be stated as factually as possible. If all the information has not been obtained within the required notification time limit, be frank and qualify your statements as not yet available. Be sure to describe all specific values stated as preliminary and approximate pending verification.

**NRC TELEPHONE REPORT INFORMATION**

**1. REPORTING PARTY (You)**

Last Name: ____________
First Name: ____________ MI ______
Contact Phones: 1-800-322-PIPE (7473)
1-800-664-5365
Company: All American Pipeline Company
Org. Type: Interstate Crude Oil Pipeline Operator
Position (yours): ____________
Address: All American Pipeline Company
10000 Ming Ave.
City: Bakersfield
State: California Zip: 93311

**2. SUSPECTED RESPONSIBLE PARTY**

Last Name: ____________
First Name: ____________ MI ______
Contact Phones: ____________
Company: ____________
Org. Type: ____________
Position: ____________
Address: ____________
City: ____________
State: ____________ Zip: ____________

Was Material Released? ☐ Yes ☐ No — Are you calling for the responsible party? ☐ Yes ☐ No

Is this telephone report to meet Federal obligation ☐ Yes ☐ No — FEDERAL CODE REQUIREMENT: 49 CFR 195.52

Crude Oil Spill release resulting in: *(check the box(es) that apply)*
☐ An explosion or fire ☐ A death or injury requiring hospitalization ☐ Property damage exceeding $50,000
☐ Polluted water ☐ Judged to be a significant incident ☐ ____________

**3. INCIDENT DESCRIPTION**

Source and/or Cause of Incident *(Check the box that apply)*
☐ External Source (Third Party) ☐ Material Failure ☐ Equipment Malfunction ☐ Operator Error
☐ Unknown (Describe) ____________

Discovery:
Date: ________ Time: (24hr.) ________

NRC Notified
Date: ________ Time (24hr.) ________

Incident Address/Location
Nearest City ____________
State ____________
County ____________
Zip ____________
T-S-R-Q ____________

Distance from city ____________
Direction from city ____________
Pipeline Milepost ____________
One Call System Area ID ____________

**4. MATERIAL**

Classification of material released ____________ Approximate Quantity ____________ BBL(s)

**5. REMEDIAL ACTION**

☐ Containment ☐ Recover ☐ Re-injection into system ☐ Evaluation ☐ Clean-up ☐ Disposal
☐ Current Status

**6. IMPACT**

Number of Injuries ______ Were there evacuations? ☐ Yes ☐ No Number Evacuated ______
Number of Fatalities ______ Was there damage? ☐ Yes ☐ No Est. Damage $ ______

**7. NOTIFICATION OF OTHER AGENCIES**

☐ CA OES ☐ NM Oil Conservation Division ☐ BLM ☐ Fire Department ☐ Highway Patrol
☐ AZ DPS ☐ Railroad Commission of Texas ☐ Sheriff ☐ ____________

Name of person reporting to NRC — NRC Watchstander taking report — Date

Retention: Life of System



103F02.EPS

*Figure 2* ◆ Telephonic notice report form.

**Classroom**

**Describe how dams, pumps, and floating booms are used to control spills. Ask trainees for reasons why ocean spills are difficult to contain.**

**Teaching Tip**

**See the Teaching Tip at the end of this module.**

**Classroom**

**If appropriate, discuss how your company handles pipeline facility waste. Stress the importance of proper waste disposal.**

**Performance Testing**

**Have each trainee perform Performance Task 4 to your satisfaction. Fill out a Performance Profile Sheet for each trainee.**

**Classroom**

**Go over the Review Questions. Encourage trainees to ask questions. Have trainees prepare for the Module Examination.**

### 8.2.0 Controlling the Spill

If the quantity of the spill is sufficient to flood the area, measures must be taken to control the spread. Dikes or dams can be thrown up quickly by bulldozers or other earth-moving equipment. Damming small, usually dry, watercourses or ditches provides catch basins. Sometimes it is heavy equipment working near the line that causes the release, and they can be used in controlling it.

Truck or trailer-mounted pumps can be used to pick up large quantities of oil and transfer it to places for safe recovery or disposal. Pneumatic or hydraulic pumps should be used instead of gasoline pumps to avoid igniting oil vapors.

When oil is spilled into and carried on the surface of water, it can be contained by floating booms or other surface barriers. Dams with low-level outlets below the oil can be used to accommodate the water flow. In large bodies of water, floating barriers are used to encircle the oil source or to trap wind-blown oil. The entrapped oil is then picked up by pumps. Spills in the oceans are complicated by wind, waves, water currents, water depths, and the great areas involved, and these require individualized solutions.

## 9.0.0 ◆ WASTE HANDLING

Waste and debris around pipeline facilities, including the right-of-way, must always be properly disposed. Any waste that may have been contaminated with petroleum liquids must receive particular attention for disposal.

Facilities may have incinerators for disposing of such items. They are designed to burn small amounts of highly flammable material at a time, including oily rags, sorbent pads, and drained oil filters. They will also burn punctured aerosol cans, but not liquids. Specific instructions will be provided for loading and burning refuse in the incinerators.

### Summary

An important part of any job on the pipeline is the ability to recognize hazards. The effects of the pipeline products can be harmful to people, wildlife, and vegetation.

Because of the hazards, company environmental manuals contain detailed procedures to teach workers how to help prevent hazards from becoming problems and how to react if problems occur. Each pipeline company is required to develop and maintain a SPCC to be ready to react immediately when necessary. Federal and state agencies, such as the EPA, have provided guidelines and regulations for pipeline operation and maintenance to help prevent spills and to contain any spills that do occur. The NRC and the Coast Guard are trained and ready to respond to pipeline emergencies as needed.

Every facet of pipeline construction, operation, and maintenance is concerned with preventing spills in the first place, but they do still sometimes happen. Maintenance personnel are likely to be the first to spot problems. During routine maintenance inspections or in the field on specific investigations, each person must remain alert to the signs that a problem exists or is possible.

When a spill occurs, the first steps are always to report and contain it if possible while ensuring the safety of the employees and the public. Containment procedures vary according to the situation, but are outlined and planned ahead of time to save time and prevent mistakes when needed. Employees are trained ahead of time on how to react for the same reason.



Instructor's Notes:

**Review Questions**

1. Company manuals will detail the _____ of the company pipeline products.
   a. chemical make-up
   b. environmental hazards
   c. user applications
   d. expected volumes

2. The _____ is a plan of action that provides decision-making criteria in the event of an oil spill.
   a. Oil Pollution Act of 1960
   b. Oil Spill Response Plan
   c. company manual
   d. Incident Command Organization

3. Employees should always respond in ways that will result in _____.
   a. the highest degree of safety
   b. meeting regulations
   c. the fastest evacuation
   d. the best public relations

4. The pipeline industry, using the _____ prevention program, has reduced the number of spills.
   a. DNR
   b. NRC
   c. EPA
   d. Coast Guard

5. In the event of marine oil spills, the Coast Guard provides _____.
   a. cleanup crews
   b. vessels
   c. pipeline repairs
   d. law enforcement presence

6. With experience, workers learn to recognize differences in the _____ of their working environment.
   a. climate
   b. sounds and smells
   c. requirements
   d. geography

7. Regulations require that maintenance of critical parts be done _____.
   a. regularly
   b. weekly
   c. when repairs are needed
   d. monthly

8. The most likely indication of ground contamination is _____.
   a. gathering of insects
   b. melted snow
   c. dead wildlife
   d. soil stain and/or odor

9. When releases occur at low points on the hydraulic profile, static pressures usually cause _____.
   a. spillage to increase
   b. continued pressure and spillage
   c. pumps to stop the flow
   d. controls to shut down

10. Waste and debris around pipeline facilities must always be _____.
    a. buried
    b. recycled
    c. burned
    d. properly disposed of



## Examination

**Administer the Module Examination. Be sure to record the results of the Exam on Craft Training Report Form 200 and submit the results to the Training Program Sponsor.**

## Performance Testing

**Ensure that all Performance Tasks have been completed and Performance Profile Sheets for each task for each trainee are filled out. Be sure to record the results of the testing on Craft Training Report 200 and submit the results to the Training Program Sponsor.**

GLOSSARY

# *Trade Terms Introduced in This Module*

**Department of Natural Resources (DNR):** Departments in state governments that supervise the use of the state's natural resources.

**Environmental Protection Agency (EPA):** Supervises and controls activities that affect the quality of the environment.



Instructor's Notes:

# MODULE 62103-02 — TEACHING TIPS

The following are suggested activities or instructional methods to help you teach the material in this AIG.

## General

When you call on someone to answer a question, the rest of the class relaxes or even tunes out because they expect that the question and answer will take place only between you and the trainee you called on. Instead, use this technique to involve more trainees in answering questions and to keep them on their toes.

1. Ask trainees to define a term or explain a concept.
2. After one trainee has answered, ask a trainee seated nearby if the answer is right. Then ask whether a trainee in the back of the room agrees.
3. Ask trainees to explain why they think an answer is right or wrong.
4. Use the session to clear up incorrect ideas and encourage trainees to learn from their mistakes.

## Section 5.0.0 *Spill Prevention*

This activity is designed to help trainees relate their reading to real-life situations. You will need blank transparency sheets and copies of news headlines about pipeline accidents/incidents. You can get them from back issues of magazines or newspapers at your local library or from the Internet. Trainees will need paper and pencils to take notes for discussion. As an alternative, you may wish to ask trainees to look up news reports of accidents/incidents as a homework assignment. In this case, have each trainee bring a copy of one article to class to report from during the safety discussion. Allow at least 60 minutes for this exercise.

1. Copy the headline excerpts onto blank transparency sheets. (Make as many transparencies as you think useful.)
2. Use the enlargement feature on a photocopier to make the headlines large enough to be seen clearly from a distance.
3. Use your knowledge of the reported accidents/incidents to do the following:
   - Underscore the importance of safety in the pipeline industry.
   - Describe real-life examples of the various government agencies involved in action.
   - Demonstrate the impact on the environment, property, and the public.
   - If appropriate, relate an accident/incident from your own experience.

## Section 8.2.0 *Controlling the Spill*

This optional activity allows you to demonstrate in a small way the difficulties of containing and cleaning up oil spills. You will need the following items:

- Plastic painter's drop cloth
- Small plastic box or cardboard box lined with heavy-duty plastic wrap
- Small bag of sand mixed with garden soil and rocks
- Garden trowel
- Wide, shallow plastic bowl
- Bottle of water
- Flexible plastic tubing
- Small bottle of automobile oil
- Plastic trash bags
- Paper towels

Trainees will need to wear appropriate clothing. You may wish to practice this exercise before demonstrating it in the classroom. Allow at least 60 minutes for this exercise.

1. Cover the demonstration table with the painter's drop cloth to protect the surface against an accidental spill. Partially fill the lined shoebox or plastic box with the sand/soil/rock mixture. Carefully pour a small pool of the automobile oil onto the soil. Point out how the oil seeps into the soil.
2. Using the garden trowel, demonstrate how a dam can be constructed to contain the pool. (The oil will likely sink into the dirt before you have a chance to construct the dam. Use this fact to demonstrate the importance of quickly limiting and containing actual spills.)
3. Partially fill the bowl with water. (Keep the water bottle for use in clean-up.) Bend the tubing into a ring large enough to contain an oil spill in the bowl. Carefully pour some of the oil into the water. Point out how the oil floats on the surface. Set the flexible tubing ring onto the water to demonstrate how a surface barrier can contain the spill. Discuss the tools and methods used to clean the oil from the water.
4. Ask trainees to observe how difficult it is to contain and clean up even a small spill. Ask them to consider the effects of a large spill on land near a populated area, on a large lake, or in the ocean.
5. Ask trainees how you should dispose of the oily water and sand. Use the paper towels and trash bag for clean-up. Dispose of the materials used in this exercise responsibly by taking them to your local hazardous waste recycling station.

## MODULE 62103-02 — ANSWERS TO REVIEW QUESTIONS

| Answer | Section |
|---|---|
| **1.** b | 2.0.0 |
| **2.** b | 2.1.0 |
| **3.** a | 2.2.0 |
| **4.** c | 3.2.0 |
| **5.** d | 4.2.0 |
| **6.** b | 5.1.2 |
| **7.** a | 5.1.5 |
| **8.** d | 6.0.0 |
| **9.** b | 8.1.0 |
| **10.** d | 9.0.0 |

## CONTREN™ LEARNING SERIES — USER UPDATES

The NCCER makes every effort to keep these textbooks up-to-date and free of technical errors. We appreciate your help in this process. If you have an idea for improving this textbook, or if you find an error, a typographical mistake, or an inaccuracy in NCCER's Contren™ textbooks, please write us, using this form or a photocopy. Be sure to include the exact module number, page number, a detailed description, and the correction, if applicable. Your input will be brought to the attention of the Technical Review Committee. Thank you for your assistance.

*Instructors* – If you found that additional materials were necessary in order to teach this module effectively, please let us know so that we may include them in the Equipment/Materials list in the Instructor's Guide.


**Write:** Curriculum Revision and Development Department
National Center for Construction Education and Research
P.O. Box 141104, Gainesville, FL 32614-1104

**Fax:** 352-334-0932

**E-mail:** curriculum@nccer.org


Craft _______________ Module Name _______________

Copyright Date _______________ Module Number _______________ Page Number(s) _______________

Description _______________

_______________

_______________

_______________

(Optional) Correction _______________

_______________

_______________

(Optional) Your Name and Address _______________

_______________

_______________

Module 62104-02

# *Tools of the Trade*

## MODULE OVERVIEW

This module introduces the tools and equipment used by pipeline maintenance technicians, such as specialized hand and power tools, welding equipment, meters and testers, nondestructive testing tools, and heavy excavating equipment. The module explains how to use, care for, and maintain such tools and also reviews inspection and safety guidelines.

## PREREQUISITES

Please refer to the Course Map in the Trainee Module. Prior to training with this module, it is recommended that the trainee shall have successfully completed the following:

Core Curriculum; Pipeline Core, Modules 66101-02 and 66102-02; Pipeline Maintenance, Level One, Module 62103-02

## OBJECTIVES

Upon completion of this module, the trainee will be able to:

1. Use and maintain hand and power tools.
2. Identify and explain the use of welding equipment.
3. Explain the use of meters and testers.
4. Explain the use of nondestructive testing tools.
5. Identify and explain the use of lifting and excavation equipment.

## PERFORMANCE TASKS

Under the supervision of the instructor, the trainee should be able to:

1. Identify a given hand tool, state its application, and describe its safe use and maintenance.
2. Demonstrate the use of a given hand tool according to standards given by the instructor.
3. Identify a given power tool, state its application, and describe its safe use and maintenance.
4. Demonstrate the use of a given power tool according to standards given by the instructor.
5. Identify a given beveling machine, state its application, and describe its safe use and maintenance.
6. Identify a given meter or tester, state its application, and describe its safe use and maintenance.
7. Identify various nondestructive examination methods and tools and state their applications.
8. Identify a given kind of lifting or excavation equipment, state its application, and describe its safe use.

## NCCER STANDARDIZED CRAFT TRAINING PROGRAM

The National Center for Construction Education and Research (NCCER) provides a standardized national program of accredited craft training. Key features of the program include instructor certification, competency-based training, and performance testing. The program provides trainees, instructors, and companies with a standard form of recognition through a National Craft Training Registry. The program is described in full in the *Guidelines for Accreditation*, published by the NCCER. For more information on standardized craft training, contact the NCCER by writing us at P.O. Box 141104, Gainesville, FL 32614-1104; calling 352-334-0911; or e-mailing info@nccer.org. More information may be found at our Web site, www.nccer.org.

## HOW TO USE THIS ANNOTATED INSTRUCTOR'S GUIDE

Each page presents two sections of information. The larger section displays each page exactly as it appears in the Trainee Module. The narrow column ties suggested trainee and instructor actions to each page and provides icons to call your attention to material, safety, audiovisual, or testing requirements. The bottom of each page includes space for your notes.

**Teaching Tip**

If you see the Teaching Tip icon, that means there is a teaching tip associated with this section. Also refer to any suggested teaching tips at the end of the module.

## SAFETY CONSIDERATIONS

Ensure that the trainees are equipped with appropriate personal protective equipment. Emphasize the importance of following all safety precautions and procedures when working with power tools.

## PREPARATION

Before teaching this module, you should review the Module Outline, Objectives, Performance Tasks, and the Materials and Equipment List. Be sure to allow ample time to prepare your own training or lesson plan and gather all required equipment and materials.

## MATERIALS AND EQUIPMENT LIST

Transparencies
Felt-tip markers/chalk
Blank acetate sheets
Transparency pens
Pencils and scratch paper
Copies of the Quick Quiz*
Module Examinations**
Performance Profile Sheets**
Overhead projector and screen
Whiteboard/chalkboard
Appropriate personal protective equipment
Probe rod
Assorted types of wrenches, including bolt-thru ratchets and hammer/slug wrenches
Banding tool and band
Rotary pipe cutter
Flange spreaders
Line plugging/stoppling equipment
Assorted pit gauges
Assorted lengths of pipe of different thicknesses
Hand drilling and tapping machines
Multi-gas monitor
Line locator
Dye penetrant

* Located at the end of this Annotated Instructor's Guide.
** Located in the Test Booklet packaged with this Annotated Instructor's Guide.

## TEACHING TIME FOR THIS MODULE

An outline for use in developing your lesson plan is presented below. Note that each Roman numeral in the outline equates to one session of instruction. Each session has a suggested time period of 2½ hours. This includes 10 minutes at the beginning of each session for administrative tasks and one 10-minute break during the session. Approximately 10 hours are suggested to cover *Tools of the Trade*. You will need to adjust the time required for hands-on activity and testing based on your class size and resources.

| Topic | Planned Time |
|---|---|
| **Session I. Using and Caring for Hand Tools** | |
| A. Probe Rods | ____________ |
| B. Banding Tool | ____________ |
| 1. Laboratory | ____________ |
| C. Wrenches | ____________ |
| D. Flange Alignment Pins | ____________ |
| E. Rotary Pipe Cutters | ____________ |
| 1. Laboratory | ____________ |
| F. Flange Spreaders | ____________ |
| 1. Laboratory | ____________ |
| G. Drilling and Tapping Machines | ____________ |
| 1. Laboratory | ____________ |
| H. Sandblasting | ____________ |
| I. Air Bags | ____________ |
| J. Bonding Clamps | ____________ |
| K. Pit Gauges | ____________ |
| L. Laboratory | ____________ |
| M. Pipeline Skids | ____________ |
| N. De-Watering | ____________ |
| O. Performance Testing | ____________ |
| **Session II. Using and Caring for Power Tools and Welding Equipment** | |
| A. Air Compressors | ____________ |
| B. Pipe Preparation Machines | ____________ |
| C. Air Movers | ____________ |
| D. Performance Testing | ____________ |
| E. Beveling Machines | ____________ |
| 1. Beveling Using Grinders | ____________ |
| 2. Thermal Beveling | ____________ |
| 3. Pipe Facing Machines | ____________ |
| 4. Band-Type Machines | ____________ |
| 5. Line-Up Clamps | ____________ |
| F. Performance Testing | ____________ |
| **Session III. Meters and Testers and Nondestructive Examination Methods and Tools** | |
| A. Multi-Gas Monitors | ____________ |
| B. Portable Gas Detectors | ____________ |
| C. Holiday Detectors | ____________ |
| D. Jeeps | ____________ |
| E. Line Locators | ____________ |
| F. Performance Testing | ____________ |
| G. Dye Penetrant Test | ____________ |

H. Magnetic Particle Test ______
I. Handheld Thickness Gauge ______
J. Radiography ______
K. Visual and Optical Inspection ______
L. Performance Testing ______

**Session IV. Hydraulic Cranes and Heavy Excavating Equipment**

A. Hydraulic Cranes ______
B. Overhead Cranes ______
C. Pipeline Rigging Equipment ______
D. Excavation Equipment ______
E. Performance Testing ______
F. Summary
    1. Summarize module ______
    2. Answer questions ______
G. Module Examination ______
    1. Trainees must score 70% or higher to receive recognition from the NCCER.
    2. Record the testing results on Craft Training Report Form 200 and submit the results to the Training Program Sponsor.
H. Performance Testing ______
    1. Trainees must perform each task to the satisfaction of the instructor to receive recognition from the NCCER.
    2. Record the testing results on Craft Training Report Form 200 and submit the results to the Training Program Sponsor.

Module 62104-02

# *Tools of the Trade*

Instructor's Notes:

Instructor's Notes:

# Introduction to the Pipeline Operator Qualification Rule, API, and NCCER

*It could be said that petroleum products keep America running. Petroleum products like oil and natural gas are the principal sources of energy in the United States and are used in many manufacturing and chemical processes. Petroleum products are primarily transported by pipeline. There are more than a million miles of liquid and natural gas pipeline in this country, ranging in diameter from two inches to four feet. These pipelines bring crude oil and natural gas from production fields to processing plants and refineries—and ultimately to the consumer.*

## Pipeline Operator Qualification

Petroleum products are considered hazardous, so pipeline personnel must take extra precautions and follow specific procedures when working with them. To help ensure the safe handling and transportation of petroleum products, the United States Department of Transportation (DOT) mandates safe practices as federal regulations.

One such regulation, the Operator Qualification (OQ) rule, requires pipeline operators to develop a written qualification plan to evaluate their personnel and their contract personnel (see *49 CFR Part 192 Subpart N—Gas,* and *49 CFR Part 195 Subpart G—Liquid*). The evaluation hinges on the operator's ability to perform certain covered tasks and to be able to recognize and respond to abnormal operating conditions that may be encountered while performing these covered tasks.

## Covered Tasks

A pipeline performance task is deemed *covered* if the activity passes the following four-part test:

1. Is it performed on a pipeline facility?
2. Is it an operations or maintenance task?
3. Is it performed as a requirement of *49 CFR Part 195 (Liquid)* or *49 CFR Part 192 (Gas)*?
4. Does it affect the operation or integrity of the pipeline facility?

If the answer to all four questions is *yes,* then the activity is considered *covered* and must figure into the evaluation process of pipeline personnel.

## American Petroleum Institute

The OQ rule and its related covered tasks provided the catalyst to develop a unified pipeline training and assessment program—so that pipeline operators and owners have ready access to the assessment and training tools they need. This effort was the direct result of the vision and leadership provided by The American Petroleum Institute (API) and API members.

Efforts to unify the industry are nothing new to API. Indeed, the impetus for forming the Institute in 1919 was the need to standardize engineering specifications for drilling and production equipment. Today, the most pressing issues revolve around public perceptions and government policies toward the industry. Speaking with one voice on these issues has become as essential as having interchangeable parts in the field. API's members determine the petroleum industry's positions on public policy as well as the standards that govern its day-to-day operations worldwide.

## Consortium on Operator Qualification

API assembled a focus group whose primary purpose was to identify covered tasks and provide compliance recommendations to pipeline operators.

The Consortium on Operator Qualification (COOQ) became the guiding force for the identification and development of covered tasks for the liquid pipeline industry. Their work product, API publication 1161, *Guidance Document for the Qualification of Liquid Pipeline Personnel,* formed the basis for the resulting Pipeliner Training and Assessment Program.

The covered task list created in API publication 1161 has now evolved to include gas covered tasks and is maintained by NCCER. The most current covered task list can be downloaded at www.nccer.org.

## National Center for Construction Education and Research

With the guidance and support of API, its members, and the Consortium on Operator Qualification, the National Center for Construction Education and Research (NCCER) was able to design and implement the Pipeliner Training and Assessment Program.

NCCER is a not-for-profit 501(c)(3) education foundation established in 1995 by the world's largest and most progressive construction and maintenance contractors and national contractor associations. NCCER was founded to address the

severe workforce shortage facing the construction and maintenance industries and to develop a standardized training process and curricula. Today, NCCER is supported by hundreds of leading construction and maintenance companies, manufacturers, and national associations and developed more than 35 construction and maintenance titles.

NCCER also maintains a National Registry, which contains the records of trainees and instructors who have received NCCER training. All completion information for every trainee is sent to the NCCER and filed within the Registry. The Registry can then confirm training and skills for workers as they move from company to company, state to state, or even within their own company.

### *Standardized Curriculum (NCCER's Contren™ Learning Series)*

NCCER curricula are developed using the expertise of industry subject matter experts (SMEs). The use of SMEs as an integral part of every new development or revision ensures that the curriculum is current, industry-driven, and indicative of industry practices. The development and/or revision process is facilitated from start to finish by NCCER project managers, who work closely with the SMEs and a select team of technical writer(s) on the development and presentation of the curriculum.

### *Pipeline: Five Training Paths*

The pipeline curriculum focuses on essential skills and competencies within five skill areas:

- Corrosion Control
- Electrical and Instrumentation
- Maintenance
- Mechanical
- Operations (Field and Control Center)

SMEs from the pipeline industry have determined the skills and competencies needed based on API's covered tasks and the consensus of the group/industry. Specific company information is generally not incorporated into the curriculum unless the SMEs determine that the information is the industry standard.

### *Training Modules and Covered Tasks*

The smallest unit of instruction in the curriculum is a *module*. Each curriculum is composed of modules (the equivalent of stand-alone chapters) that are logically sequenced into complete *levels* of instruction. Every module is task-oriented and requires the successful completion of a written (knowledge-based) examination and a performance (hands-on) test. Modules may be individually selected and/or combined for efficient task training. Modules in the pipeline curricula, to the extent possible, will align with API-developed covered tasks. While not all modules will relate to covered tasks, all covered tasks will relate to at least one module.

### *Benefits of Training to the Pipeline Marketplace*

Applying NCCER's standardized training process allows pipeline owners and contractors in the United States to establish the benchmark for acceptable skill and competency levels within the pipeline industry. This process will assist training organizations in developing and/or improving standardized pipeline training programs that effectively match their resources and operating conditions while responding to regulatory demands and to the needs of the industry. The process will also recognize training organizations that meet the minimum standards of competence, warranting regulatory, public, and professional confidence. This in turn will foster national unity within the pipeline industry for the mutual benefit of individual technicians, owners, contractors, and ultimately pipeline-users and society.

## Additional Resources

To review the Pipeliner Training and Assessment Program Implementation and Resource Guide and the latest version of pipeline covered tasks, visit NCCER's Web site at www.nccer.org, or contact:

**National Center for Construction Education and Research**
PO Box 141104
Gainesville, FL 32614-1104
(888) 622-3720
Web site: www.nccer.org

For more information on API, please contact:

**American Petroleum Institute**
1220 L Street, Northwest
Washington, D.C. 20005
(202) 682-8000
Web site: www.api.org

To order *API Pub 1161*, contact Global Engineering Documents by calling 1-800-854-7149 or visiting their Web site at www.global.his.com.

Additional sources of information on the pipeline operator qualification regulation include:

**DOT – Office of Pipeline Safety**
www.ops.dot.gov

**Transportation Safety Institute**
www.tsi.dot.gov



Instructor's Notes:

## COURSE MAP

This course map shows all of the modules in the first level of the Pipeline Maintenance curriculum. The suggested training order begins at the bottom and proceeds up. Skill levels increase as you advance on the course map. The local Training Program Sponsor may adjust the training order.

**PIPELINE MAINTENANCE LEVEL ONE**

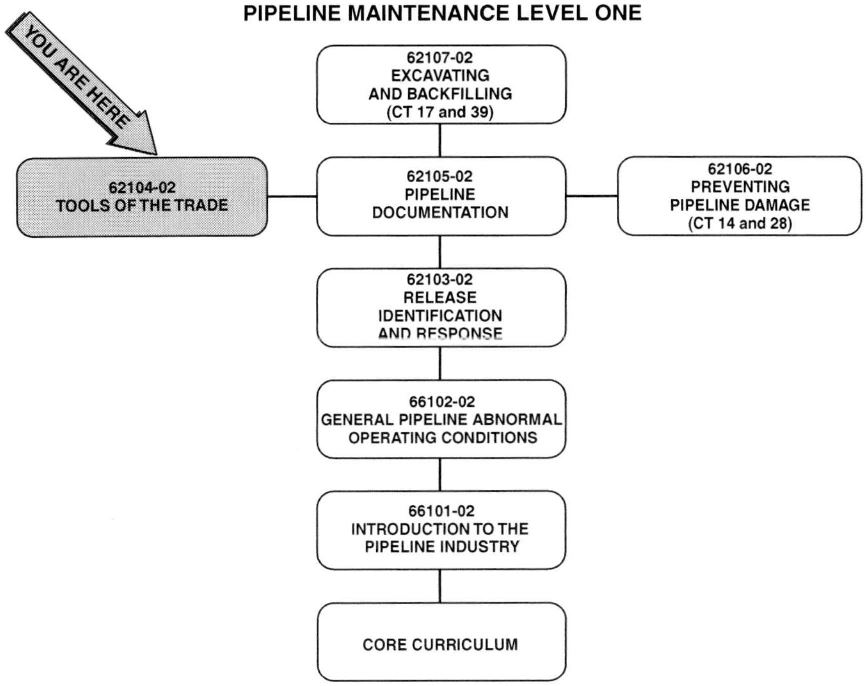


104CMAP.EPS





**Homework**

**Assign reading of Module 62104-02.**

Instructor's Notes:

## MODULE 62104-02 CONTENTS

## MODULE 62104-02 CONTENTS CONTINUED





Instructor's Notes:

## Figures

**Table**





Instructor's Notes:

MODULE 62104-02

# *Tools of the Trade*

## Objectives

When you have completed this module, you will be able to do the following:

1. Use and maintain hand and power tools.
2. Identify and explain the use of welding equipment.
3. Explain the use of meters and testers.
4. Explain the use of nondestructive testing tools.
5. Identify and explain the use of lifting and excavation equipment.

## Prerequisites

Before you begin this module, it is recommended that you successfully complete the following: Core Curriculum; Pipeline Core, Modules 66101 and 66102; Pipeline Maintenance Level One, Module 62103.

## Required Trainee Materials

1. Pencil and paper
2. Appropriate personal protective equipment

## 1.0.0 ◆ INTRODUCTION

This module focuses on the use, care, and maintenance of hand and power tools, welding equipment, meters and testers, nondestructive testing tools, and heavy excavating equipment. Use and care of basic hand tools and power tools are explained in the *Core* curriculum, and this module describes the more specialized tools used by pipeline maintenance technicians.

Using the correct type and size of tool that has been properly maintained can prevent accidents and damage to your equipment, system components, and tools. The money spent on the tools required to maintain a pipeline represents a significant cost to your employer. Performing the required maintenance and using the tools properly will protect you and your co-workers, your employer's investment, and will also make the job run more smoothly.

## 2.0.0 ◆ USING AND CARING FOR HAND TOOLS

Occupational Safety and Health Administration (OSHA) *CFR 1926.301-Hand Tools* provides four very simple, common-sense rules that must be followed when using hand tools. It is the pipeline maintenance technician's responsibility to ensure that these rules are strictly followed:

- Employers shall not issue or permit the use of unsafe hand tools.
- Wrenches, including adjustable, pipe, end, and socket wrenches, shall not be used when the jaws are sprung to the point at which slippage occurs.
- Impact tools, such as drift pins, wedges, and chisels, shall be kept free of mushroomed heads.
- The wooden handles of tools shall be kept free of splinters or cracks and shall be kept tight in the tool.

**NOTE**

OSHA writes these rules directed at the employer. However, it is the employee's responsibility to not use defective tools and to notify the employer of any defective tools on the job site.



**Materials**

**Ensure that you have everything necessary to teach the course. Check the Materials and Equipment List at the front of this module.**

**Classroom**

**Explain that the terms shown in bold are defined in the Glossary at the back of this module.**

**Audiovisual**

**Show Transparency 1, Course Objectives.**

**Show Transparency 2, Performance Tasks.**

**Teaching Tip**

**See the general Teaching Tip at the end of this module.**

**Safety**

**Review the four OSHA rules for using hand tools. Ask trainees to explain what would happen if the rules were ignored.**

**Demonstration**

**Show trainees how to use a probe rod. Answer any questions trainees have.**

**Safety**

**Invite trainees to recall the do's and don'ts of using probe rods. Ensure that trainees know that probes cannot be used on certain kinds of pipelines.**

**Laboratory**

**Show trainees how to use the banding tool. Pair trainees and ask each pair to review the steps in using the tool. Have trainees practice loading the band and tightening it.**

**Audiovisual**

**Show Transparency 3 (Figure 2). Ask trainees to identify the most common uses of banding tools.**

**Classroom**

**Display different types of wrenches, including bolt-thru ratchets and hammer/slug wrenches. Ask trainees how each wrench makes a task easier.**

The following hand tools are described in this section:

- Probe rods
- Banding tools
- Wrenches
- Flange alignment pins
- Rotary pipe cutters
- Flange spreaders
- Drilling and **tapping** machines
- Sandblasting
- Air bags
- Bonding clamps
- Pit gauges
- Pipeline **skids**

### 2.1.0 Probe Rods

Probe rods (*Figure 1*) made of metal or fiberglass are used to locate existing utilities. The probe rod is pushed into the soil and used to probe for the utility. When the utility is located, its exact placement can be determined and marked by measuring the amount of buried probe rod in the **excavation**. Follow these guidelines when using probe rods:

- Always use a fiberglass or insulated rod in the vicinity of electric, phone, or television cables.
- Never use a probe rod to locate electric utilities.
- The probe rod should always be stored when not in use to prevent damage.
- Do not use excessive force when using the probe rod.
- Always use proper lifting procedures.
- The probe should have blunt end.

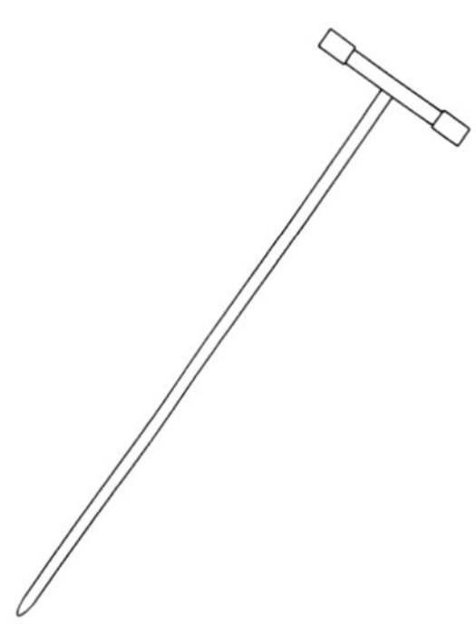


*Figure 1* ◆ Probe rod.

### 2.2.0 Banding Tool

The banding tool (*Figure 2*) is used to perform a number of tasks on the job site. It is commonly used for repairing suction and discharge hoses and for banding together stacks of pipe or bundles of wood. The banding tool is also used to install diapers, to band skids to carrier pipe for installation in casings, and occasionally for strengthening concrete forms. The banding tool operates by using a ratchet action to tighten the banding material. Once the band is tightened, it is held in place by striking the anvil. A small protrusion on the back side of the anvil places a dimple on the band clip, preventing the band from loosening.

Follow these steps to use a banding tool:

*Step 1* Place the band around the material to be banded.

*Step 2* Load the band into the receiver.

*Step 3* Depress the tension handle repeatedly until the band is tight.

*Step 4* Strike the tool anvil sharply with a hammer to dimple the band clip.

*Step 5* Release the tool.

*Step 6* Clean and store the banding tool.

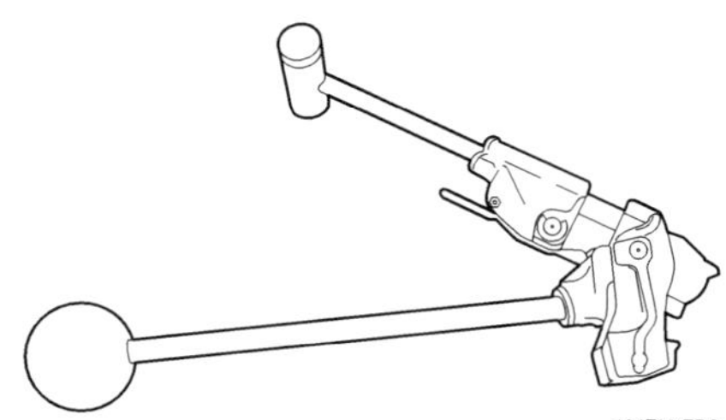


*Figure 2* ◆ Banding tool.

### 2.3.0 Wrenches

Several specialty wrenches are used in pipeline maintenance. These tools are designed to make a specific task easier. Examples of wrenches include the valve wrench, T-handle torque wrench, bolt-thru ratchets, hydrant wrenches, and hammer/slug wrenches.



*Instructor's Notes:*

### 2.3.1 Bolt-Thru Ratchets

Bolt-thru ratchets (see *Figure 3*) are commonly used to fasten the nuts on a variety of pipe joints. The unique aspect of these wrenches is the opening in the head. This opening allows the bolt or rod to extend through the wrench so that the nut can be threaded any distance required. The wrench is available with a variety of interchangeable sockets and is reversible.

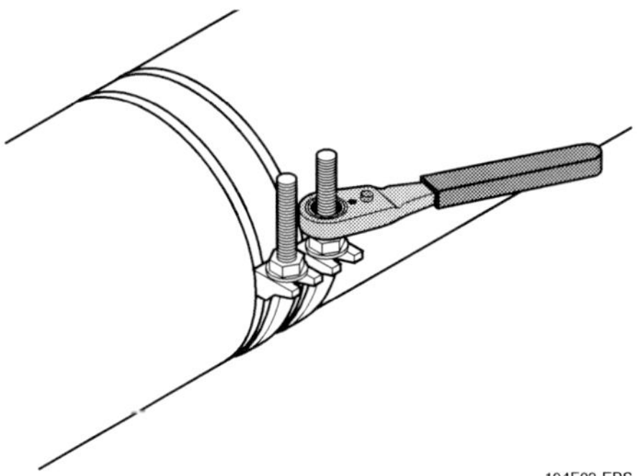


*Figure 3* ◆ Bolt-thru ratchet.

### 2.3.2 Hammer/Slug Wrench

A hammer/slug wrench is designed with a heavy box end on one end that completely surrounds the bolt head or nut and a large striking area on the other end for striking with a hammer or mallet (*Figure 4*). The slug wrench is the only wrench that is designed to be hit with a hammer.

> **WARNING!**
> Never hammer or strike any other type of wrench.

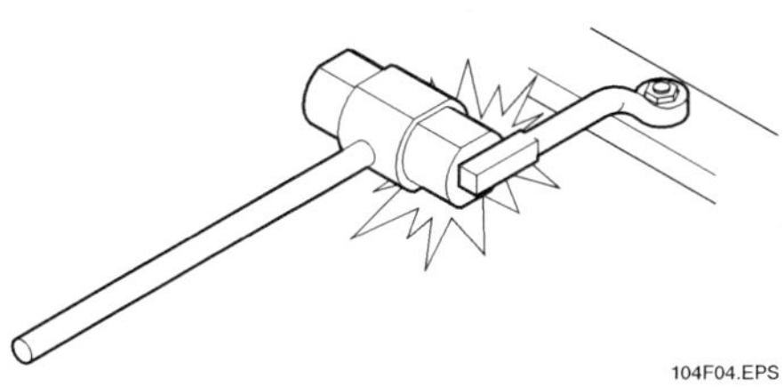


*Figure 4* ◆ Hammer/slug wrench.

## 2.4.0 Flange Alignment Pins

A variety of flange alignment pins are manufactured for the pipeline industry. A drift pin, also commonly referred to as a line-up pin, is a round steel piece that is tapered on one end. A barrel pin (*Figure 5*) is a drift pin that is tapered on both ends. Drift pins and barrel pins are used to align bolt holes when connecting flanges, structural steel, and other holes in mating surfaces. Drift pins and barrel pins are available in various sizes to fit different size bolt holes. The pin selected for a specific job should fit all the way through both holes being aligned and still be large enough and strong enough to pull the mating pieces together.

Two-hole pins (see *Figure 6*) are used to level or plumb two adjacent bolt holes on flanges. The pins line up with the centers of each of the holes when tightened, and then a level is placed across the top of the pins to assist in aligning the flange.

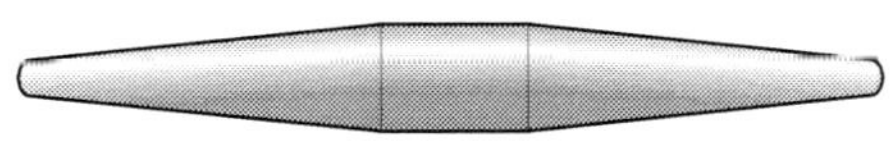


*Figure 5* ◆ Common barrel pin.

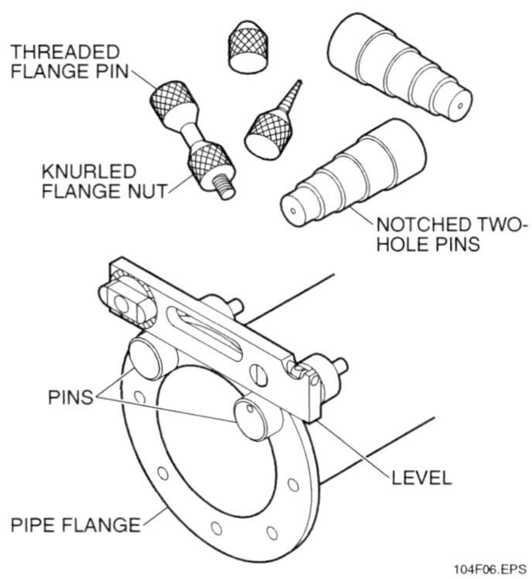


*Figure 6* ◆ Two-hole pins.



**Audiovisual**

**Show Transparency 4 (Figure 3). Explain the unique aspects of bolt-thru ratchets.**

**Show Transparency 5 (Figure 4). Explain the dangers of striking a wrench with a hammer.**

**Show Transparencies 6 and 7 (Figures 5 and 6). Have trainees describe drift pins, barrel pins, and two-hole pins. Ask other trainees to review the uses of each type.**

**Classroom**

**Stress that trainees should know the correct pipe diameters for each tool. Ask a trainee to identify the diameter of pipe that can be cut with a rotary pipe cutter.**

**Laboratory**

**Pair trainees. Ask one to review the steps in cutting while the other practices using a rotary pipe cutter. Ensure that trainees are wearing appropriate personal protective equipment. Review their work.**

**Safety**

**Stress the importance of releasing pressure on the pipeline before you cut.**

**Audiovisual**

**Show Transparency 8 (Figure 7). Ask trainees what materials can be cut with rotary pipe cutters. Ensure that trainees know that different materials require specific blades.**

**Demonstration**

**Demonstrate how each of the five types of flange spreaders is used. Have a trainee identify the types of flange spreaders.**

### 2.5.0 Rotary Pipe Cutters

Rotary pipe cutters come in a wide range of sizes (see *Figure 7*) and can be used to cut steel, stainless steel, ductile iron, and cast-iron pipe in tight spaces. The tool usually needs only about 4 to 8 inches of clearance to operate, and will cut pipe up to 34 inches in diameter with wall thicknesses to about ½ inch. The cutting blades are interchangeable so that blades for a specific material can easily be installed.

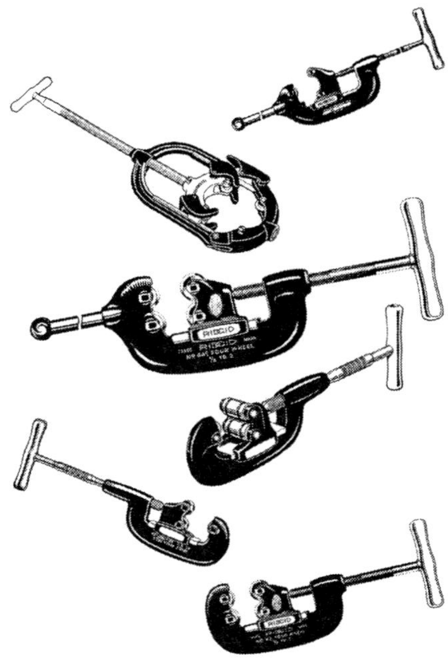
104F07.EPS

*Figure 7* ◆ Rotary pipe cutters.

Follow these steps to cut pipe with a rotary cutter.

> **WARNING!**
> Always make sure that all pressure on the pipeline has been released before working on the line.

*Step 1* Place the cutter around the pipe and tighten it in place.

*Step 2* Using a ratchet action, move the pipe cutter back and forth around the pipe.

*Step 3* Feed the blade further into the cut by tightening the tool around the pipe.

*Step 4* Repeat Steps 2 and 3 until the pipe is cut.

> **CAUTION**
> A pipe may snap off before the cut is all the way through the pipe. To reduce the risk of injury, be sure that you are standing clear of the pipe should it unexpectedly snap off.

### 2.6.0 Flange Spreaders

Flange spreaders are used to separate flanges when disassembling flanged piping systems for removal or maintenance purposes.

#### *2.6.1 Hex Turn Flange Spreaders*

Flange spreaders are small mechanical devices used to part flanges quickly and easily. These tools are typically made of steel alloys to withstand the severe forces encountered during use. Spreaders are also heat treated to add to their ability to undergo heavy use without distortion or breakage.

Hex turn flange spreaders (see *Figure 8*) are easy to set up and use on flanges that do not easily separate because of corrosion or coating that might be binding them together. The user removes one bolt from the flange face and inserts the flange spreader support. Next, the remaining flange bolts are removed. Finally the hex nut is turned with the appropriate tool to drive the spreader between the flange faces.

> **NOTE**
> A spreader is required on each side of the flange for even distribution of pressure.

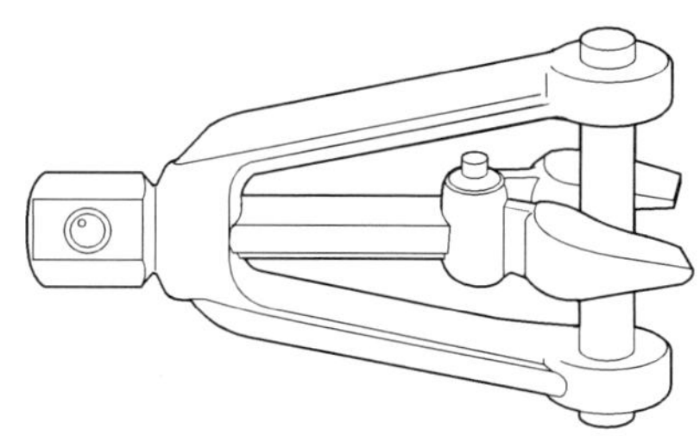
104F08.EPS

*Figure 8* ◆ Hex turn flange spreader.



*Instructor's Notes:*

### *2.6.2 T-Handle Flange Spreaders*

T-handle flange spreaders (see *Figure 9*) are another variation on the flange spreader line of tools. Used in pairs to keep the pressure equal on both sides of a flange, they allow the user to part flanges quickly for small jobs like gasket replacement. T-handle spreaders do not require additional tools to exert pressure on tight flanges.

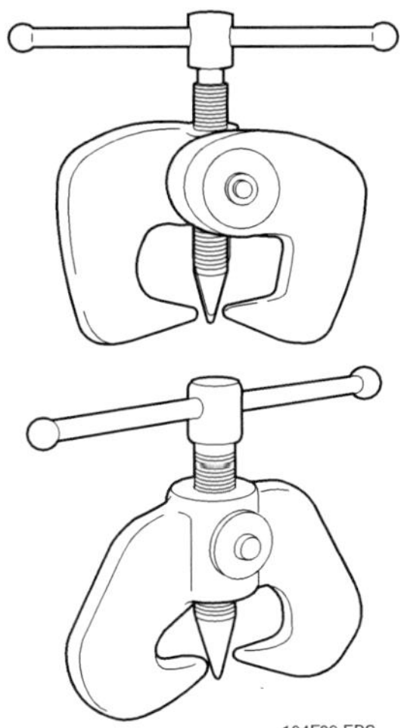


*Figure 9* ◆ T-handle flange spreaders.

### *2.6.3 Ratchet Action Flange Spreaders*

Ratchet action flange spreaders (*Figure 10*) are also small and lightweight but offer the user a way of applying greater force to separate flange faces through the use of a ratcheting tool that advances the spreader point into the gap between the flange faces. These flange spreaders are typically used to separate pipe connections to permit insertion of blinds or gaskets.

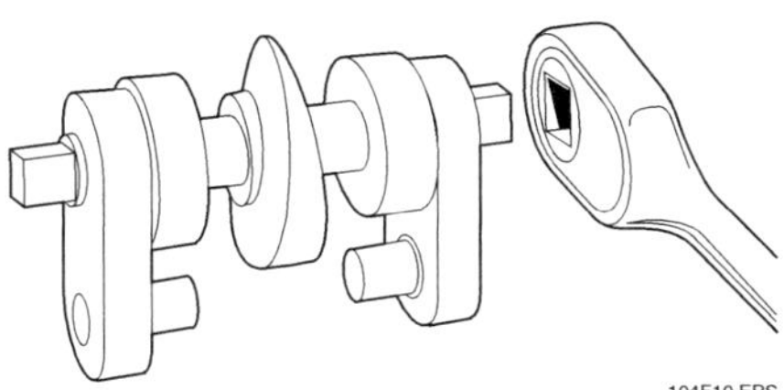


*Figure 10* ◆ Ratchet action flange spreader.

### *2.6.4 Hydraulic Flange Spreader*

Hydraulic flange spreaders (see *Figure 11*) are very effective tools for separating flange faces where abnormally high spreading loads are required, such as on heat exchangers, pressure vessels, and other applications in the marine, oil, and gas industries. Different models of flange spreaders can exert pressures as high as 10 tons and can separate flanges that are up to 6⅝ inches in thickness. These tools require minimal clearance between flanges for insertion.

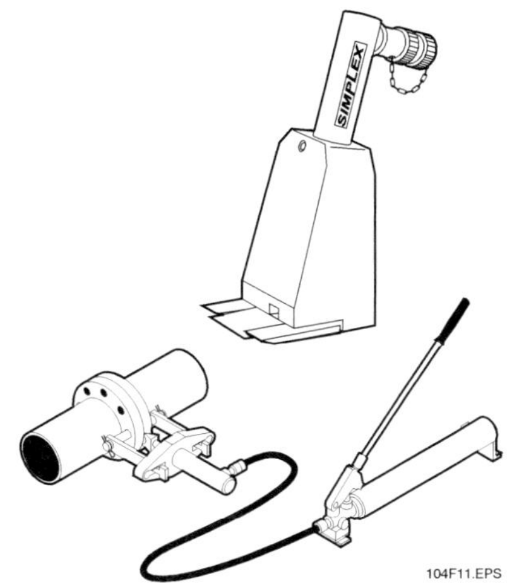


*Figure 11* ◆ Hydraulic flange spreaders.

## 2.7.0 Drilling and Tapping Machines

Drilling machines are used to make connections to pipelines, tanks, and plant piping without shutdown and are used to make hot **taps** in preparation for plugging machine applications. With the proper attachments, large drilling machines can also be used to install line stops and valves into main lines. Most taps and connections can be made with the main line pressurized and in service.

The machines are widely available for cuts or taps ranging from ¼ inch to 12 inches. Much larger equipment is available, but these are not common tools and are usually operated by crews who specialize in that type of work. Drilling and tapping machines can cut most pipe types including cast iron, cement-lined cast iron, ductile and cement-lined ductile iron pipe, steel, concrete, reinforced concrete, asbestos cement, and plastic.



**Laboratory**

**Have trainees practice using various flange spreaders. Discuss with trainees any difficulties they may have encountered.**

**Audiovisual**

**Show Transparencies 9 through 12 (Figures 8 through 11). Ask trainees to recall the uses of different types of flange spreaders.**

**Classroom**

**Review the uses and limits of drilling and tapping machines.**

**Have trainees state the four functions of stoppling. Ensure that trainees understand each of the functions.**

**Show Transparency 13 (Figure 12). Ask trainees which of the 13 steps apply to each of the stages of stoppling.**

**Have trainees practice stoppling. Check their work after each step and answer any questions.**

### *2.7.1 Line Plugging/Stopping*

When maintenance must be done on a section or part of the pipeline without shutting down the operation, equipment can be installed to provide a temporary bypass line for the product. This allows the section being worked on to be drained and removed. This is accomplished by using valves, line plugging or stopping machines, and an appropriate bypass line. Once repairs are made, the plugs and bypass are removed, and the product flow restored to its original path.

*Stoppling* is the term applied to the use of TD Williamson (TDW) equipment and fittings to make either permanent or temporary changes while performing the following:

- Connecting new lines
- Removing, replacing, or repairing damaged or corroded lines
- Relocating lines
- Completing other maintenance or repair tasks

This general procedure represents a typical stoppling operation to clear an obstruction or make a modification to a section of line. *Figure 12* shows the process described in the step-by-step procedure.

*Step 1* Install the required fittings on the line.

*Step 2* Install temporary valves.

*Step 3* Install a temporary bypass.

*Step 4* Make taps through the valves.

*Step 5* Install STOPPLE® plugs.

*Step 6* Close the plugs to direct product through the bypass.

*Step 7* Complete the task on the isolated section of pipe.

*Step 8* Retract the plug heads.

*Step 9* Close valves and remove the temporary bypass.

*Step 10* Install LOCK-O-RINGS® in the fittings.

*Step 11* Remove all equipment and install blind flanges to complete the job.

*Step 12* Notify the control center that the line is returned to full service.

*Step 13* Complete all documentation as required by your company.

Always follow your company's policies and procedures when performing this type of task.

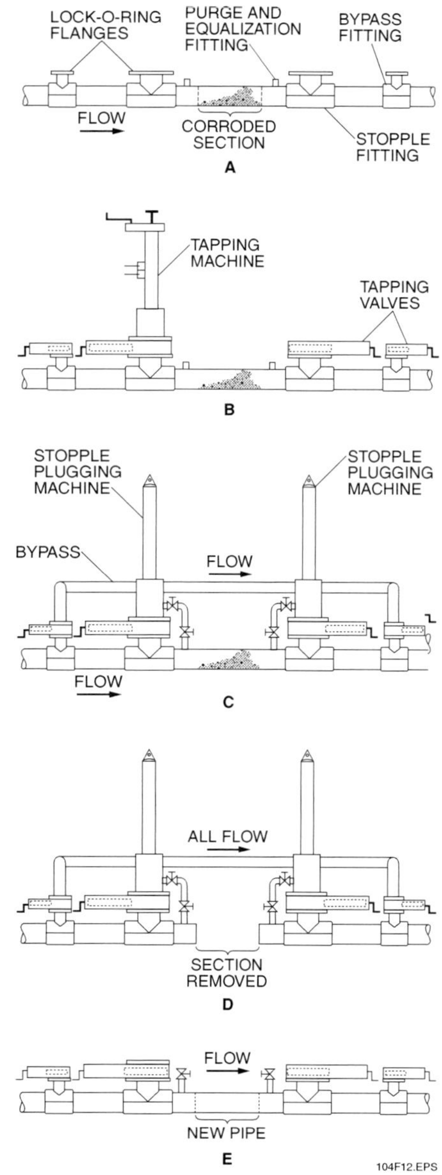


*Figure 12* ◈ Stoppling process.



Instructor's Notes:

### 2.7.2 Hand Drilling and Tapping Machines

Hand drilling and tapping machines are available as manual or power-driven tools that allow the pipeliner to tap into pipes for repairs or adding lines while the pipe is under operating pressure and in service. These tools come in a wide range of models and sizes for making cuts or taps from ½ inch to 4 inches, and larger, in diameter. Hand-operated cranks are replaced by handheld air motors for taps of 6 inches or larger. These smaller tapping machines are also used to install completion plugs once tapping work is complete.

One widely used small machine, the TD Williamson Model T-101, is available in two versions: the standard T-101B with 18" of travel and the XL Model with 28" of travel. These and other commonly used drilling machines meet NACE standards for oil field equipment. *Figure 13* shows TDW T-101 series drilling machines.

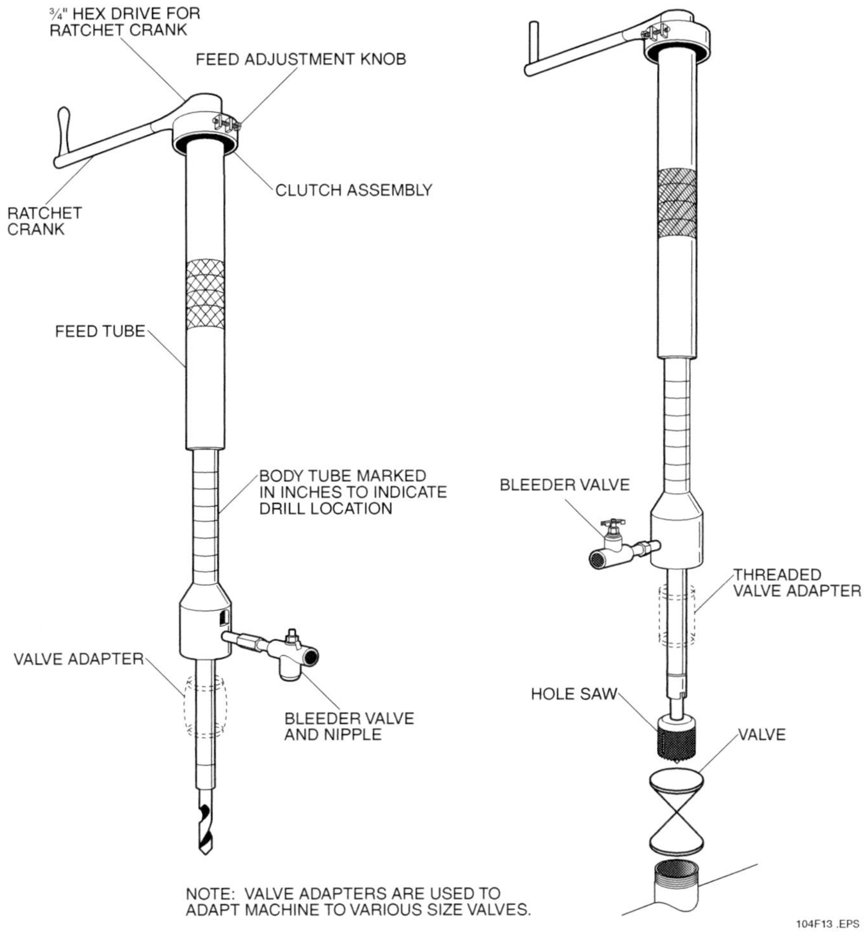


*Figure 13* ◆ T-101 series drilling machines.



**Teaching Tip**

**See the Teaching Tip at the end of this module.**

**Audiovisual**

**Show Transparency 14 (Figure 13). Ask trainees to point out parts of the T-101 drilling machine and functions of each part.**

**Audiovisual**

**Show Transparency 15 (Figure 14). Show trainees how to set up to tap into a pipeline. Explain the advantages of hydraulic powered tapping machines.**

**Classroom**

**Explain the importance of sandblasting in pipeline maintenance.**

**Audiovisual**

**Have trainees explain the advantages and disadvantages of using airbag jacks. Show Transparencies 16 and 17 (Figures 15 and 16). Review how cribbing is used.**

**Classroom**

**Explain why airbags are effective lifting tools. Answer trainee questions.**

#### *2.7.3 Hydraulic Powered Tapping Machines*

Larger tapping machines are more suited for making hot taps into main lines, storage tanks and plant piping. These taps can be made without shutdown of the system.

The TD Williamson Model 1200H tapping machine is widely used throughout the petroleum industry where larger taps are necessary. It is hydraulically operated and can make taps ranging from 12 inches to 36 inches in size. It is used to prepare for plugging operations, in support of maintenance requirements, or to allow the addition of new lines. Using this equipment, taps may be made in a vertical or horizontal position or at any other angle where there is adequate space for equipment installation and safe work practices.

This model, like other large machines, is equipped with an access ladder and work platform for operator use. The Model 1200H meets NACE standards for oil field equipment. See *Figure 14* to view a typical setup for tapping into a pipeline with a large tapping machine.

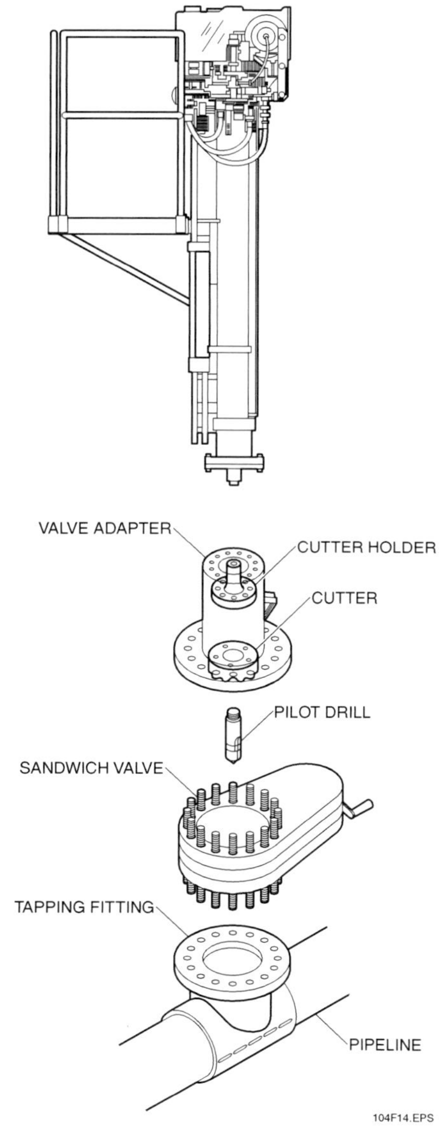


*Figure 14* ◆ Typical setup for tapping.

### 2.8.0 Sandblasting

Sandblasting, also referred to as abrasive blast cleaning, is used to prepare pipe surfaces for coating applications and/or repairs. Sand, plastic pellets, and walnut hulls are widely used for treating different surface materials. The abrasive material used impacts the surface at a high velocity and cuts or knocks the old paint, rust, and dirt from the pipe while creating a good surface profile for the coating material on which to adhere.

Follow your company's procedure for use of personal protective equipment and review applicable sandblasting guidelines prior to starting the process.

### 2.9.0 Air Bags

Air bags, or airbag jacks, were developed to meet the need for a tool that could meet three critical requirements. Those are:

- Low height
- Fast inflation
- Large lifting capacity

Some models of airbag jacks are capable of lifting more than 70 tons and even larger loads when used in sets of two or more. The gas and oil pipeline industry is but one of the industries that find airbag jacks a very useful tool for lifting heavy loads. *Figure 15* shows multiple airbags being used to lift a piece of heavy equipment for maintenance.



Instructor's Notes:

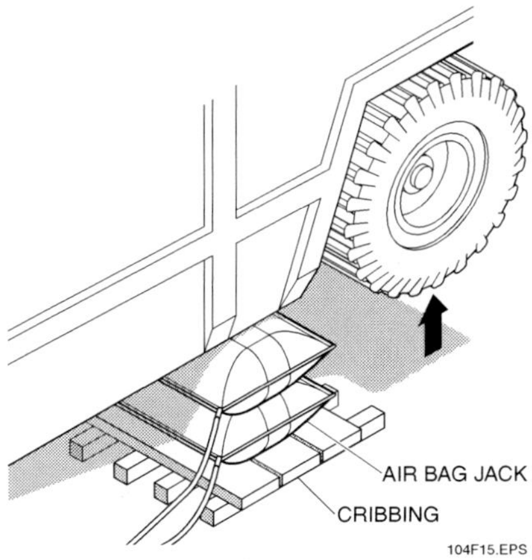


*Figure 15* ◆ Multiple airbag jacks under a load.

The one disadvantage when using airbag jacks is the loss of lifting power as the bags inflate. To offset this, the user places cribbing between the airbag and the load to shorten the distance the bag must inflate before contacting the load. Refer to *Figure 15* to see how cribbing was used to shorten the distance the bag must be raised before coming into contact with the load.

Because of variables in load size and contact area, it is a good practice to place a steel plate (or other stabilizer material) on top of the bag. This increases the area of surface contact between the bag and the load (plate) and offers a more stable support for oddly shaped loads while spreading the load over a greater part of the bag surface. For pipe, cradles can be used in conjunction with the plates to prevent the pipe from rolling off to the side while being elevated. *Figure 16* shows this concept.

These illustrations are very general in nature. Always refer to your company's policy on airbag jack use before attempting to raise a load with these tools.

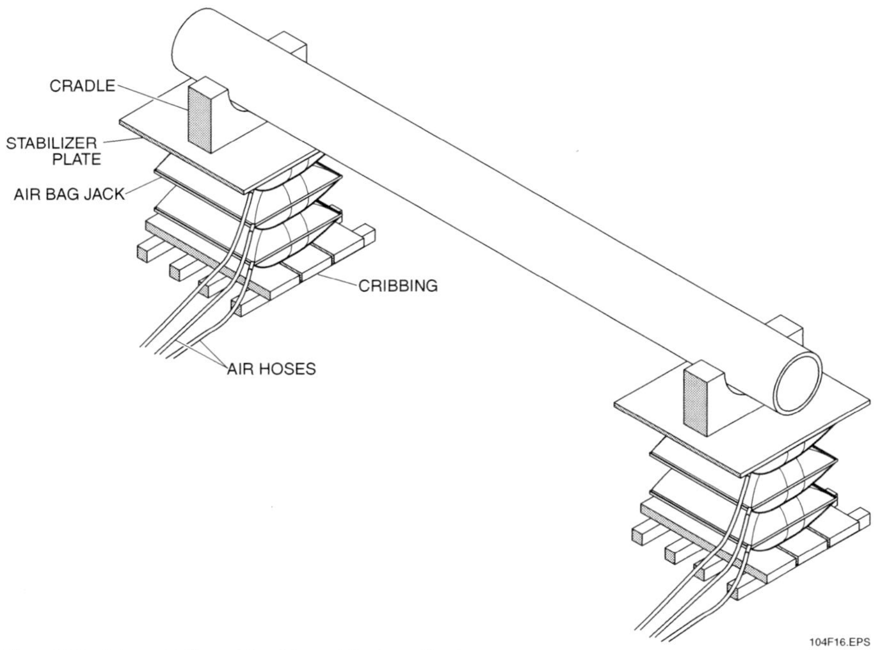


*Figure 16* ◆ Using a stabilizer plate with an oddly shaped load.

**Classroom**

**Review how to ensure the protective function of bonding and grounding clamps.**

**Audiovisual**

**Show Transparency 18 (Figure 17). Point to a clamp and ask a trainee to identify it. Ask another trainee to identify how it is used. Stress the importance of knowing how to choose the correct type of clamp.**

**Laboratory**

**Have trainees practice using pit gauges to measure the thickness of pipe walls. Review their techniques and suggest ways to improve.**

**Classroom**

**Ask trainees to explain the functions and limitations of the common pit gauge. Answer trainee questions.**

> **NOTE**
> Airbags are effective lifting tools because each pound of air pressure (psi) pumped into the bag creates a lifting force that is multiplied over the entire lifting surface of the bag, resulting in a much greater lifting force that can be used to lift the load.

## 2.10.0 Bonding Clamps

Bonding and grounding are acceptable controls for static electricity that can build up in a variety of ways. Static electricity can be produced when any of the following occurs:

- Liquid flows through a pipeline or hose
- Tanks or drums are filled
- Dry material passes through a chute or a pneumatic conveyor

Bonding is the term used to indicate the connection of two or more objects with a conductor. This conductor, such as a copper wire, equalizes but does not eliminate the potential electrical charge (static electricity) between the connected objects, reducing the chance that sparks will jump the gap between objects. Grounding, on the other hand, connects one or more of the objects to the earth using ground rods or steel building components, allowing the static to drain away.

Clamps used for bonding and grounding must provide a good conductive path in order to perform their protective function. To ensure this:

- Remove all dirt, paint or corrosion from the area where the clamp is placed.
- Use strong connectors that will maintain a firm grip, keeping the flow path intact.
- Use the appropriate size and type of connector based on the location and conditions.
- Always connect the metal clamp to other metal.

*Figure 17* shows various types of clamps and connectors used for bonding and grounding.

## 2.11.0 Pit Gauges

One of the most common tools used to check the extent of corrosion to a pipe is the pit gauge. Most pit gauges are easy to use and to carry. These gauges, if properly maintained, can provide accurate measurements.

When corrosion or damage is discovered, the corrosion technician or maintenance team member needs to determine the remaining wall thickness of the pipe in order to set the priority for repair or replacement of the corroded section of pipe. The mechanical pit gauge is a small, sturdy tool and easy to use in making a quick determination of the extent of corrosion damage present. These gauges are usually made of polished stainless steel and can easily be carried in a tool kit. Care must be taken to use these gauges properly and to recognize and understand their limitations.

Inaccurate readings can result if the gauge is not aligned properly or is resting on an uneven surface. Where severely pitted or partially coated pipe and associated structures are involved, these handheld pit gauges will provide the user with only an approximation of pit depth. The depth measurement scale on the mechanical pit gauge is marked in sixteenths of an inch and in mils (thousands of an inch). Most mechanical pit gauges include formulas, tables, and data giving outside diameter (OD), wall thickness, and weight per foot for various sizes of pipe.

When pit depth measurements are critical and accuracy is a priority, dial pit gauges, which are designed for a more accurate measurement of corrosion pit depth, may be used. Most dial pit gauges are capable of measuring depth to extremely close tolerances. On the down side, these indicators are more fragile than mechanical gauges and should be handled with care to ensure their repeated accuracy. *Figure 18* shows a dial pit gauge positioned on a pipe.

The accuracy of any pit measurement is only as reliable as the instrument that is used to take the measurement. Before a pit gauge is used for measurement, it must be inspected for any damage and any worn points. During this inspection, always closely inspect the point of the instrument for wear. If a pit gauge does have a worn point, the pit gauge should be replaced. Only use an instrument that is in excellent condition; otherwise, your inspection may not be accurate. Prior to using the pit gauge, clean the pit in accordance with your company's approved policy.

## 2.12.0 Pipeline Skids

Pipeline skids are timbers used to support the pipe off the ground. These timbers, which typically measure 4" × 6" × 36", are usually cut from oak or other hardwood for strength and durability. They are built up one level at a time until the structure is sufficient for the task at hand.

For larger jobs, pipeline skids can be used in multiples, in a cross-buck style, to support larger sections of pipe for construction or maintenance and repair jobs.

*Figure 19* shows examples of both single-skid and multiple-skid, or cross-buck, applications.



Instructor's Notes:

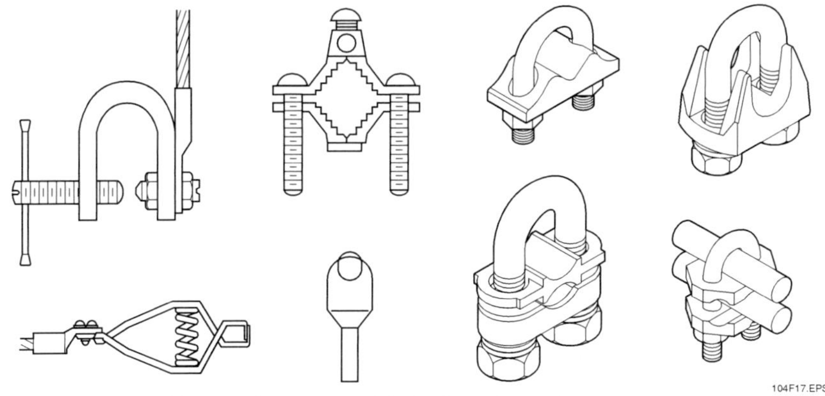

*Figure 17* ◈ Various types of clamps and connectors used for bonding and grounding.

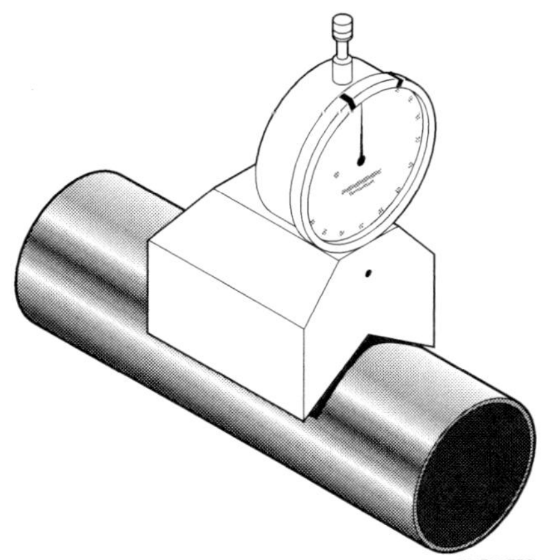

*Figure 18* ◈ Dial pit gauge positioned on a pipe.

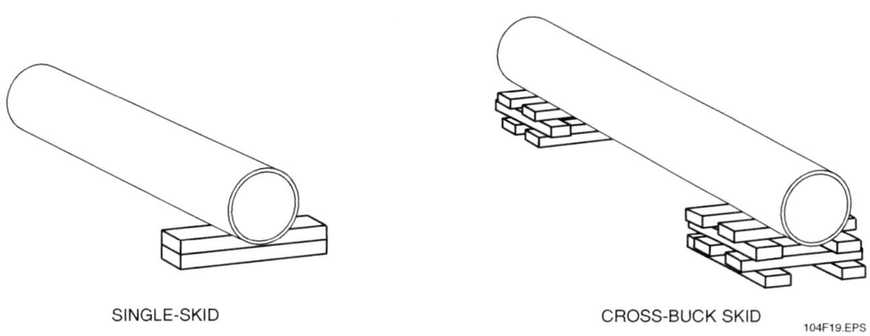


*Figure 19* ◈ Single-skid and cross-buck skid arrangements.



**Audiovisual**

**Show Transparency 19 (Figure 19). Explain single-skid and cross-buck arrangements.**

**Performance Testing**

**Have each trainee demonstrate Performance Tasks 1 and 2 to your satisfaction. Fill out a Performance Profile Sheet for each task for each trainee.**

## Homework

**Assign reading of Sections 3.0.0–4.1.5 for the next class session. Answer any questions trainees have.**

## Materials

**Ensure that you have everything required for testing during this session.**

## Safety

**Review the seven safety checks that must be performed before power tools can be used. Stress the importance of performing safety checks when using a power tool.**

## Audiovisual

**Show Transparency 20 (Figure 20). Explain the principle behind air compressors. Ask trainees to give the operating pressure for portable air compressors.**

### 2.13.0 De-Watering

Before any excavation, **trenching**, or grading can begin on a construction project, such as laying or relining pipelines, groundwater must be removed. Material that contains too much moisture will slow or prevent construction and leave the project site without the necessary preparation for building or compaction. Optimum construction project conditions can be maintained with the proper use of de-watering pumps. Once a trench has been dug, or grading has been done and open bodies of groundwater have collected, de-watering is needed to evacuate the moisture and return the material to a drier state for further work.

De-watering pumps are self-contained units typically mounted on a frame or trailer. The power source turns a shaft connected to the pump end. Inside the pump end is an impeller, a fan-like piece that rotates at a very high speed and forces the air out of the pump end. This creates a vacuum inside the pump end and draws water into the pump. The water is then discharged through piping or hose and emptied at another location.

## 3.0.0 ◆ USE AND CARE OF POWER TOOLS

Pipeline maintenance technicians work with and around various types of powered hand tools throughout their careers. Care and maintenance of these tools is a critical factor in completing job tasks in a safe and timely manner. Before you operate any power tool, make sure you understand and follow the safety precautions for that tool. The following checklist should be reviewed before operating any power tool:

- Are grinders, saws, and similar equipment provided with appropriate safety guards?
- Are power tools used with the correct shield, guard, or attachment that is recommended by the manufacturer?
- Are portable circular saws equipped with guards above and below the base shoe? Are circular saw guards checked to assure they are not wedged up, leaving the lower portion of the blade unguarded?
- Are rotating or moving parts of equipment guarded to prevent physical contact?
- Are all cord-connected, electrically operated tools and equipment effectively grounded or of the approved double-insulated type and plugged into a ground fault interrupter (GFI) outlet?
- Are effective guards in place over belts, pulleys, chains, and sprockets on equipment such as concrete mixers, generators, and air compressors?
- Are pneumatic and hydraulic hoses on power-operated tools checked regularly for deterioration or damage?

### 3.1.0 Air Compressors

Compressors are machines that are designed to compress air from an initial intake pressure and release it at a higher discharge pressure. Portable air compressors consist of a compressor and drive motor that can be readily moved as a unit. Portable air compressors are driven by either gas, diesel, or electric motors. A portable air compressor can be a small engine compressor mounted on two air tanks that can be handled by one person or a larger engine compressor mounted to a trailer that can be towed by a truck or a tractor. The larger types are more widely used on job sites.

The principle behind air compressors is to force air into an area smaller than it normally occupies so that the pressure increases significantly. This compressed air is then released and allowed to expand to its normal free state, thus producing work. *Figure 20* shows portable air compressors.

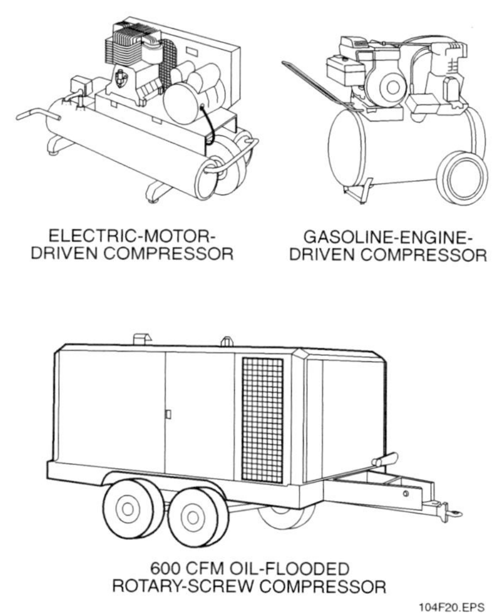


*Figure 20* ◆ Portable air compressors.



Instructor's Notes:

The rated operating pressure for most portable air compressors is 100 psig (pounds per square inch gauge) with a maximum allowable pressure of 120 or 125 psig. Portable air compressors have a safety valve setting, usually 10 psig above the maximum allowable pressure. The safety valve automatically opens to release the pressure if the pressure reaches the safety valve setting. The actual free air delivery is measured in cubic feet per minute (cfm). The smaller air compressors deliver air between 2.5 cfm and 6 cfm, while the larger, trailer-mounted compressors usually deliver air between 100 and 200 cfm.

Portable air compressors are classified as positive-displacement compressors, and positive-displacement compressors are classified as either reciprocating or rotary.

### 3.2.0 Pipe Preparation Machines

Line travel pipe preparation machines include cleaning/priming machines, cleaning/priming/tape wrapping machines, and coating/wrapping machines.

#### *3.2.1 Powered Pipe Wheel Cutters*

Pipe wheel cutters were developed for fast and easy pipe cutting. Different models of wheel cutters come in a variety of sizes capable of cutting pipe ranging from 2 to 36 inches in diameter. The Wachs wheel cutter (see *Figure 21*) is used for special pipe cutting situations where it is important that there is no metal chip contamination. The Wachs tool uses three cutting wheels that are fed equally into the cut using Wachs' starwheel feed system. As the cutting wheels are fed into the cut line, the pipe wall is pierced and displaced by the cutting wheels.

The combination of a parting blade and a beveling blade coupled to the Wachs LCSF/3 split frame is a more widely used tool in the oil and gas pipeline industry. Using this tool set to bevel the pipe after cutting may still produce some chips, but much of the time the metal is removed in ribbons, making it easier to control chip contamination.

#### *3.2.2 Cleaning/Priming Machines*

Cleaning/priming machine (CPL) units clean the full pipe circumference with wire cup brushes and/or scraper knives and, simultaneously, apply the primer coating. Three- or four-speed forward travel control is available. These are available for pipe sizes 3 to 60 inches.

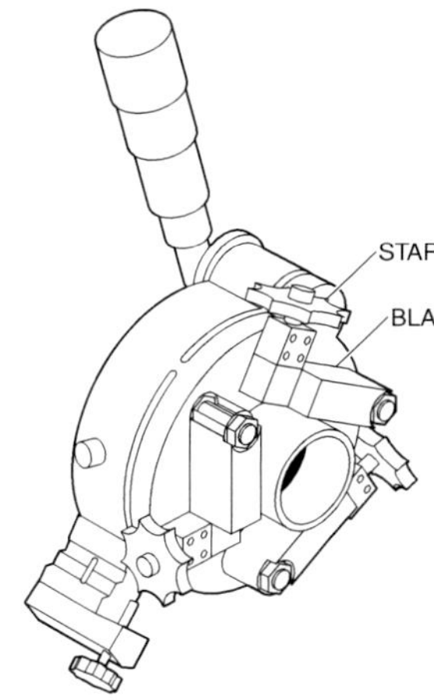


*Figure 21* ◆ Wachs powered pipe wheel cutter.

#### *3.2.3 Cleaning/Priming/Tape Wrapping Machines*

Line-traveling cleaning/priming/tape wrapping (CPT) machines perform pipe cleaning, primer application, and tape wrapping in a single pass. These units are equipped with powered pressure rollers, three- or four-speed forward travel, tape tension adjustment, tape overlap adjustment, and exhaust fan(s). These are available for pipe sizes 3 to 60 inches.

#### *3.2.4 Coating/Wrapping Machines*

Line-traveling coating/wrapping (CWL) machines apply coal tar enamel to $\frac{3}{32}$" thickness, a wrap of reinforcing material, and an outer wrap in a simultaneous operation. Independent clutches govern travel speed, coating, and wrapping. Double coating and wrapping machines are also available. These are available for pipe sizes 4 to 60 inches. Hand coating/wrapping equipment is also used. These types of wrappers are common in remote areas of the pipeline. *Figure 22* shows a hand coating/wrapping machine.

### 3.3.0 Air Movers

Air movers and fans are used in pipeline facilities for ventilation and air removal. Many areas require explosion-proof air movers and fans due to volatile fumes in the working environment. Two main types of air movers are air horns and axial fans.



**Classroom**

**Explain how powered pipe wheel cutters work and list their advantages.**

**Stress the functions of CPL, CPT, and CWL machines.**

**Classroom**

**Ask trainees to explain the functions and advantages of pneumatic air horns and axial fans.**

**Performance Testing**

**Have each trainee demonstrate Performance Tasks 3 and 4 to your satisfaction. Fill out a Performance Profile Sheet for each task for each trainee.**

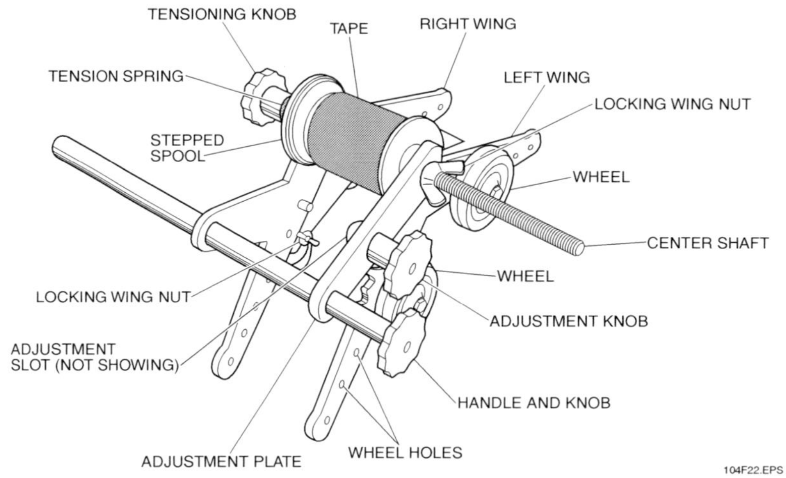


*Figure 22* ◆ Hand coating/wrapping machine.

### 3.3.1 Pneumatic Air Horns

Pneumatic, explosion-proof air horns (see *Figure 23*) are compact, quiet, and highly efficient for moving large volumes of air. Operating on compressed air, they are ideal for ventilation in hazardous and/or volatile areas. Having no moving parts, they are safe and easy to maintain.

### 3.3.2 Axial Fans

Pneumatic axial fans (see *Figure 24*) that are spark- and explosion-proof are available. The pneumatic fans are driven by a rugged, low-maintenance turbine directly from compressed air. These fans are especially suitable for ventilation, exhaust, and moving large air volumes through ductwork. They can be used both in normal atmospheres and where an explosion hazard is present in pump stations, refineries, and tank farms. Features available include protective grills on the intake side and provisions to mount flanges for installation.

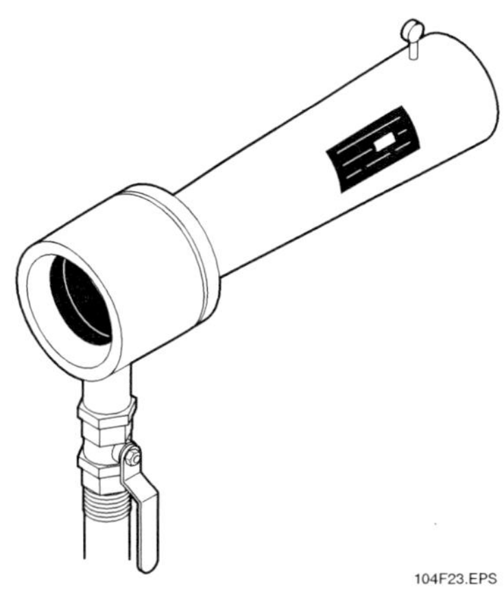


*Figure 23* ◆ Pneumatic air horn.

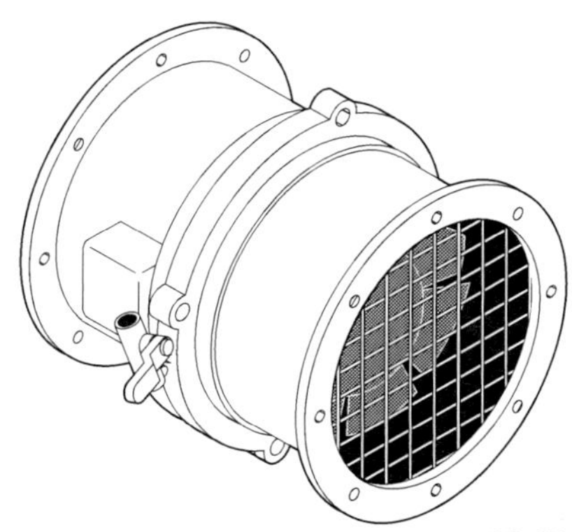


*Figure 24* ◆ Pneumatic axial fan.



Instructor's Notes:

## 4.0.0 ◈ WELDING EQUIPMENT

Several different types of welding equipment and accessories are on the market to help pipeline maintenance technicians and fabricators weld pipe to the high quality standards required in the pipeline industry. You must follow company policies and procedures, which dictate the certifications required to make welds on pipe. *Figure 25* shows an engine-driven welding machine.

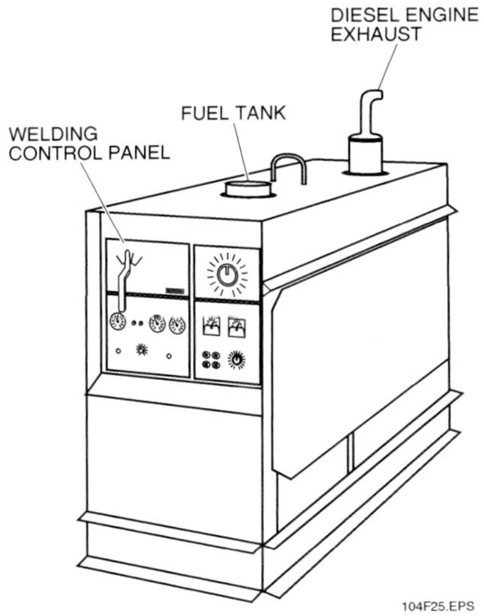


*Figure 25* ◈ Engine-driven welding machine.

### 4.1.0 Beveling Machines

Various types of beveling machines are used to prepare pipe ends before a joint can be fitted up for welding. Pipe ends can be beveled using grinders, small-diameter pipe bevellers, nibblers and cutters, thermal beveling machines, pipe facing machines, and band-type beveling machines.

#### *4.1.1 Beveling Using Grinders*

Portable hand grinders can be used to prep bevel the pipe, but great care must be taken to make the bevel at the specified angle and keep the bevel square prior to welding.

#### *4.1.2 Thermal Beveling (Beveling Machine)*

Thermal joint preparation using an oxyacetylene torch is a quick and easy way to bevel pipe ends; however, the cutting torch leaves an oxide coating on the pipe that must be ground away before welding. The oxyacetylene pipe-beveling machine is fitted over the pipe end and tightened to the pipe (*Figure 26*). The torch is then adjusted to cut the correct angle. An electric drive motor moves the cutting torch around the pipe end to cut the preset bevel.

There are many different types of oxyacetylene pipe-beveling machines on the market. The best way to learn how to operate the pipe-beveling machine is through on-the-job training with qualified personnel or the equipment manufacturer.

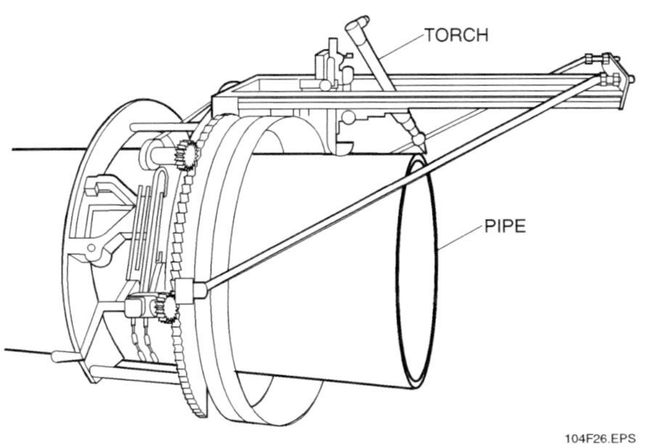


*Figure 26* ◈ Cutting/beveling machine.



**Classroom**

**Ask trainees to name five ways to bevel a pipe end before they fit a joint. Ask them to list the pros and cons of various beveling machines.**

**Audiovisual**

**Show Transparency 21 (Figure 28). Emphasize the advantages of band-type machines in cutting pipes.**

**Classroom**

**Have a trainee explain the workings of external and internal clamps to position pipe ends for welding. Correct any errors. Answer trainee questions.**

**Performance Testing**

**Have each trainee demonstrate Performance Task 5 to your satisfaction. Fill out a Performance Profile Sheet for this task for each trainee.**

**Homework**

**Assign reading of Sections 5.0.0–6.5.0 for the next class session. Answer any questions trainees have.**

**Materials**

**Ensure that you have everything required for demonstrations and testing during this session.**

#### *4.1.3 Pipe Facing Machines*

Pipe facing machines (PFMs) (see *Figure 27*) ensure a precise and consistent narrow-gap bevel for high quality welds. The machine mounts to the pipe end, square with the pipe axis, by means of hydraulically operated shoes. A rotating face-plate, fitted with as many as six cutters, machines the desired bevel on the pipe end.

Major projects normally require the use of two PFMs. The units are mounted on sideboom tractors or A-frames. A facing operation is completed in two to five minutes, depending on pipe wall thickness and skill of the operator. Hydraulic power for each unit can be supplied by diesel or electric drivers. They can be tractor-mounted on the ground in close proximity to the workpiece.

Pipe facing machines are available for pipe diameters up to 60 inches.

#### *4.1.4 Band-Type Machines*

Band-type machines (see *Figure 28*) feature a rugged design for accurate cuts and long life. The transmission/torch unit, with an appropriate band, cuts any size pipe. A 9-inch clearance requirement allows for minimum space use, such as small bell holes.

#### *4.1.5 Line-Up Clamps*

Prefabricated pipe welding clamps are available for both external and internal use to position pipe ends to be joined by welding. Internal alignment clamps are fitted with various types of wheels, depending on whether the pipe is coated internally or not, to allow the unit to easily be rotated and positioned inside the pipe. External clamps are constructed to overlap the two pipe faces to be joined by fitting around the exterior surface. Both types of clamps hold the pipe faces firmly in place to permit welding around the complete circumference of the pipe without having to adjust the clamp position.

Different types of external clamps match up and secure the two pipe ends using a hand-actuated levering device or, in the case of some larger models, with hydraulic jack assistance. Some internal clamp units have on-board compressed air tanks to retract or extend wheels and to move the unit through the pipe.

When required, line-up clamps can be site-fabricated to hold pipe in place for welding. Site-fabricated clamps are typically made up from iron bars, usually 6 to 16 inches in length. The clamps are welded together to form a circle around the pipe and hold the two pieces in place for welding.

### 5.0.0 ◈ METERS AND TESTERS

Meters and testers frequently used by pipeline maintenance technicians include multi-gas detectors, **holiday** testers, jeeps, and line locators.

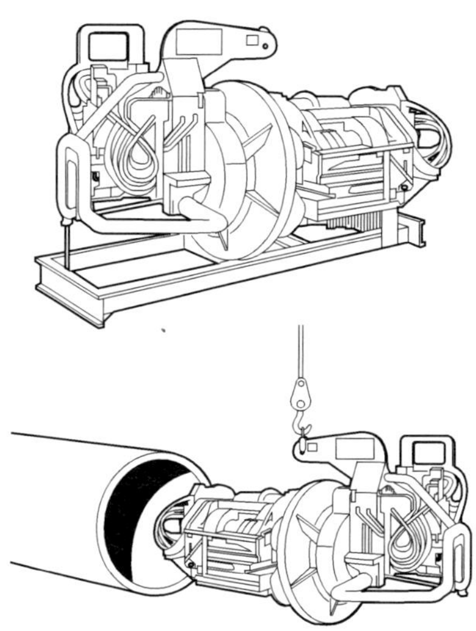


*Figure 27* ◈ Pipe facing machine.

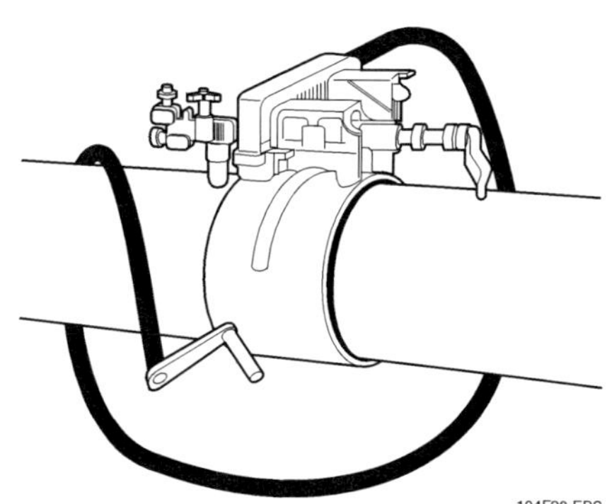


*Figure 28* ◈ Band-type machines.



*Instructor's Notes:*

### 5.1.0 Multi-Gas Monitors

Multi-gas monitors (see *Figure 29*) are used to sample the air for **hazardous atmosphere**. The most common gas detector, the Draeger tube, consists of a disposable direct reading detector tube and a piston-type volumetric pump.

To use this type of gas detector, simply break off the ends of the detector tube and place the tube in the pump. When the pump handle is squeezed, a predetermined 50 cc or 100 cc sample of air is collected. The absorbing media immediately reacts with the gas or vapor being analyzed, and a constant color stain is produced. The stain will vary in length to show the concentration being measured.

It is imperative when using this type of gas detector that you use the proper type of detection tube for the vapor you are analyzing. After taking the reading from the detection tube, the tube is disposed.

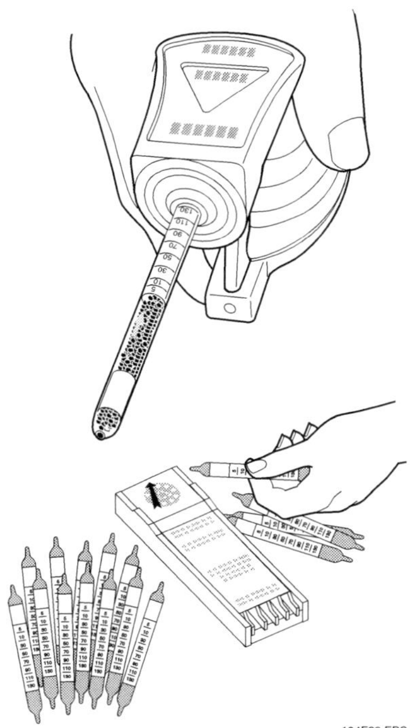

104F29.EPS

*Figure 29* ◆ Multi-gas monitor.

Other models of multi-gas monitors include those designed to detect combustible gases, carbon monoxide, hydrogen sulfide, and low levels of oxygen. Some of these detectors also are equipped with dataloggers and modems that allow them to communicate directly with a base unit that can receive real-time readings, a function that can be very valuable in emergency situations. Selected models include lithium ion battery packs that enable the monitor to be used for up to 20 hours of continuous operation without requiring a re-charge.

### 5.2.0 Portable Gas Detectors

Many models of portable gas detectors can monitor as many as five atmospheric conditions simultaneously, including oxygen, combustible gas, and selected toxic gases. Toxic gas monitoring capability typically covers both carbon monoxide and hydrogen sulfide sensors. Most portable gas detectors are powered by interchangeable nickel-cadmium or alkaline battery packs that provide up to 12 hours of continuous operation. *Figure 30* shows a widely used portable gas detector, the Biosystems PhD5.

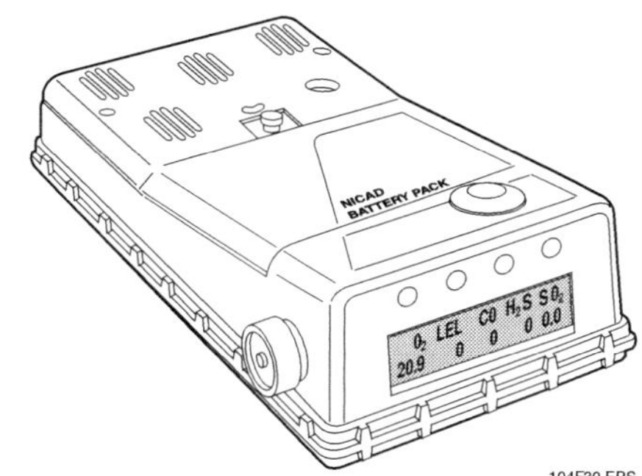


104F30.EPS

*Figure 30* ◆ Portable gas detector.

### 5.3.0 Holiday Detectors

Pipeline maintenance technicians are sometimes called upon to perform conductivity tests, or holiday tests, on piping systems. Conductivity testing ensures that there is no unprotected piping that is not covered with coating. Conductivity tests use amperage to test the conductivity of a material. The holiday testers have variable amperage ratings, and the amperage rating used must comply with the specifications of the piping system.



**Classroom**

**Ask trainees to define hazardous atmosphere. Give examples from your own experience of using multi-gas monitors, portable gas detectors, and holiday detectors.**

**Demonstration**

**Show trainees how a multi-gas monitor works. Let them practice putting in detector tubes and taking readings. Review their techniques and suggest ways to improve.**

**Classroom**

**Ask trainees to explain how a holiday detector works. Ask if they have any questions.**

**Review conductivity testing procedures. Ask trainees to explain the principles behind this technique. Review how to ensure accuracy.**

**Audiovisual**

**Show Transparency 22 (Figure 31). Ask trainees to explain the function of the metallic ring, the wand, and the alarm unit. Make sure they know what the term holiday means.**

**Classroom**

**Emphasize the importance of matching frequencies of transmitter and receiver to maximize effectiveness.**

**Safety**

**Ask trainees how a damp accessory improves safety.**

**Demonstration**

**Demonstrate how a line locator works. Ask trainees to explain how locators improve their ability to trace and analyze a pipeline.**

The basic principle behind this technique is based on establishing an electric potential between the pipe and an external electrode with the help of the holiday detector. The holiday detector (see *Figure 31*)consists of a metallic ring that encircles the pipe; a handheld voltage source, known as a wand; and an alarm unit. The amount of voltage applied to the ring can be adjusted according to the type of coating.

Thicker coatings, such as coal tars, generally require a higher setting. Too much voltage, however, can damage the thinner coatings, such as tapes. The pipeline maintenance technician pushes or rolls the holiday detector along the pipe. When the detector passes over a gap, or holiday, in the coating, an electric spark completes the circuit between the electrode and the pipe. An alarm on the unit is then activated, and the exact location of the flaw is marked.

### 5.4.0 Jeeps

The output voltage of the holiday detector is subject to influence by weather conditions, moisture on the pipe, improper grounding, and soil conditions. To ensure the absolute highest accuracy and performance from the holiday detectors, special jeeps are used to **calibrate** detector voltage output. These jeeps are designed to accurately measure the output voltage of any holiday detector.

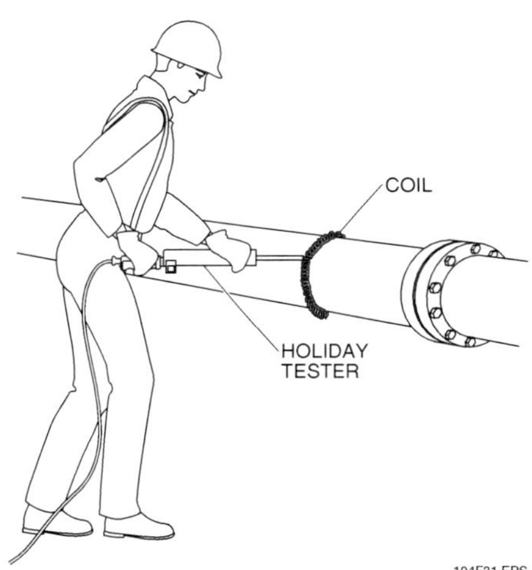


*Figure 31* ◆ Holiday detector.

The digital jeep provides high accuracy and versatility. It features a digital display of measured voltages. The precision measurement display minimizes interoperation errors and is good for daily tests. Frequent calibration is necessary to ensure that holiday detectors are actually giving the output voltage they were set to give.

Other types of jeeps measure both pulse and DC voltage outputs. Pulse-type holiday detectors that are used to inspect shrink sleeves, tapes, and coal tars will measure up to 40 kilovolts (kV). DC holiday detectors that are used to inspect thin film epoxy coatings have a DC range up to 5 kV. *Figure 32* shows a digital jeep.

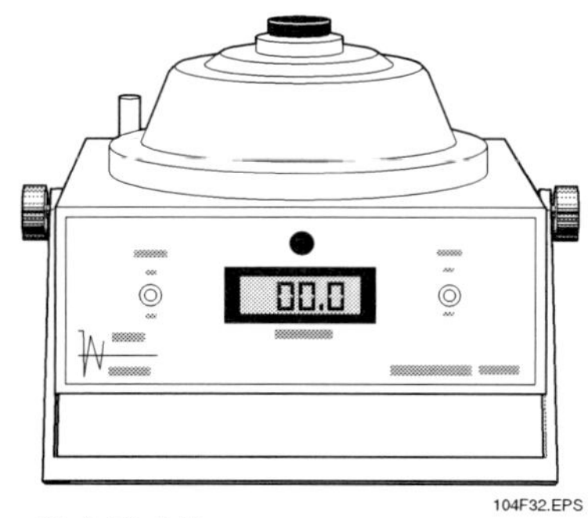


*Figure 32* ◆ Digital jeep.

### 5.5.0 Line Locators

It is sometimes necessary to determine the location of buried items such as the pipeline to be worked on and the location of foreign utilities, ground bed cables, or other concealed metallic structures. Pipe and cable locators are used when trying to locate existing pipeline utilities. Pipe and cable locators impress an electrical signal onto the target line, and this permits a handheld receiver to detect their location.

The type of signal that is transmitted to the pipe or cable can vary from manufacturer to manufacturer and can even be variable within a single locator. Alteration of the signal frequency allows the operator to set the most appropriate signal for the application. For example, poorly coated pipes, well-coated pipes, shorted pipelines, and insulated cables might each be best located using different frequencies. It is imperative to remember to set both the transmitter and the receiver to matching frequencies.

Other locator features can be useful in cathodic protection work. Some transmitters have a damp accessory that can be dipped around a conductor



Instructor's Notes:

to induce the signal, such as when tracing a live electrical cable. This technique saves time as well as enhances safety. The damp can be placed around an electrical conduit leading to a rectifier, for instance, so that unlocking and opening the rectifier enclosure becomes unnecessary.

Advances in locator technology have also enabled a variety of attributes to be incorporated into pipe and cable locators Depth measurement capability has greatly improved and allows the operator to measure, with increasing confidence, the depth of the target. Other pipe locators have the ability to measure and trace the amount of signal current being conducted by the pipeline or cable. These locators can trace the pipeline as well as improve the operator's ability to analyze and evaluate the pipe's current-carrying characteristics.

## 6.0.0 ◆ NONDESTRUCTIVE EXAMINATION METHODS AND TOOLS

Nondestructive examinations (NDE) are any tests that are used to test equipment and components without damaging them. The approach of nondestructive examination is a system of one or more types of tests that quantify defects or detect future failures.

With NDE, it is possible to detect defects or changes that will lead to premature failure. The results can be used to find the real cause of the change or abnormality and correct the problem, not just the symptom. *Table 1* lists the types of NDE.

**Table 1** Types of NDE

| Fluid | Type of Reading | Advantages | Limitations | Examples of Uses |
|---|---|---|---|---|
| Liquid penetrant | Detects surface openings due to cracks, porosity, seams, or folds. | Is inexpensive, easy to use, readily portable, and sensitive to small surface flaws. | Flaw must be open to the surface. Is not useful for porous materials. | Turbine blades for surface cracks or porosity. |
| Magnetic particles | Detects leakage of magnetic flux caused by surface or near-surface cracks, voids, inclusions, and material or geometry changes. | Is inexpensive and sensitive to both surface and near-surface flaws. | Is limited to ferromagnetic material; surface preparation and postinspection demagnetization may be required. | Railroad wheels for cracks. Detection of weld defects. |
| Ultrasonics | Detects changes in acoustic impedance caused by cracks, nonbonds, inclusions, or interfaces. | Can penetrate thick materials; excellent for crack detection; can be automated. | Normally requires coupling either by contact to surface or immersion in a fluid. Orientation, detection, or interpretation of the defect can present problems. | Adhesive assemblies for bond integrity. Detection of cracks. |
| Radiography | Detects changes in material density from voids, inclusions, material variations, and placement of internal parts. | Can be used to inspect a wide range of materials and thicknesses; is versatile; film provides a record of inspection. | Radiation safety requires precautions, is expensive, and detection of cracks can be difficult. | Pipeline welds for penetration, inclusions, and voids. Verification of parts in assemblies. |
| Visual-optical | Detects surface characteristics, such as finish, scratches, cracks, or color, and strain in transparent material. | Is often convenient and can be automated. | Can be applied only to surfaces, through surface openings, or to transparent material. | Paper, wood, or metal for surface finish and uniformity. |

104T01.EPS



**Performance Testing**

**Have each trainee demonstrate Performance Task 6 to your satisfaction. Fill out a Performance Profile Sheet for this task for each trainee.**

**Audiovisual**

**Show Transparency 23 (Table 1). Cover the right-hand side of the transparency. Ask trainees to identify the function of each nondestructive examination method. Review the answers for accuracy.**

**Classroom**

**Group trainees into threes. Ask one trainee in each group to list the five types of nondestructive examination methods and tools. Ask another to list the advantages of each type. Have the third trainee in the group list the disadvantages.**

**Teaching Tip**

**See the Teaching Tip at the end of this module.**

**Audiovisual**

**Show Transparency 24 (Figure 33). Ask one trainee to explain the principle behind the magnetic particle test. Ask another how to obtain maximum sensitivity.**

**Classroom**

**Describe the advantages of ultrasonic monitors. Ask trainees if they have questions.**

### 6.1.0 Dye Penetrant Test

Liquid penetrant inspection is a method of highlighting cracks and gaps that are not visible to the naked eye in the surface of smooth or nonporous samples. The liquid penetrant seeps into surface cracks, porosity, delaminations, or shrinkage areas by capillary action. It is also used to highlight wear patterns and measure stress cracking.

Liquid penetrant inspection involves wetting the surface of a test object so that the liquid flows over that surface to form a continuous and uniform coating and migrates into cracks and surface defects. The liquid is then washed off the surface, and a developer is placed on the surface. As the developer draws the penetrant out of the defects, it is stained by the penetrant and highlights the cracks and surface irregularities to form a visual image of the flaw.

With liquid dye penetrants, two types of indications are revealed: true indications and false indications. True indications are caused by the penetrant bleeding out from actual discontinuities in the metal. The standard defects indicated are fatigue cracks, underbead cracks, pits, and porosities. Large fatigue cracks are indicated by wide lines that become apparent quickly after the developer is applied. Underbead cracks are undersurface cracks that bleed through the surface. Underbead cracks are indicated by a line of dots that appear a few minutes after the developer is applied. Porosity is indicated by dots that come to the surface almost immediately.

The major reasons for false indications are failure to apply the penetrant correctly or rough, irregular surfaces of the test metal. The procedure for liquid dye penetrant testing must be carefully performed in order to achieve the proper results.

### 6.2.0 Magnetic Particle Test

When a magnetic field is generated in and around a part made of a ferromagnetic material, such as steel or iron alloys, and the lines of magnetic flux are intersected by a defect such as a crack, magnetic poles are generated on either side of the defect. Magnetic particle inspection works only for materials that can be magnetized, such as steel and nickel alloys.

In a magnetic particle test, the sample is magnetized by an electrical current, and magnetic particles are applied over the surface. These particles can be applied either dry or in a liquid carrier such as oil or water. The flaws on or just under the surface of the sample modify the magnetic field so that the particles outline the defect in a pattern and highlight the shape and location of the flaw.

This method is very sensitive to minor surface cracks and flaws that are near the surface. The disadvantages are the limitation to magnetic samples and the inability to read deeper flaws in the sample.

Maximum sensitivity is obtained when the part is oriented perpendicular to the magnetic field and the strength of the field saturates the entire part. The magnetic particle equipment uses either AC or DC and can be applied to the part with coils, probes, or flexible conductors that can wrap around a test piece.

Due to the variety of shapes of samples, the inductor for the test must come in a variety of sizes and shapes. Some sample parts, such as gears, balls, and ring-shaped parts, can be placed in the magnetic field without contact from a probe. Other samples use coils, clamps, central conductors, or other forms of direct contact with the sample. A welded joint would use a yoke to induce a magnetic field around the weld seam and show any discontinuities as extra particles scattered outside the weld seam. *Figure 33* shows the electromagnetic yokes of a magnetic particle tester.

### 6.3.0 Handheld Thickness Gauge

Handheld thickness gauges use high-frequency sound waves to detect the noise generated by plant equipment. Handheld thickness gauges measure thickness in materials or listen for turbulence in pipes to detect internal leaks and malfunctions. Other ultrasonic test instruments send focused, amplified sound waves into solid materials and compare them with the return wave to find flaws below the surface, such as corrosion, bad welds, or casting flaws. The frequency range for ultrasonic monitoring is from 20,000 Hz to 100 kiloHz.

A beam of ultrasound is similar to a beam of light. Both are waves and travel in a constant velocity in a consistent medium. The velocity depends on the medium, not on the properties of the wave form. Like a beam of light, ultrasonic waves are bent, reflected from surfaces, refracted when they cross boundaries of a different surface, and diffracted at edges or around obstacles. This beam, when reflected back, creates a signature that is interpreted as an image by the ultrasonic monitor.

Ultrasonic monitors can penetrate deep and with high sensitivity and accuracy. The output is readily digitized, and operation of the monitor is entirely nonhazardous. Some of the different types of ultrasonic techniques are pulse-echo, pulse-transmission, ultrasonic attenuation, continuous wave resonance, and ultrasonic spectroscopy. Ultrasonic monitors are very portable



Instructor's Notes:

but require experienced technicians to monitor equipment. Reading an ultrasonic flaw detector (see *Figure 34*) requires certification because the screen image shows the peaks of back reflection.

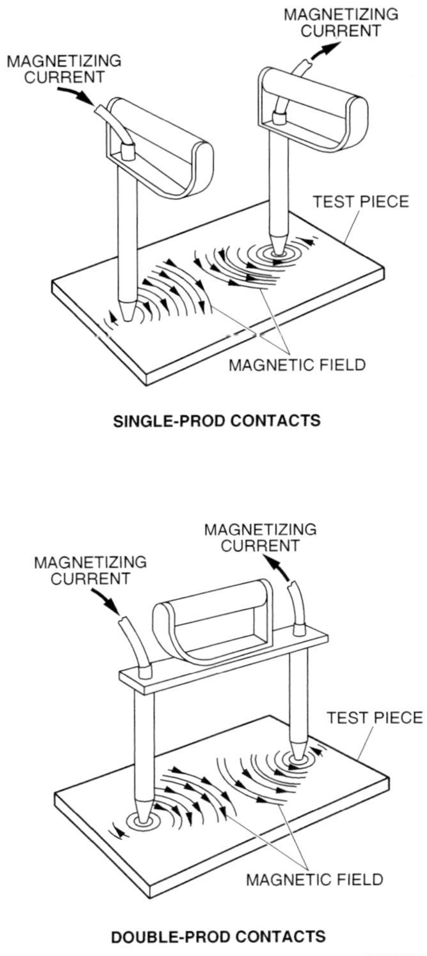


*Figure 33* ◈ Electromagnetic yokes of a magnetic particle tester.

*Figure 34* ◈ Ultrasonic flaw detector.

### 6.4.0 Radiography

Radiography is a nondestructive examination method that uses a beam of X-ray radiation that penetrates the weldment. X-rays are very sensitive and can be used to inspect the density of almost any ferrous, nonferrous, organic, or inorganic material. Some of the radiation energy is absorbed, and the intensity of the beam is reduced. The variations in the beam intensity are recorded on film. These variations are seen as differences in shading that are typical of the types and sizes of any discontinuities present.

X-rays can detect both surface and subsurface discontinuities. Because of the extremely powerful nature of gamma rays, much radiographic work is now done with radium or Cobalt 60 sources. The principal advantages of gamma ray inspection are the low cost and portability of the source. Safety is an essential part of radiography, and special precautions must be taken when working with X-ray materials to protect yourself from overexposure to radiation.

The X-ray method is used to detect internal defects and to check for proper alignment of assembled parts. X-rays are also used extensively to examine welds, check wall thickness, and examine the quality of casting and forging. *Figure 35* shows how an X-ray is made using gamma rays.



**Audiovisual**

**Explain the advantages of X-rays in detecting pipeline flaws. Show Transparency 25 (Figure 35). Ask trainees if they have questions.**

**Safety**

**Emphasize the importance of visual and optical inspection.**

**Performance Testing**

**Have each trainee demonstrate Performance Task 7 to your satisfaction. Fill out a Performance Profile Sheet for this task for each trainee.**

**Homework**

**Assign reading of Sections 7.0.0–7.4.3 and completion of the Review Questions for the next class session.**

**Materials**

**Ensure that you have everything required for testing during this session.**

**Safety**

**Ensure that trainees learn the applications and safety hazards of heavy excavating equipment.**

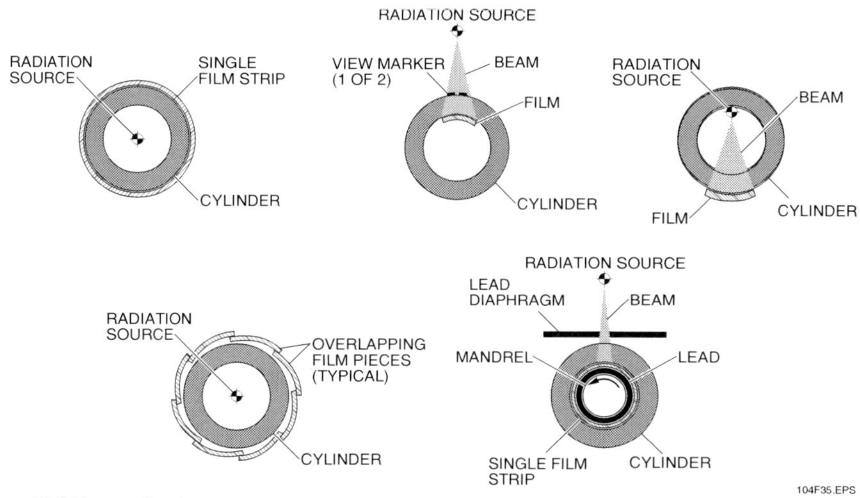


*Figure 35* ◆ X-ray made using gamma rays.

### 6.5.0 Visual and Optical Inspection

Visual and optical inspection of equipment is the easiest to perform but is the most likely to be overlooked as an inspection technique. There should be a systematic approach to visual inspection so that no details or events are missed. Make a detailed observation while the equipment is running and follow the path of the product, carefully watching every step and noting any abnormalities. Then, with the equipment off or in a safe position, look for any external detail that could provide a clue to potential trouble. Excessive corrosion, bent shafts, leaking seals, and any sign of abnormality should be documented in detail.

Hard-to-reach, internal, or hot parts of the equipment can be inspected with optical borescopes and fiberscopes. The optical borescope (see *Figure 36*) is designed to reach deep inside a machine, magnifies like a microscope, and has its own light source. The fiberscope is flexible because of its fiber-optic cable and can provide a close look in areas that can normally be seen only after the equipment is disassembled.

## 7.0.0 ◆ HYDRAULIC CRANES AND HEAVY EXCAVATING EQUIPMENT

Pipeline maintenance personnel must understand the types, applications, and safety hazards of cranes and heavy excavating equipment, although they are not required to operate this equipment because operators must be licensed. Pipeline maintenance personnel must also be familiar with the special types of rigging harness used to carry pipe.

### 7.1.0 Hydraulic Cranes

The basic operational characteristics of all mobile cranes are the same. These include the following:

- Adjustable boom lengths
- Adjustable boom angles
- Ability to lift and lower loads
- Ability to swing loads
- Ability to travel about the job site

The most common types of mobile cranes that a pipeline maintenance technician is exposed to are boom trucks, cherry pickers, carry decks, and pipeline side boom tractors.

#### *7.1.1 Boom Trucks*

Boom trucks (see *Figure 37*) are unlike any of the other mobile cranes in that they are mounted on carriers that are not solely designed for crane service. These cranes are mounted onto a commercial truck chassis that has been strengthened to carry the crane. The boom truck can also be used to haul loads and be driven on the highway system when the boom is in the transport position. Boom trucks typically range in capacity from 6 to 22 tons.



Instructor's Notes:

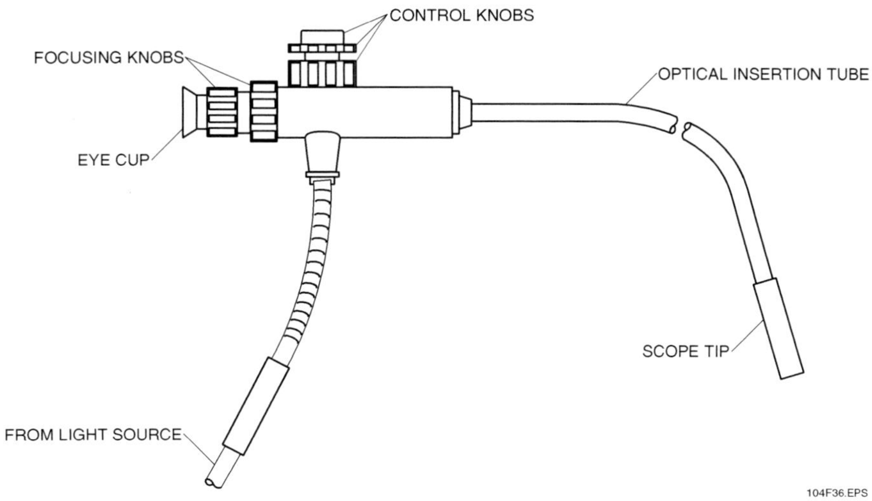


*Figure 36* ◆ Optical borescope.

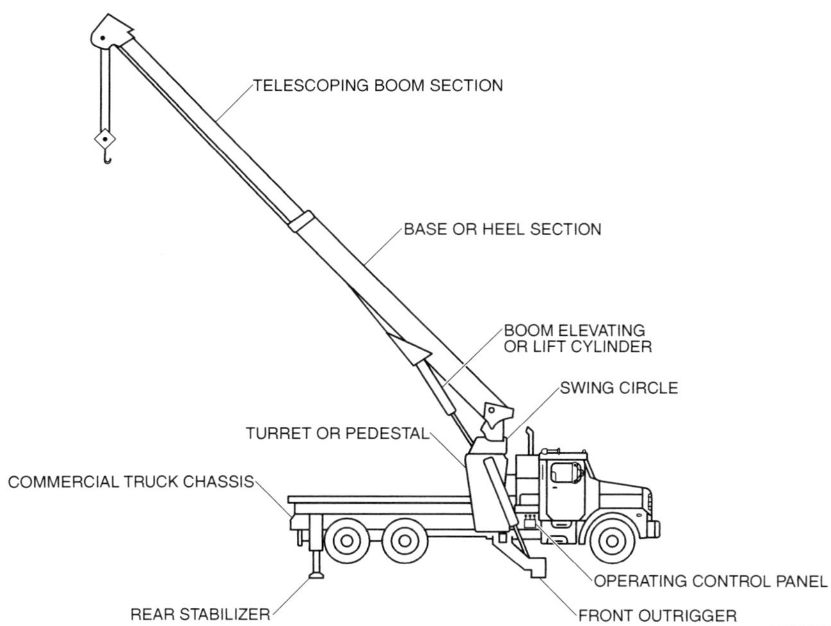


*Figure 37* ◆ Typical boom truck.



**Teaching Tip**

**See the Teaching Tip at the end of this module.**

**Classroom**

**Discuss the operational characteristics of mobile cranes.**

**Audiovisual**

**Show Transparency 26 (Figure 37). On an acetate sheet, list the characteristics and uses of boom trucks as trainees call them out.**

Audiovisual

**Show Transparency 27 (Figure 39). Stress how side boom tractors work and how to ensure safety during their use.**

Classroom

**Ask trainees to define backfilling. Describe how trackhoes function in backfilling.**

### *7.1.2 Cherry Pickers*

Cherry pickers (see *Figure 38*) are rough-terrain mobile cranes. They have oversized tires that allow them to move across the rough terrain of construction sites and other broken ground. Two types of cherry pickers are the fixed cab and the rotating cab.

### *7.1.3 Pipeline Side Boom Tractors*

A pipeline side boom tractor (see *Figure 39*), also known as a pipe layer, lays pipe down a trench while moving along the side of the trench. The pipeline side boom tractor is driven by a diesel engine. It moves on tracks and has a winch-driven boom mounted to one side. The boom lifts pipe and places it in the trench. A counterweight is mounted on the other side of the tractor and prevents the tractor from tipping over when lifting the pipe.

### *7.1.4 Trackhoes*

Trackhoes (see *Figure 40*) come equipped with a hydraulically powered bucket with a hydraulically operated closing lower "jaw" attached to help with grasping.

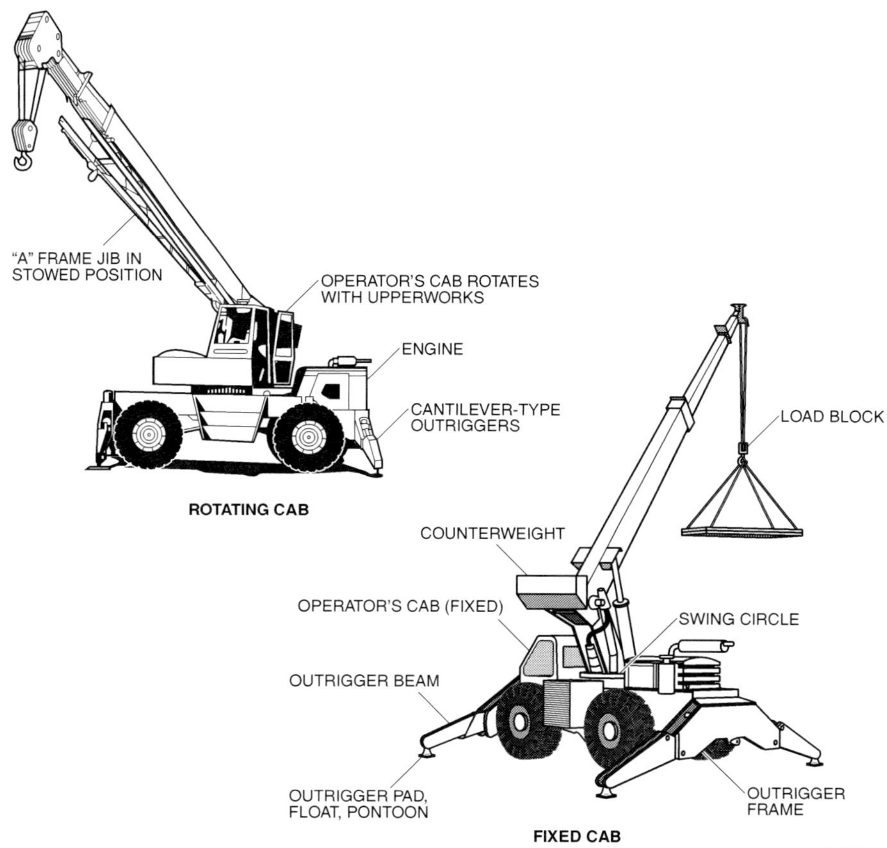


*Figure 38* ◆ Cherry pickers.



Instructor's Notes:

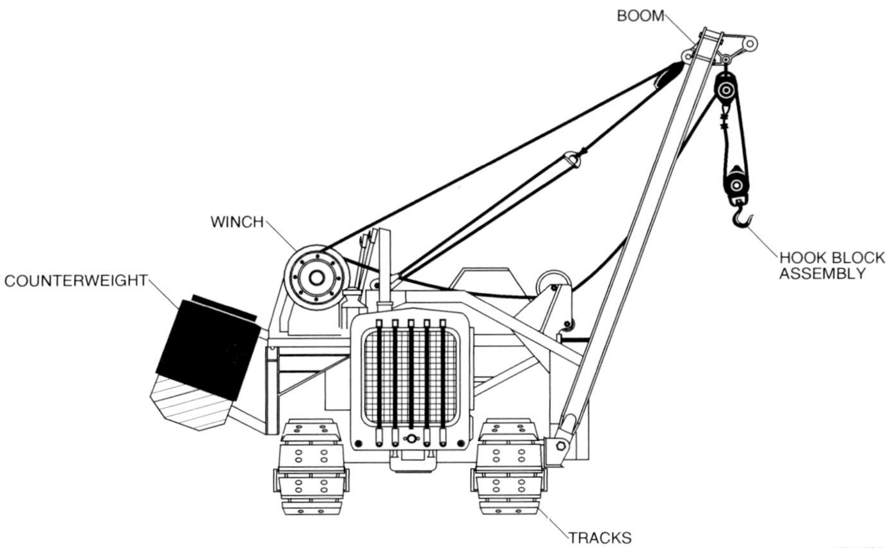


*Figure 39* ◆ Pipeline side boom tractor.

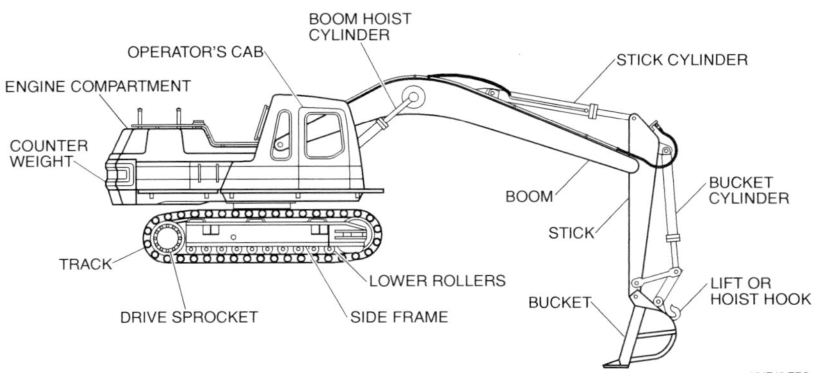


*Figure 40* ◆ Trackhoe.

Trackhoes are used for various applications such as:

- Digging flowline
- Digging dugouts
- **Backfilling** line leaks
- Loading trucks with clay
- Pulling subsoil/topsoil material back on leases to allow other equipment to be able to get behind material on fenced locations

### 7.2.0 Overhead Cranes

Electrically powered overhead cranes, with a wide range of lifting capacities, have become a mainstay throughout industry. Overhead cranes are available in both free-standing (self-supporting frame) and building-supported configurations. Overhead cranes employed in industrial settings range from small, 1-ton capacity light cranes serving a shop or work center to heavy-duty, rail-supported models



**Safety**

**Review the OSHA guidelines that apply to safe crane operation. Stress the importance of adhering to these guidelines.**

**Classroom**

**Describe the advantages of fabric slings and lowering-in belts. Answer trainee questions.**

**Refer to Figure 42. Explain how to choose the proper equipment depending on the diameter of a pipe.**

**Audiovisual**

**Show Transparency 28 (Figure 43). Ask trainees to list the advantages of choker belts.**

used on production lines, in maintenance shops, and in warehouses capable of lifting 50 tons or more.

Overhead cranes are made up of the rail or frame system, the trolley that moves along the rails, electric trolley drive unit, electric brakes, hoist motor and lifting mechanism, and controls. The largest cranes have an on-board cabin for the operator while others are controlled from the work floor by pendant-mounted controls or through the use of a remote radio-signal control unit.

Crane operators are certified in accordance with OSHA guidelines. Never use an overhead crane to move equipment or materials before reviewing your company's policies and procedures covering the training and certification of crane operators.

### 7.3.0 Pipeline Rigging Equipment

Pipeline rigging equipment ensures secure control in placement of the pipe and also prevents damage to the **pipe coating** and structural integrity of the pipe. Some of the more common pipeline rigging equipment is explained below.

#### *7.3.1 Lowering-In Belts*

Lowering-in belts are made up of a steel lifting iron and a fabric **sling** formed of several bands, or layers, of high-grade rip-resistant nylon. Each band should be manufactured to the same length, width, and stretching capacity as other bands in the sling.

The slings are flexible and easy to handle and conform to the pipe being lifted, even at very low temperatures. Lowering-in belts (see *Figure 41*) are available to handle a range of pipe sizes from 12 to 66 inches in diameter. Larger sizes may be special ordered. They are designed to permit easy hooking and unhooking by one person. Each model can be disassembled for storage and transport to and from the job site.

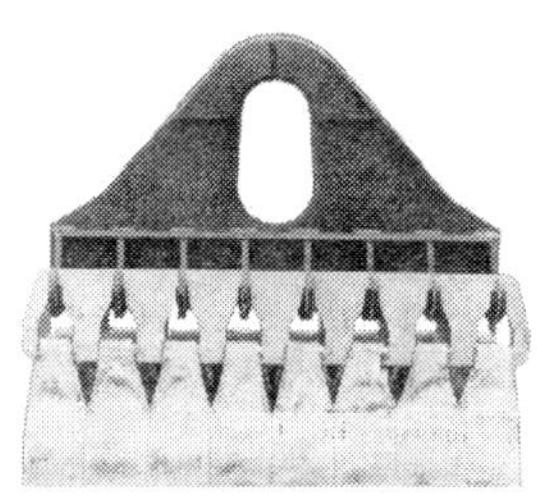


*Figure 41* ◆ Lowering-in belt.

#### *7.3.2 Roller Cradles and Wheel Cradles*

Roller cradles and wheel cradles (see *Figure 42*) are designed to lower-in and lift-out pipe during pipeline construction and maintenance activity. These cradles easily meet a broad range of lifting and other operational requirements. Cradle design is such that the a range of pipe sizes can be handled with the same size cradle. The pipe is protected from damage if loaded properly using the required number of cradle units to evenly distribute the load.

Roller cradles and wheel cradles are available for lifting and handling pipe from 4 to 60 inches in diameter. Some models of cradles are fitted with polyurethane-coated rollers that have high load capacity, good wear resistance, and elasticity under load that minimize breakage during use. Some models of cradles can be used with a matching strongback to couple two cradles together.

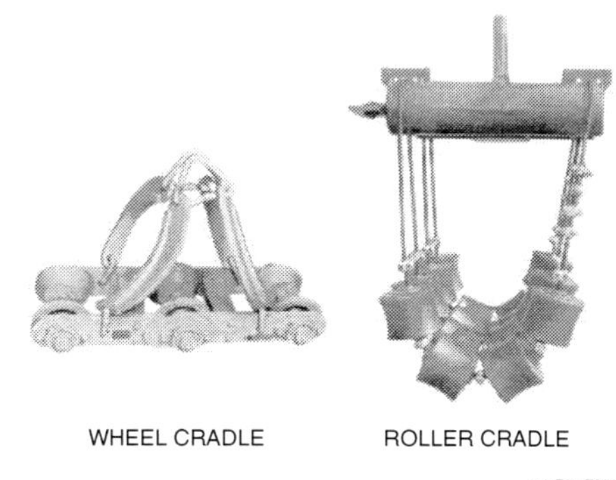


*Figure 42* ◆ Roller cradles and wheel cradles.

#### *7.3.3 Lined Steel Choker Belts*

Lined steel choker belts (see *Figure 43*) can be used to align pipe precisely for tasks like bending or welding. A distinct advantage to using choker belts is that these belts allow you to roll the pipe, pull the pipe horizontally, and secure extreme bends in the pipe without damaging the pipe or any applied coating.

Choker belts are adjustable to accommodate differences in coating material thickness. Choker belts are available in sizes that will handle pipe diameters up to 60 inches and lift loads up to 27,000 pounds.



Instructor's Notes:

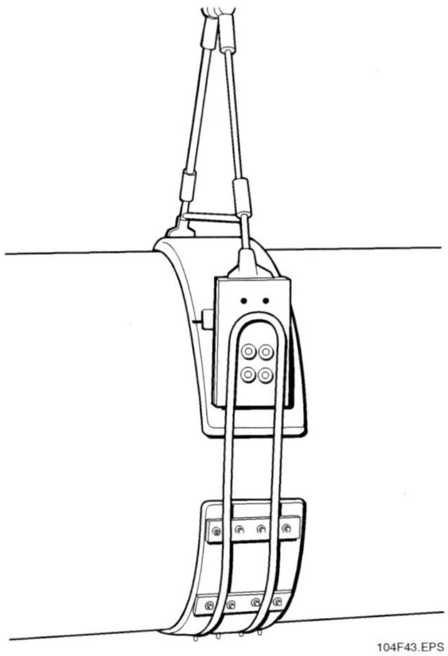

*Figure 43* ◆ Lined steel choker belts.

### *7.3.4 Pipe Handling Calipers*

Pipe handling calipers, sometimes referred to as pipe handling tongs, are used along with lowering-in belts, cradles, and slings to lift and position pipe for storage, construction, and maintenance tasks. Calipers are usually available to handle pipe up to 38 inches in diameter. Larger calipers can be special ordered when required.

Some calipers are designed so that a single individual can position a section of pipe and then release the calipers, eliminating the need for a second person to be stationed in a position to release the pipe.

## 7.4.0 Excavation Equipment

This section provides a general overview of heavy excavating equipment used to install, replace, and maintain pipelines and pipeline facilities. Types of excavating equipment explained include backhoes, loaders, and bulldozers.

### *7.4.1 Backhoes*

A backhoe (see *Figure 44*) is a hydraulically operated unit mounted on the back of a wheel-type tractor (or sometimes on a track-type tractor). The most common activities that backhoes are used for include:

- Digging and cleaning trenches
- Excavating hard to reach areas
- Hoisting light loads

**Safety**

**Refer to Figure 45. Stress the importance of understanding the operation of a loader and its accessories. Emphasize knowing what material is being carried and the effect on safety.**

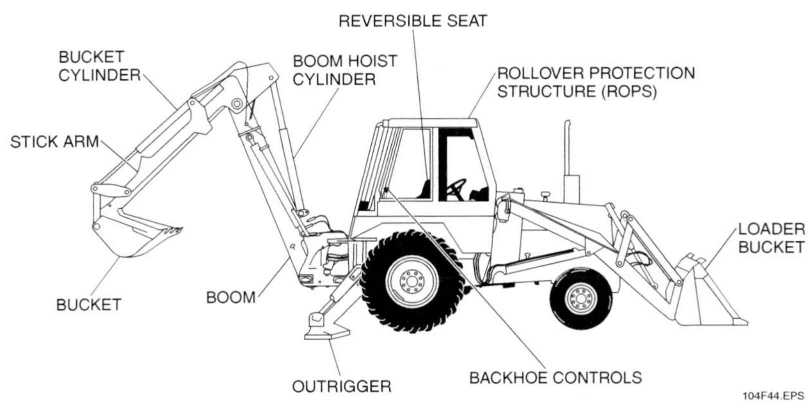


*Figure 44* ◆ Backhoe.

There are numerous type of backhoes such as:

- Rubber-tired tractor-type, hydraulically operated
- Truck-mounted, cable-operated
- Track-type, hydraulically operated
- Trenching machine with backhoe accessory

The tractor-mounted backhoe is smaller than most other hydraulic excavating equipment. It has a limited swing compared to the swing of the tracked excavator or telescoping excavator and is less powerful. The tractor-mounted backhoe is a useful piece of equipment because it is very mobile and can be maneuvered easily.

The backhoe attachment is usually mounted on a rubber-tired two- or four-wheel drive tractor. It may also be fitted onto a crawler tractor or a truck. The tractor usually has permanent mountings, such as a three-point hitch, to which the backhoe can be attached with pins. Most models are backhoe front-end loader combinations with permanent attachments.

### 7.4.2 Loaders

Backhoe loaders are used for a wide range of jobs. They include:

- Loading trucks
- Stockpiling
- Rough grading
- Digging
- Hauling materials and equipment
- Removing snow
- Backfilling

Understanding the operation of a backhoe loader and use of the bucket and other accessories is very important. The operator not only needs to know how to operate the machine and its accessories effectively, but also must understand the material being handled. This includes knowing the general properties of such materials as clay, sand, mud, and different types of aggregate.

The two categories of backhoe loaders are the wheel-type and the track-type. The wheel-type loader has rubber tires that are driven by either a gasoline or diesel engine. Early models of wheel-type backhoe loaders were on a rigid frame with only the rear wheels being driven. *Figure 45* shows the basic parts of a wheel loader.

Newer wheel-type backhoe loaders are articulated and allow for tighter turning and better maneuvering.

The track-type backhoe loader (see *Figure 46*) is mounted on tracks and looks very much like a bulldozer with a loading bucket instead of a blade. The engine for this loader is usually diesel. This type of backhoe loader comes in many sizes with bucket capacities up to four cubic yards.

### 7.4.3 Bulldozers

The bulldozer (see *Figure 47*) is designed to move large amounts of earth and provide maximum power and stability at speeds of up to approximately 8-10 miles per hour. Although it is not very fast, its good traction, stability, and lower center of gravity make it especially suited to work on irregular terrain and steep slopes.

Bulldozers may also be called crawler tractors, track-type tractors, dozers, or crawlers. They come in many different configurations and sizes and are made by a number of manufacturers. However, they all have the same basic features and perform the same operations.

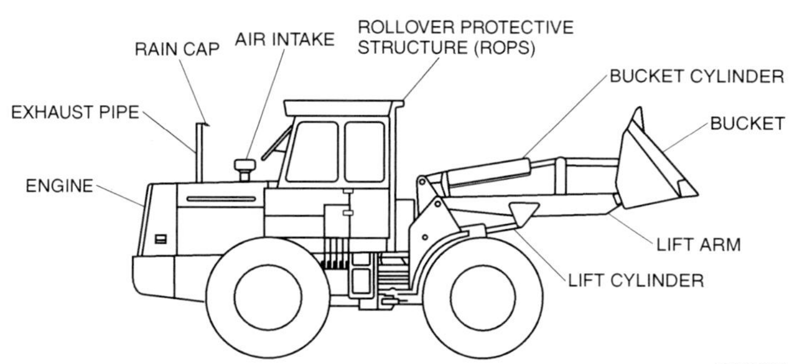


*Figure 45* ◈ Basic parts of a wheel loader.



Instructor's Notes:

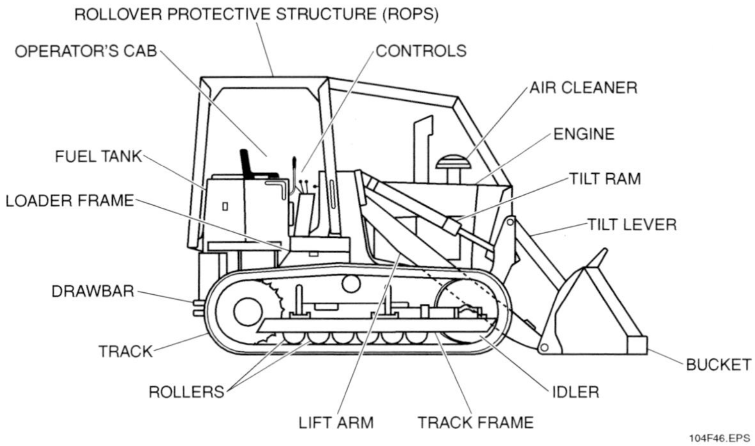


*Figure 46* ◈ Track-type backhoe loader.

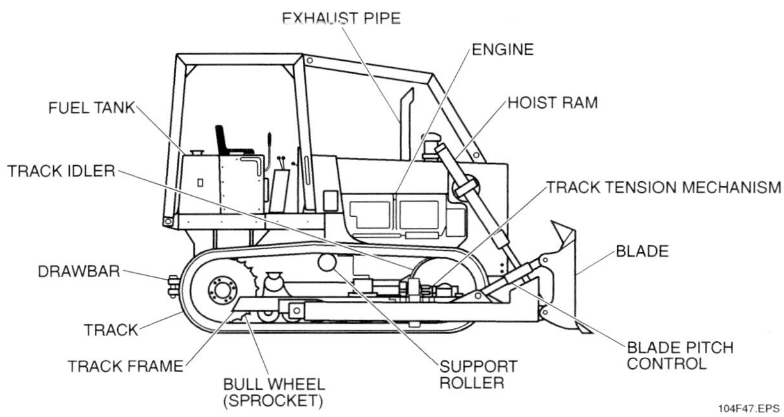


*Figure 47* ◈ Bulldozer.

Understanding how a bulldozer operates and how to set up the blade and other accessories for proper operation is very important. The operator must know how to operate the machine and its accessories effectively, aso well as the characteristics of the material being moved. This includes understanding the general properties of materials such as clay, sand, mud, different types of rock, and different types of manufactured materials like asphalt and concrete.

Because the bulldozer is equipped with tracks instead of wheels, it provides the maximum amount of traction. The primary uses of the bulldozer are as follows:

- Excavate, cut, fill, backfill, and spread material
- Maintain haul roads and temporary roads through rocky or rough terrain
- Clear construction sites of debris, clear land, and demolish stumps and other buried items
- Tow accessory equipment or push other heavy equipment



**Performance Testing**

**Have each trainee demonstrate Performance Task 8 to your satisfaction. Fill out a Performance Profile Sheet for this task for each trainee.**

## Summary

The purpose of this training module is not to teach you how to use all of the equipment presented here but to introduce you to the types of equipment that you may encounter on the job as a pipeline maintenance technician. Before using any piece of equipment on the job, you must be properly trained in the correct use of the tool as to prevent injury to yourself or to your co-workers. Before using any tool, the tool must be inspected to ensure that it is in proper working condition and free of defects, excessive wear, and other damage that might make the tool inoperable or dangerous.



Instructor's Notes:

### Review Questions

1. Never use a probe rod to locate a(n) _____ line.
   a. water
   b. electrical
   c. drain
   d. gas

2. The only wrench designed to be hit with a hammer is a _____ wrench.
   a. standard
   b. T-handle
   c. slug
   d. bolt-thru

3. A drift pin that is tapered on both ends is called a _____ pin.
   a. hole
   b. flange
   c. alignment
   d. barrel

4. Rotary pipe cutters will cut pipe up to inches in diameter.
   a. 24
   b. 32
   c. 34
   d. 36

5. T-handle flange spreaders make _____ easier.
   a. pipe bending
   b. pipe removal
   c. gasket replacement
   d. maintenance

6. Hydraulic flange spreaders operate at pressures of _____ tons.
   a. 2 to 4
   b. 4 to 5
   c. 4 to 8
   d. 5 to 10

7. Most taps and connections are made with the main line _____.
   a. shut down
   b. in service
   c. removed
   d. disassembled

8. The T-101 series hand drilling and tapping machine is used to make _____ inch taps.
   a. ½ to 4
   b. 1 to 6
   c. 2 to 8
   d. 6 to 10½

9. Portable compressors have a safety value that is usually set _____ psig above the maximum allowable pressure.
   a. 5
   b. 10
   c. 15
   d. 20

10. Cleaning/priming machines are available for pipe up to _____ inches in diameter.
    a. 20
    b. 40
    c. 60
    d. 80

11. Air horns operate on _____.
    a. electricity
    b. compressed air
    c. gasoline
    d. hydraulics

12. To square the ends of a pipe, a _____ is used.
    a. beveling machine
    b. grinder
    c. pipe facing machine
    d. bandsaw

13. Handheld thickness gauges use _____ to check pipe wall integrity.
    a. light beams
    b. vibration
    c. sound waves
    d. infrared beams

14. Pipeline side boom tractors are sometimes called _____.
    a. trenches
    b. trackhoes
    c. pipe layers
    d. excavators

15. Bulldozers are capable of top speeds from _____ miles per hour.
    a. 8–10
    b. 10–12
    c. 13–14
    d. 14–16



**Classroom**

**Go over the Review Questions. Encourage trainees to ask questions. Have trainees prepare for the Module Examination.**

**Examination**

**Administer the Module Examination. Record the results of the exam on Craft Training Report Form 200 and submit the results to the Training Program Sponsor.**

**Performance Testing**

**Ensure that all Performance Tasks have been completed to your satisfaction. Fill out a Performance Profile Sheet for each task for each trainee.**

**Be sure to record the results of the testing on Craft Training Report Form 200 and submit the results to the Training Program Sponsor.**

GLOSSARY

# Trade Terms Introduced in This Module

*Backfilling:* The technique for covering a completed pipeline so that adequate fill material is underneath and over the pipe to protect it from loose rocks, abrasion, shifting, and washouts.

*Calibrate:* To determine, by measurement or comparison with a standard, the correct value of each scale reading on a meter or other device.

*Excavation:* A man-made cavity or depression in the earth's surface formed by earth removal. This includes a trench, deep foundation, tunnel, shaft, or open excavation, but does not include borrow pits, gravel pits, and quarries unless specified by the local regulations or safety and health officer/inspector.

*Hazardous atmosphere:* An atmosphere that is identified as being explosive, flammable, poisonous, corrosive, oxidizing, irritating, oxygen deficient, toxic, or otherwise harmful, which may cause illness, injury, or death.

*Holiday:* An imperfection in the pipe coating that exposes unprotected surface to the environment.

*Pipe coating:* Dielectric coating on pipe to provide electrical insulation from other materials.

*Skid:* Pipeline skids are timbers used to support and stabilize the pipe off the ground during construction or maintenance work. These timbers, which typically measure 4" × 6" × 36", are usually cut from oak or other hardwood for strength and durability. Skids are built up one level at a time until the structure is sufficient for the task. Skids can be used as a single support or in multiples for larger jobs.

*Sling:* A wide rubber and fabric sling at the end of the boom cat's hoisting lines that is used for lowering in or handling coated or wrapped pipe.

*Tap:* A hole or opening in a line into which a gauge or valve may be inserted and screwed tight.

*Tapping (hot):* To tap into a pipeline that is in service and contains a hazardous liquid and/or vapors.

*Trench:* An excavation having a depth that exceeds its width measured at the bottom.



Instructor's Notes:

ACKNOWLEDGMENT

# *Figure Credit*

| | |
|---|---|
| **Ridge Tool Company** | 104F07 |

# MODULE 62104-02 — TEACHING TIPS

The following are suggested activities or instructional methods to help you teach the material in this AIG.

## General

When you call on someone to answer a question, the rest of the class relaxes or even tunes out because they expect that the question and answer will take place only between you and the trainee you called on. Instead, use this technique to involve more trainees in answering questions and to keep them on their toes.

1. Ask trainees to define a term or explain a concept.
2. After one trainee has answered, ask a trainee seated nearby if the answer is right. Then ask whether a trainee in the back of the room agrees.
3. Ask trainees to explain why they think an answer is right or wrong.
4. Use the session to clear up incorrect ideas and encourage trainees to learn from their mistakes.

## Section 2.0.0 *Using and Caring for Hand Tools*

This quick quiz reviews the proper and effective use of hand tools. You will need photocopies of the quiz provided on the following page. Trainees will need pencils. Allow 15 to 20 minutes for this quiz. If you allow trainees to use the Trainee Guide, decrease the amount of time to complete the quiz.

1. Make as many photocopies as you will need of the quiz on the following page.
2. Give trainees 15 to 20 minutes to complete the quiz.
3. Go over the answers to the quiz.
4. Ask trainees if they have questions about the procedures, tools, or equipment.

### *Answers to Quick Quiz*

1. c. hex turn flange spreader
2. a. repair suction and discharge hoses
3. c. ratcheting the blade into the pipe little by little
4. b. t-handle flange spreader
5. d. making taps of 36 inches in size
6. a. ratchet action flange spreader

**Quick Quiz** ***Using and Caring for Hand Tools***

Circle the correct answer.

**1.** To spread a flange that is corroded or stuck, you would use a _____.
  a. hydraulic flange spreader
  b. t-handle flange spreader
  c. hex turn flange spreader
  d. ratchet action flange spreader

**2.** The banding tool can _____.
  a. repair suction and discharge hoses
  b. locate exact placement of a pipeline
  c. hit a slug wrench safely
  d. fasten the nuts on pipe joints

**3.** The proper technique for a rotary pipe cutter involves _____.
  a. standing close to the pipe you are cutting
  b. feeding the blade into the pipe in one full motion
  c. ratcheting the blade into the pipe little by little
  d. keeping full pressure on the pipeline

**4.** The only type of flange spreader that is used in pairs is the _____.
  a. hex turn flange spreader
  b. t-handle flange spreader
  c. ratchet action flange spreader
  d. hydraulic flange spreader

**5.** Hand drilling and tapping machines can be used for all of the following except _____.
  a. making hot taps in small lines
  b. making connections to pipelines and tanks
  c. cutting cement-lined cast iron
  d. making taps of 36 inches in size

**6.** To insert a blind or gasket into a pipe connection, you would use a _____.
  a. ratchet action flange spreader
  b. t-handle flange spreader
  c. hex turn flange spreader
  d. hydraulic flange spreader

## Section 2.7.2 *Hand Drilling and Tapping Machines*

This exercise will give trainees an opportunity to observe hand drilling and tapping machines being used. Arrange a field trip to a nearby job site and obtain permission to either observe personnel or perform these tasks yourself.

You will need these items:

- Hand drilling and tapping machines
- Appropriate personal protective equipment

Trainees will need these items:

- Pencils and notebooks to take notes
- Appropriate personal protective equipment

To complete this exercise, do the following:

1. Ensure that trainees are wearing appropriate personal protective equipment.
2. Tell trainees to observe the procedures carefully and to take notes.
3. Observe pipeline personnel hand drilling and using tapping machines or perform the demonstration yourself.
4. If time permits and if appropriate, have trainees practice on the hand drilling and tapping machines. Have trainees practice on the T-101 series if available.
5. Review the trainees' work and provide guidance as necessary.
6. If time permits, show trainees how to tap into a pipe while it is under operating pressure.
7. After the trip, ask trainees if they have questions about procedures and ask about any difficulties they encountered.

## Section 6.1.0 *Dye Penetrant*

This experience will help trainees find pipeline defects that are not visible to the naked eye.

1. Demonstrate dye penetrant tests.
2. Stress exactness in the procedure for best results. What happens if the test is not exact?
3. Have trainees practice dye penetrant tests. Tell them to look for examples of fatigue cracks, underbead cracks, pits, and porosities. Provide guidance as necessary.
4. Help trainees note the condition that indicates each defect.
5. Explain the difference between true indications and false indications.
6. Ask trainees what safety advantages the dye penetrant test provides.

## Sections 7.0.0–7.4.3 *Hydraulic Cranes and Heavy Excavating Equipment*

Hearing from an outside expert will help trainees understand the applications and safety hazards of cranes and heavy excavating equipment. Although pipeline maintenance personnel do not operate this equipment (operators must be licensed), they must be familiar with the applications. Invite a representative from an excavating company to explain how the company works with heavy excavating equipment.

1. Have trainees prepare questions for the visiting expert. Encourage them to pinpoint what conditions call for the use of a certain type of unit, such as a backhoe, loader, or bulldozer.
2. Ask the company representative to stress safety features and share insights from actual incidents.

## MODULE 62104-02 — ANSWERS TO REVIEW QUESTIONS

| Answer | Section |
|---|---|
| **1.** b | 2.10 |
| **2.** c | 2.3.2 |
| **3.** d | 2.4.0 |
| **4.** c | 2.5.0 |
| **5.** c | 2.6.2 |
| **6.** d | 2.6.4 |
| **7.** b | 2.7.0 |
| **8.** a | 2.7.2 |
| **9.** b | 3.1.0 |
| **10.** c | 3.2.2 |
| **11.** b | 3.3.1 |
| **12.** c | 4.1.3 |
| **13.** c | 6.3.0 |
| **14.** c | 7.1.3 |
| **15.** a | 7.4.3 |

## CONTREN™ LEARNING SERIES — USER UPDATES

The NCCER makes every effort to keep these textbooks up-to-date and free of technical errors. We appreciate your help in this process. If you have an idea for improving this textbook, or if you find an error, a typographical mistake, or an inaccuracy in NCCER's Contren™ textbooks, please write us, using this form or a photocopy. Be sure to include the exact module number, page number, a detailed description, and the correction, if applicable. Your input will be brought to the attention of the Technical Review Committee. Thank you for your assistance.

*Instructors* – If you found that additional materials were necessary in order to teach this module effectively, please let us know so that we may include them in the Equipment/Materials list in the Instructor's Guide.


**Write:** Curriculum Revision and Development Department
National Center for Construction Education and Research
P.O. Box 141104, Gainesville, FL 32614-1104

**Fax:** 352-334-0932

**E-mail:** curriculum@nccer.org


Craft ____________________ Module Name ____________________

Copyright Date __________ Module Number __________ Page Number(s) __________

Description

______________________________________________

______________________________________________

______________________________________________

(Optional) Correction

______________________________________________

______________________________________________

(Optional) Your Name and Address

______________________________________________

______________________________________________

Module 62105-02

# *Pipeline Documentation*

# Pipeline Documentation
# Annotated Instructor's Guide

**Module 62105-02**

## MODULE OVERVIEW

This module introduces trainees to the various reports and methods used to document the location and condition of pipelines. Topics covered include pipeline maps and drawings, documentation, and documentation management.

## PREREQUISITES

Please refer to the Course Map in the Trainee Module. Prior to training with this module, it is recommended that the trainee shall have successfully completed the following:

Core Curriculum; Pipeline Core, Modules 66101-02 and 66102-02; Pipeline Maintenance, Level One, Module 62103-02

## OBJECTIVES

Upon completion of this module, the trainee will be able to:

1. Read alignment sheets.
2. Identify and explain documentation processes and forms.

## PERFORMANCE TASKS

Under the supervision of the instructor, the trainee should be able to:

1. Read and interpret alignment sheets.
2. Using a company's manuals or procedure books, identify and explain some of the company's required documentation and records.
3. Fill out, properly submit, and file a company maintenance document according to the company's documentation procedures (or simulate doing so).

## NCCER STANDARDIZED CRAFT TRAINING PROGRAM

The National Center for Construction Education and Research (NCCER) provides a standardized national program of accredited craft training. Key features of the program include instructor certification, competency-based training, and performance testing. The program provides trainees, instructors, and companies with a standard form of recognition through a National Craft Training Registry. The program is described in full in the *Guidelines for Accreditation*, published by the NCCER. For more information on standardized craft training, contact the NCCER by writing us at P.O. Box 141104, Gainesville, FL 32614-1104; calling 352-334-0911; or e-mailing info@nccer.org. More information may be found at our Web site, www.nccer.org.

## HOW TO USE THIS ANNOTATED INSTRUCTOR'S GUIDE

Each page presents two sections of information. The larger section displays each page exactly as it appears in the Trainee Module. The narrow column ties suggested trainee and instructor actions to each page and provides icons to call your attention to material, safety, audiovisual, or testing requirements. The bottom of each page includes space for your notes.

**Teaching Tip**

If you see the Teaching Tip icon, that means there is a teaching tip associated with this section. Also refer to any suggested teaching tips at the end of the module.

## SAFETY CONSIDERATIONS

Ensure that the trainees are equipped with appropriate personal protective equipment. Emphasize the importance of following all safety precautions and procedures when working with power tools.

## PREPARATION

Before teaching this module, you should review the Module Outline, Objectives and Performance Tasks, and the Materials and Equipment List. Be sure to allow ample time to prepare your own training or lesson plan and gather all required equipment and materials.

## MATERIALS AND EQUIPMENT LIST

Transparencies and blank transparencies
Markers/chalk
Highlighter
Pencils and scratch paper
Masking tape or pushpins
Graph paper
Module Examinations*
Performance Profile Sheets*

Overhead projector and screen
Whiteboard/chalkboard
Elevation scale
Samples of topographic maps, strip maps, piping and instrumentation diagrams (P&IDs), and electrical blueprints
Appropriate personal protective equipment

* Located in the Test Booklet packaged with this Annotated Instructor's Guide.

## TEACHING TIME FOR THIS MODULE

An outline for use in developing your lesson plan is presented below. Note that each Roman numeral in the outline equates to one session of instruction. Each session has a suggested time period of 2½ hours. This includes 10 minutes at the beginning of each session for administrative tasks and one 10-minute break during the session. Approximately 15 hours are suggested to cover *Pipeline Documentation*. You will need to adjust the time required for hands-on activity and testing based on your class size and resources.

| Topic | Planned Time |
|---|---|
| **Session I. Pipeline Maps and Global Positioning Systems (GPS)** | |
| A. Maps | ________ |
| B. Survey Maps/Alignment Sheets | ________ |
| C. Topographic Maps | ________ |
| D. Profile Maps | ________ |
| E. Strip Maps | ________ |
| F. GPS | ________ |
| **Session II. P&IDs and Drawings** | |
| A. P&IDs | ________ |
| B. Electrical Drawings | ________ |
| 1. Electrical Blueprints | ________ |
| 2. Electrical Schematics | ________ |
| **Session III. Documentation, Part I** | |
| A. Valve Inspection Reports | ________ |
| B. Exposed Pipe Reports | ________ |
| C. Aerial Patrol Reports | ________ |
| D. Aerial Patrol Followups | ________ |
| E. Tank Inspection Reports | ________ |

**Session IV. Documentation, Part II**

A. Pipe Repair/Replacement Reports ____________

B. Safety-Related Condition Reports ____________

C. Accident Reports ____________

D. Work Orders ____________

E. Permits ____________

**Session V. Document Management, Part I**

A. Retention Rates

1. Construction Records ____________
2. Hydrostatic Testing Records ____________
3. Operation and Maintenance Maps and Records ____________
4. Daily Operating Records ____________
5. Miscellaneous Records ____________

**Session VI. Document Management, Part II**

A. Document Control ____________

B. Liquid Inspection Intervals ____________

C. Gas Inspection Intervals ____________

D. Summary

1. Summarize module ____________
2. Answer questions ____________

E. Module Examination ____________

1. Trainees must score 70% or higher to receive recognition from the NCCER.
2. Record the testing results on Craft Training Report Form 200 and submit the results to the Training Program Sponsor.

F. Performance Testing ____________

1. Trainees must perform each task to the satisfaction of the instructor to receive recognition from the NCCER.
2. Record the testing results on Craft Training Report Form 200 and submit the results to the Training Program Sponsor.

Module 62105-02

# *Pipeline Documentation*

Instructor's Notes:

Instructor's Notes:

# Introduction to the Pipeline Operator Qualification Rule, API, and NCCER

*It could be said that petroleum products keep America running. Petroleum products like oil and natural gas are the principal sources of energy in the United States and are used in many manufacturing and chemical processes. Petroleum products are primarily transported by pipeline. There are more than a million miles of liquid and natural gas pipeline in this country, ranging in diameter from two inches to four feet. These pipelines bring crude oil and natural gas from production fields to processing plants and refineries—and ultimately to the consumer.*

## Pipeline Operator Qualification

Petroleum products are considered hazardous, so pipeline personnel must take extra precautions and follow specific procedures when working with them. To help ensure the safe handling and transportation of petroleum products, the United States Department of Transportation (DOT) mandates safe practices as federal regulations.

One such regulation, the Operator Qualification (OQ) rule, requires pipeline operators to develop a written qualification plan to evaluate their personnel and their contract personnel (see *49 CFR Part 192 Subpart N—Gas,* and *49 CFR Part 195 Subpart G—Liquid*). The evaluation hinges on the operator's ability to perform certain covered tasks and to be able to recognize and respond to abnormal operating conditions that may be encountered while performing these covered tasks.

## Covered Tasks

A pipeline performance task is deemed *covered* if the activity passes the following four-part test:

1. Is it performed on a pipeline facility?
2. Is it an operations or maintenance task?
3. Is it performed as a requirement of *49 CFR Part 195 (Liquid)* or *49 CFR Part 192 (Gas)*?
4. Does it affect the operation or integrity of the pipeline facility?

If the answer to all four questions is *yes*, then the activity is considered *covered* and must figure into the evaluation process of pipeline personnel.

## American Petroleum Institute

The OQ rule and its related covered tasks provided the catalyst to develop a unified pipeline training and assessment program—so that pipeline operators and owners have ready access to the assessment and training tools they need. This effort was the direct result of the vision and leadership provided by The American Petroleum Institute (API) and API members.

Efforts to unify the industry are nothing new to API. Indeed, the impetus for forming the Institute in 1919 was the need to standardize engineering specifications for drilling and production equipment. Today, the most pressing issues revolve around public perceptions and government policies toward the industry. Speaking with one voice on these issues has become as essential as having interchangeable parts in the field. API's members determine the petroleum industry's positions on public policy as well as the standards that govern its day-to-day operations worldwide.

## Consortium on Operator Qualification

API assembled a focus group whose primary purpose was to identify covered tasks and provide compliance recommendations to pipeline operators.

The Consortium on Operator Qualification (COOQ) became the guiding force for the identification and development of covered tasks for the liquid pipeline industry. Their work product, API publication 1161, *Guidance Document for the Qualification of Liquid Pipeline Personnel,* formed the basis for the resulting Pipeliner Training and Assessment Program.

The covered task list created in API publication 1161 has now evolved to include gas covered tasks and is maintained by NCCER. The most current covered task list can be downloaded at www.nccer.org.

## National Center for Construction Education and Research

With the guidance and support of API, its members, and the Consortium on Operator Qualification, the National Center for Construction Education and Research (NCCER) was able to design and implement the Pipeliner Training and Assessment Program.

NCCER is a not-for-profit 501(c)(3) education foundation established in 1995 by the world's largest and most progressive construction and maintenance contractors and national contractor associations. NCCER was founded to address the

severe workforce shortage facing the construction and maintenance industries and to develop a standardized training process and curricula. Today, NCCER is supported by hundreds of leading construction and maintenance companies, manufacturers, and national associations and developed more than 35 construction and maintenance titles.

NCCER also maintains a National Registry, which contains the records of trainees and instructors who have received NCCER training. All completion information for every trainee is sent to the NCCER and filed within the Registry. The Registry can then confirm training and skills for workers as they move from company to company, state to state, or even within their own company.

### *Standardized Curriculum (NCCER's Contren™ Learning Series)*

NCCER curricula are developed using the expertise of industry subject matter experts (SMEs). The use of SMEs as an integral part of every new development or revision ensures that the curriculum is current, industry-driven, and indicative of industry practices. The development and/or revision process is facilitated from start to finish by NCCER project managers, who work closely with the SMEs and a select team of technical writer(s) on the development and presentation of the curriculum.

### *Pipeline: Five Training Paths*

The pipeline curriculum focuses on essential skills and competencies within five skill areas:

- Corrosion Control
- Electrical and Instrumentation
- Maintenance
- Mechanical
- Operations (Field and Control Center)

SMEs from the pipeline industry have determined the skills and competencies needed based on API's covered tasks and the consensus of the group/industry. Specific company information is generally not incorporated into the curriculum unless the SMEs determine that the information is the industry standard.

### *Training Modules and Covered Tasks*

The smallest unit of instruction in the curriculum is a *module*. Each curriculum is composed of modules (the equivalent of stand-alone chapters) that are logically sequenced into complete *levels* of instruction. Every module is task-oriented and requires the successful completion of a written (knowledge-based) examination and a performance (hands-on) test. Modules may be individually selected and/or combined for efficient task training. Modules in the pipeline curricula, to the extent possible, will align with API-developed covered tasks. While not all modules will relate to covered tasks, all covered tasks will relate to at least one module.

### *Benefits of Training to the Pipeline Marketplace*

Applying NCCER's standardized training process allows pipeline owners and contractors in the United States to establish the benchmark for acceptable skill and competency levels within the pipeline industry. This process will assist training organizations in developing and/or improving standardized pipeline training programs that effectively match their resources and operating conditions while responding to regulatory demands and to the needs of the industry. The process will also recognize training organizations that meet the minimum standards of competence, warranting regulatory, public, and professional confidence. This in turn will foster national unity within the pipeline industry for the mutual benefit of individual technicians, owners, contractors, and ultimately pipeline-users and society.

## Additional Resources

To review the Pipeliner Training and Assessment Program Implementation and Resource Guide and the latest version of pipeline covered tasks, visit NCCER's Web site at www.nccer.org, or contact:

**National Center for Construction Education and Research**
PO Box 141104
Gainesville, FL 32614-1104
(888) 622-3720
Web site: www.nccer.org

For more information on API, please contact:

**American Petroleum Institute**
1220 L Street, Northwest
Washington, D.C. 20005
(202) 682-8000
Web site: www.api.org

To order *API Pub 1161*, contact Global Engineering Documents by calling 1-800-854-7149 or visiting their Web site at www.global.his.com.

Additional sources of information on the pipeline operator qualification regulation include:

**DOT – Office of Pipeline Safety**
www.ops.dot.gov

**Transportation Safety Institute**
www.tsi.dot.gov

5.iv PIPELINE MAINTENANCE LEVEL ONE — TRAINEE MODULE 62105-02

Instructor's Notes:

## COURSE MAP

This course map shows all of the modules in the first level of the Pipeline Maintenance curriculum. The suggested training order begins at the bottom and proceeds up. Skill levels increase as you advance on the course map. The local Training Program Sponsor may adjust the training order.

### PIPELINE MAINTENANCE LEVEL ONE

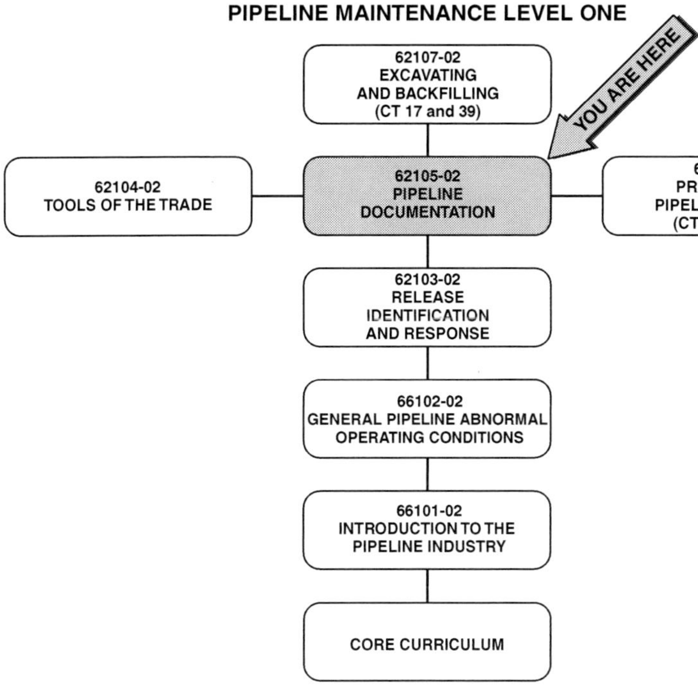


105CMAP.EPS

**Homework**

**Assign reading of Module 62105-02.**

Instructor's Notes:

## MODULE 62105-02 CONTENTS

## Figures





Instructor's Notes:

MODULE 62105-02

# Pipeline Documentation

### Objectives

When you have completed this module, you will be able to do the following:

1. Read alignment sheets.
2. Identify and explain documentation processes and forms.

### Prerequisites

Before you begin this module, it is recommended that you successfully complete the following: Core Curriculum; Pipeline Core, Modules 66101 and 66102; Pipeline Maintenance Level One, Module 62103.

## 1.0.0 ◆ INTRODUCTION

The generation and maintenance of maps and records associated with the pipeline is necessary for a number of reasons. Below are just some of these:

- Maps are used in the design of the pipeline.
- Maps are used to show the location of the pipeline.
- Drawings and schematics are used to provide greater detail on the parts of the pipeline, such as pump stations and tank facilities.
- Records are used to document what has been done on the pipeline and to equipment.
- Records are required to plan personnel training and to document personnel qualifications.
- Records and maps are required by federal, state, and local regulations.

Specific forms and maps are needed for specific purposes. The forms needed for different purposes are identified in a company's procedure manual. Some of the most common ones are discussed here.

For documents to be useful, they must be controlled and maintained, so that they are current, accurate, and available. Document management makes this possible. Although each company may have its own form designs and document management system, they all have a common purpose. An overview of these is presented in this module. This module is divided into the following sections:

- Pipeline maps and drawings
- Documentation
- Document management

## 2.0.0 ◆ PIPELINE MAPS AND DRAWINGS

Maps of the pipeline route showing how the route is aligned with true north and its surroundings are called alignment maps. They include several types of these maps as well as diagrams and schematics:

- Maps
- Global positioning system (GPS)
- **Piping and instrumentation diagrams (P&IDs)**
- **Electrical schematics**

### 2.1.0 Maps

Maps are important documents throughout a pipeline's existence. Initially, maps are used to determine what the pipeline route should be. Maps showing area highways, railroads, political boundaries, and topographic features are studied. The locations of oil fields, refineries, other pipelines, river ports, and facilities that could be used are identified. Crossing certain topographic features, such as mountains, lakes, or major rivers must be analyzed for feasibility, and the most favorable sites chosen.



**Materials**

**Ensure that you have everything required to teach the course. Check the Materials and Equipment List at the front of this module.**

**Classroom**

**Explain that the trade terms shown in bold are defined in the Glossary at the back of this module.**

**Teaching Tip**

**See the general Teaching Tip at the end of this module.**

**Audiovisual**

**Show Transparency 1, Course Objectives.**

**Show Transparency 2, Performance Tasks.**

**Classroom**

**Ask trainees to give reasons for generating and maintaining pipeline maps and records. Record answers on the board.**

**Classroom**

**Refer trainees to Figure 1. Discuss right-of-way and station numbers.**

**Performance Testing**

**Have each trainee perform Performance Task 1 to your satisfaction. Fill out a Performance Profile Sheet for each trainee.**

**Classroom**

**Refer trainees to Figures 2 and 3. Explain the relationship between topographic features and contour lines.**

**Teaching Tip**

**See the Teaching Tip at the end of this module.**

The pipeline route is imposed on the maps, and these are used throughout the pipeline's operation for locating the pipe when necessary and for defining the area along its route. Current maps of the pipeline are required by regulations as well.

Maps that may be used include the following:

- Survey maps/**alignment sheets**
- Topographic maps
- Profile maps
- Strip maps

### *2.1.1 Survey Maps/Alignment Sheets*

When the route is decided, the proposed **right-of-way (ROW)** is laid out and permits are obtained. At that point, the proposed ROW is surveyed using design maps or aerial photographs, and alignment maps of the route are produced. As the pipeline is built, the line is imposed on the survey map, which includes angles of direction changes, property crossings, land survey lines, and features legally defining property limits (see *Figure 1*). The position of each feature such as fences, direction changes, property lines, ditches, creeks, and others is shown by its distance along the route from the company-designated starting point.

**Station numbers** are the number of feet from the company-designated starting or ending point. Special features such as rivers, highways, and railroads require additional information and planning. These maps/alignment sheets will be maintained and updated throughout the life of the pipeline.

### *2.1.2 Topographic Maps*

Topographic maps, also called topo maps, show the ground's physical features and their elevations above sea level. The earth's shapes and forms are shown by contour lines spaced to show the most useful contours of features. Spacing may be different in different areas, depending on the rate of elevation changes. Each contour line connects points of equal elevation and contour lines cannot cross other lines or divide. Closely spaced lines indicate steep slopes, while contours spaced far apart represent gentle slopes. Contours trend up valleys, cross the streambed and return down the valley on the other side, forming a V which points upstream. *Figure 2* shows the relationship between topographic features and contour lines.

Numbers giving the elevation are incorporated into lines at appropriate intervals, generally every fifth line. These are called index contours and are shown with darker lines. Lines between the index contours represent specific increments that are indicated on the map. These increments can be from 5 to 200 feet, depending on the topography being mapped (see *Figure 3*). The elevation of a particular point is determined by its location relative to these lines.

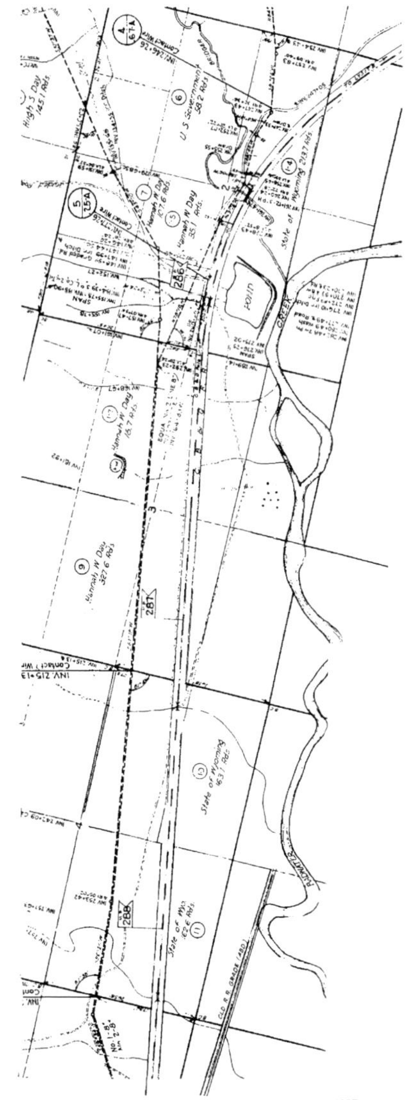

105F01.EPS

*Figure 1* ◈ Pipeline survey map.



*Instructor's Notes:*

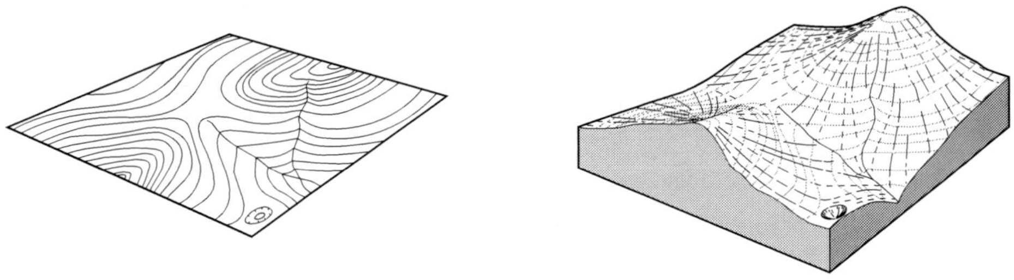


*Figure 2* ◈ Relationship between topographic features and contour lines.

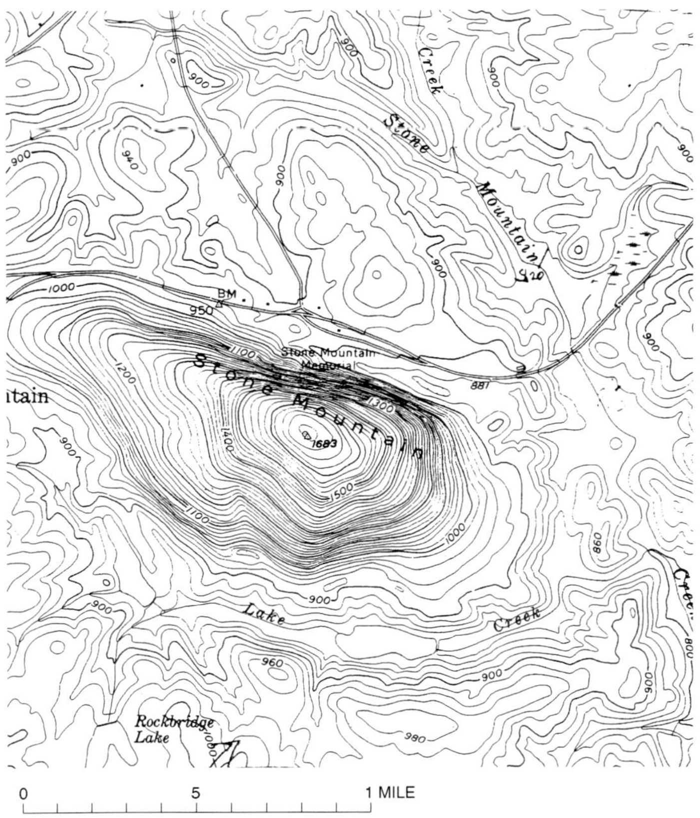


*Figure 3* ◈ Sample topographic map.

**Demonstration**

**Show trainees how to construct a simple profile map like the one in the Trainee Guide.**

**Teaching Tip**

**Refer trainees to Figure 5. See the Teaching Tip at the end of this module.**

**Classroom**

**Ask trainees if they have heard of or used GPS. Describe its use in the pipeline industry.**

**Ask trainees if they have questions about the material covered so far.**

**Homework**

**Ask trainees to read Sections 2.3.0–2.4.2 for the next class session.**

Topographic maps are important for pipeline planning because they indicate extreme elevation changes, slope of the ground, and features that should be avoided. They can also be useful in field work by helping to identify locations from physical features. Gradients of land areas or streams can be calculated from the maps by measuring the distance, converting it to miles, and dividing it into the difference in elevation.

Topographic maps look down on the ground as from an aerial photograph; in fact, most modern ones have been made from photographs taken by planes or satellites. Topo maps are available for every area on the earth, including the ocean floors. The U.S. Geological Survey is a good resource for topographic maps.

### *2.1.3 Profile Maps*

Profile maps show the elevation of the ground and pipeline from the side view or as a cross section. Hills and valleys are seen from the side as they normally would be. To produce a profile, a graph is drawn with an elevation scale on the vertical axis and the distance scale on the horizontal axis. The elevation readings along the line on the topo map are transferred to this graph and plotted at specific points along the line distances indicated by the horizontal axis (see *Figure 4*).

The pipeline does not always run in a straight line. Sections that change direction can be shown in profile by separate sections, or usually the profile is shown as a straight line all the way across. The vertical scale of the profile is usually exaggerated to provide better definition of land features. The depth of the pipeline is also plotted on the profile in the same way. Profile maps give a good visual indication of where the pipe is located both vertically and horizontally.

### *2.1.4 Strip Maps*

Strip maps are usually combinations of different types of maps or drawings over the same area; in this case, the area is the pipeline route. The strips of each type of map or drawing can be laid out side by side on the same sheet so that they line up along the route. There may be views shown of the ground from an aerial or plan view and side or elevation views, showing the ground profile along a specific line. The pipeline may be superimposed on all views. Strip maps help to coordinate surface features with topography and pipeline depth, as well as other pipeline data at any point on the route. *Figure 5* shows a sample strip map.

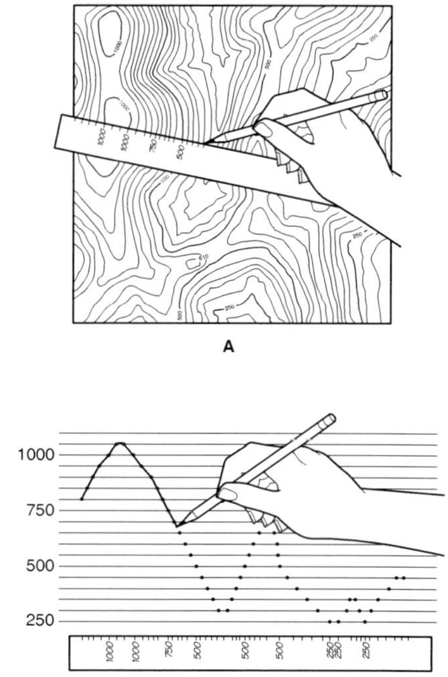


*Figure 4* ◆ Simple method for constructing a profile.

## 2.2.0 GPS

The Global Positioning System (GPS) is a worldwide navigation system that uses radio signals broadcast by satellites. The GPS is included here because it is being used to make maps of the earth that are more accurate than any ever made and can be used with other maps to precisely determine one's location.

NAVSTAR, or NAVigation System using Time and Ranging, is the name of the navigational satellite system. The system uses 24 satellites in six orbits with an average height of about 12,550 miles. As many as eight of these satellites may be above the horizon when viewed from any point on earth. *Figure 6* shows the arrangement of these satellites.

For navigation, a computerized radio receiver in an airplane, ship, or other vehicle uses the satellite signals to calculate its own location. Hikers and others on foot may use small, portable receivers. The system has both civilian and military users.



Instructor's Notes:

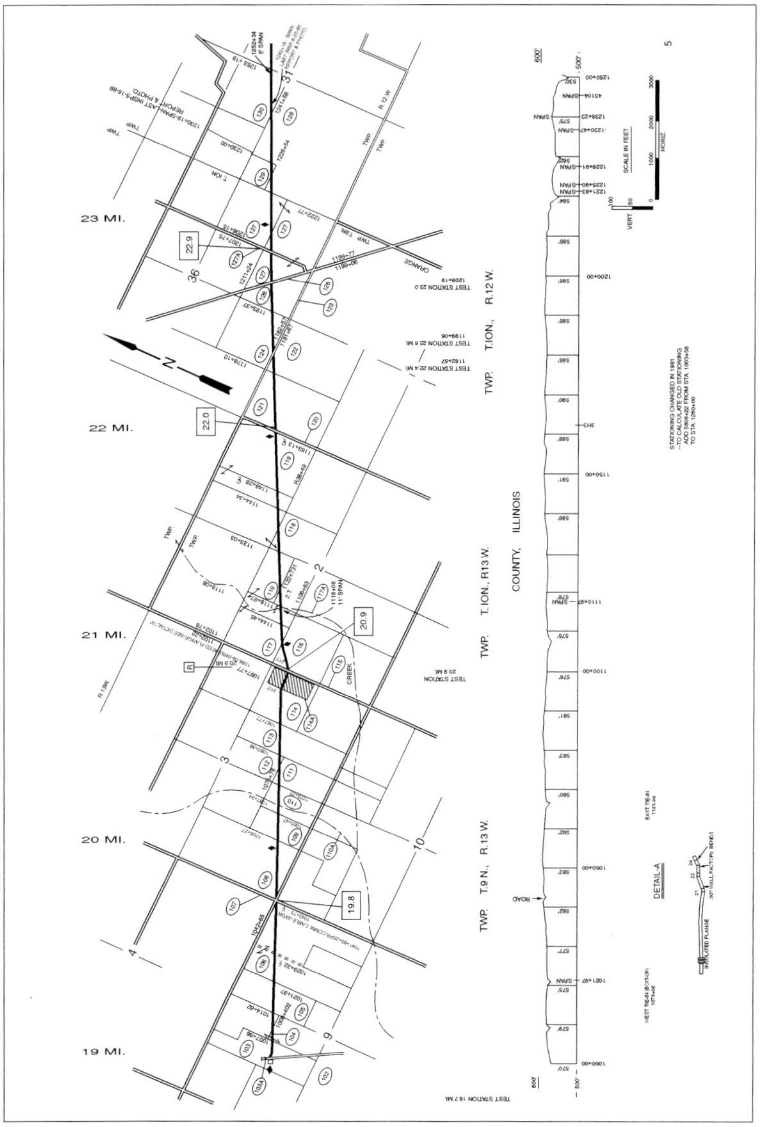


105F05.EPS

*Figure 5* ◈ Sample strip map.

**Materials**

**Ensure that you have everything required for teaching this session.**

**Teaching Tip**

**Refer trainees to Figures 7, 8, and 9. See the Teaching Tip at the end of this module.**

**Refer trainees to Figures 10 and 11. See the Teaching Tip at the end of this module.**

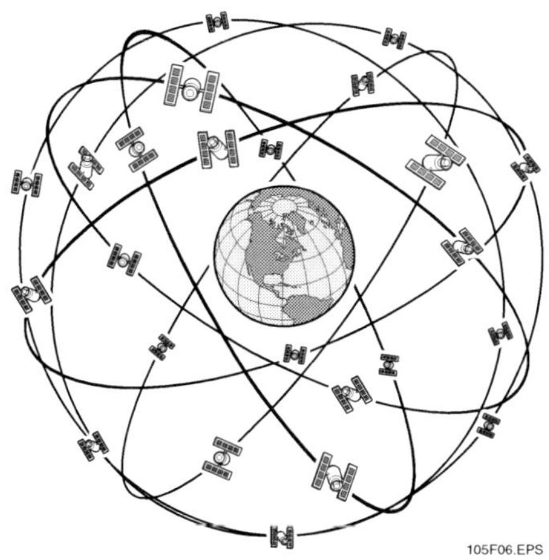


*Figure 6* ◆ Arrangement of navigational satellites.

A large percentage of the world's base information maps (including topographic maps) display both geographic coordinates (latitude and longitude) and the Universal Transverse Mercator Grid (UTM). The portable receivers are designed to report positions in both of these coordinate systems, so users have maps showing where they want to go and receivers that give them the location where they are. With this and a compass, users should be able to find their target. For military use, some receivers are designed to use the Military Grid Reference System (MGRS) as well as the other coordinate systems. MGRS is also available on some topographic maps.

A receiver uses signals from at least four and often fourteen satellites. It will lock onto the best four signals. The signal indicates the precise time and the coordinates of the satellite. From these, the receiver determines its latitude and longitude. GPS is used for surveying, exploring, and tracking ground movements. In the newest systems, GPS can be used to locate positions within inches.

### 2.3.0 P&IDs

A P&ID depicts the functional layout of a liquid or gas system and its piping, valves, and instrumentation (see *Figure 7*). It is accurate to the extent that all components are connected to each other as shown in relation to flow path orientation. A P&ID does not always represent the actual physical layout of equipment; that is, a valve that may appear to be right of the discharge of a pump can actually be located some distance away or even on another floor. Many times a P&ID will use a broken line around groups of equipment to indicate that they are all in the same building.

The P&ID depicts all components and details of a particular system, including pipe sizes, flange sizes, valve sizes, flow direction, and references to other diagrams. Standardized symbols are used for these components instead of pictures. Industry standards for these symbols have been developed and adopted by organizations such as ANSI, ISA, and ASME to aid interpretation. *Figure 8* shows standard P&ID piping symbols, and *Figure 9* shows P&ID instrumentation symbols.

### 2.4.0 Electrical Drawings

Electrical drawings include several types, which are used for different applications. The most common of these are the blueprint and schematic.

#### *2.4.1 Electrical Blueprints*

Electrical **blueprints** are layouts of electrical circuits and their related components shown to scale on the structure (see *Figure 10*). The lines and symbols used on the blueprint may vary from one designer to another, so the legend must be reviewed in order to interpret it correctly.

ANSI has published and recommends standard symbols to be used on electrical drawings. An electrical symbol is a figure or mark that stands for a component used in the system. Many of the symbols use a similar form, but with minor differences which change their meaning. Each category in a system is usually represented by a basic distinguishing symbol. *Figure 11* shows standard electrical symbols.



Instructor's Notes:

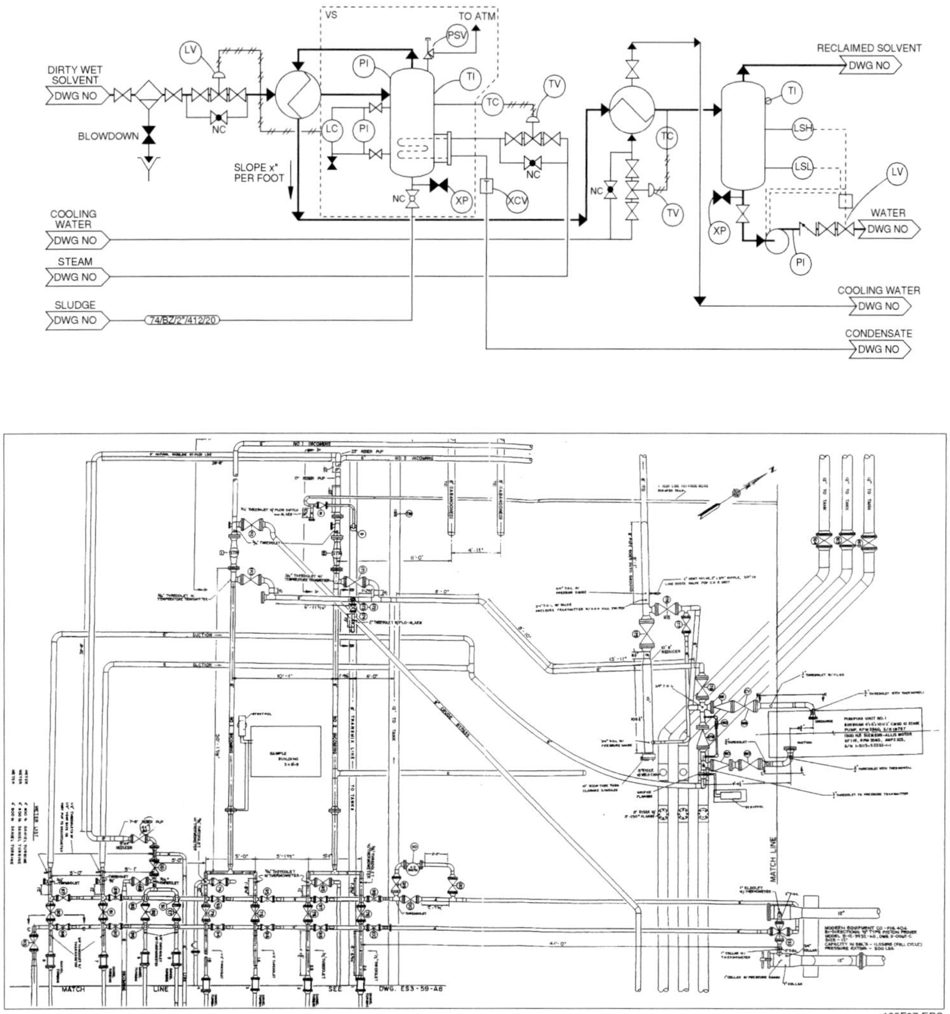


105F07.EPS

*Figure 7* ◈ Sample P&IDs.

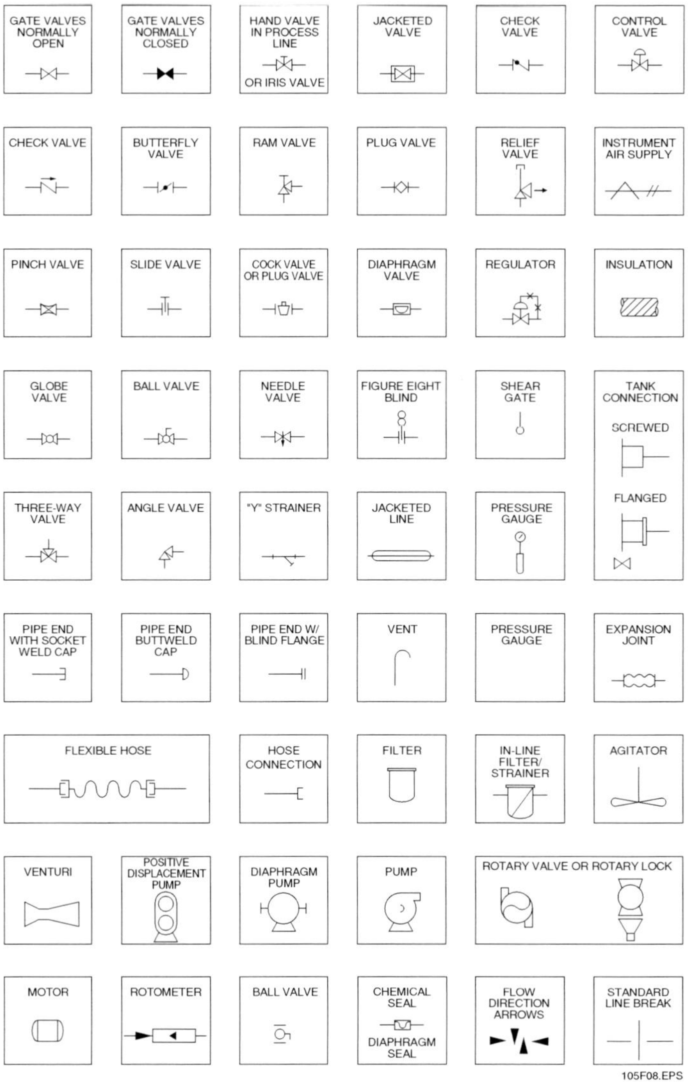


*Figure 8* ◆ Standard P&ID piping symbols.



Instructor's Notes:

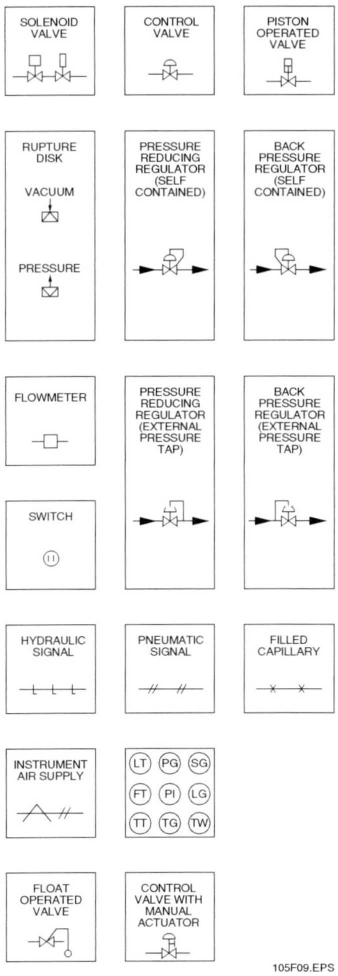


*Figure 9* ◆ P&ID instrumentation symbols.

### *2.4.2 Electrical Schematics*

Unlike an electrical blueprint, the schematic is seldom drawn to scale (see *Figure 12*). Its purpose is to show electrical components and their related connections. Schematics are normally used only in complicated electrical systems, such as control circuits. Components are represented by symbols and every wire is shown by itself or included in an assembly of wires, which appear as one line. Each wire is numbered when it enters an assembly and has the same number when it comes out again to be connected to some component.

## 3.0.0 ◆ DOCUMENTATION

Documentation is an important part of pipeline operations. Many documents are required by regulations, but many are also generated to ensure smooth operations.

Documents that may be required include the following:

- Valve inspection reports
- Exposed pipe reports
- Aerial patrol reports
- Aerial patrol followups
- Tank inspection reports
- Pipe repair/replacement reports
- Safety-related condition reports
- Accident reports
- Work orders
- Permits

### 3.1.0 Valve Inspection Reports

All valves that are necessary for safe operation of the pipeline shall be maintained in **good working order** at all times. In addition, valves must be protected from vandalism and unauthorized operation. Minimum inspection and maintenance schedules and general procedures for performing this work are given in the operations and maintenance manual.

Valve inspections are divided into main-line valves and station valves. Main-line valve maintenance and inspections are to be documented on a main-line valve inspection report form. Station valve maintenance and inspections are to be documented on a station valve inspection report form.

Both of these forms ask for similar information. This information to be provided includes service performed, condition of valve and site, deficiencies, action required, and action taken. This is documented with the date, valve tag number, serial number, and location. A main-line valve form might also include the site inspection checklist. This checklist includes markers, soil, fencing security, connections, trash and vegetation, component fittings, and other items connected with the condition of the site. The form is dated and signed by the inspector. *Figure 13* shows a valve inspection report.



**Classroom**

**Refer trainees to Figure 12. Ask trainees to state the purpose of electrical schematics.**

**Ask trainees if they have questions about the material covered so far.**

**Homework**

**Ask trainees to read Sections 3.0.0–3.5.0 for the next class session.**

**Audiovisual**

**Show Transparency 3. Leave the transparency up for your discussion of Sections 3.0.0–3.5.0. Use it again during the next session for discussion of Sections 3.6.0–3.10.0.**

**Classroom**

**Ask a trainee to define good working order. Refer trainees to Figure 13 and discuss the valve inspection report.**

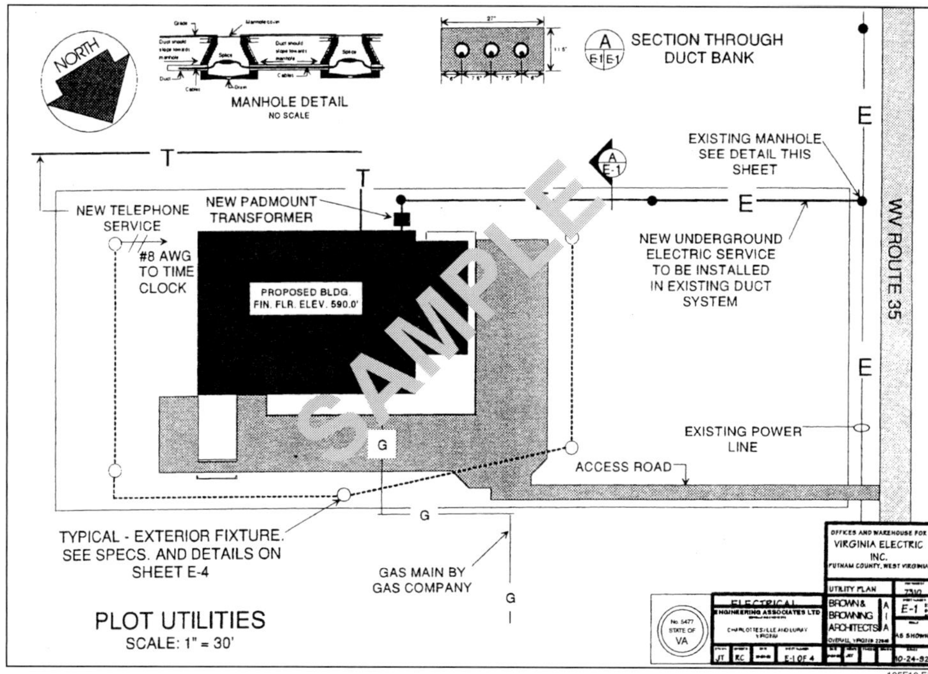


105F10.EPS

*Figure 10* ◆ Electrical blueprint.

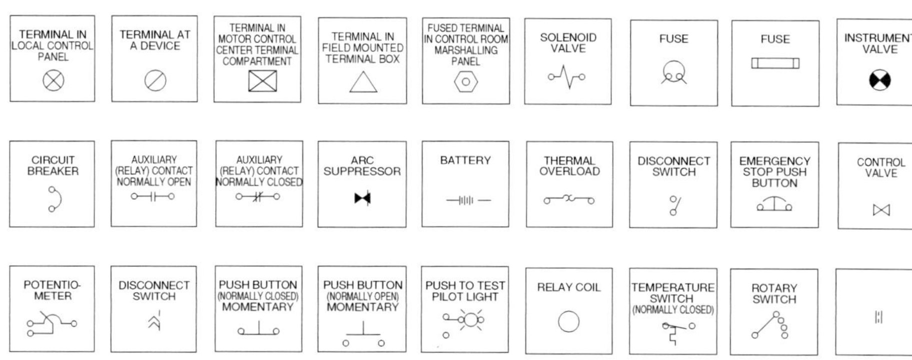


105F11.EPS

*Figure 11* ◆ Standard electrical symbols.



Instructor's Notes:

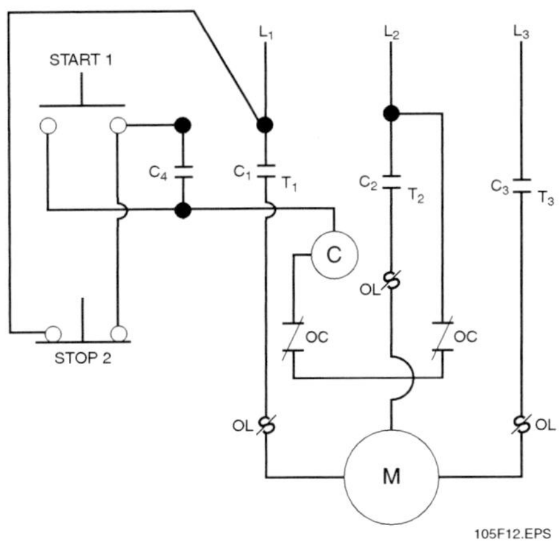


*Figure 12* ◆ Electrical schematic.

Rev. 12/97

| Service Performed | Condition of Valve/Site | Deficiency | Action Required | Action Taken |
|---|---|---|---|---|
| VI - Visual Inspection<br>MO - Mechanical Operation<br>LO - Local Mode Operation<br>RO - Remote Mode Operation<br>RM - Routine Maintenance<br>AS - All Services Listed | AC - Acceptable Condition<br>NR - Needs Repair / Work<br>RR - Replacement Required<br>OS - Out of Service | DO - Does Not Operate<br>LK - Leaking<br>PM - Part Missing<br>DP - Defective Paint<br>NS - Security<br>WO - Wash Out<br>SU - Site Upkeep<br>OT - Other (describe) | RP - Replace Part<br>RR - Repair Required<br>PR - Paint Required<br>SS - Secure Site<br>DW - Dirt Work Required<br>SC - Site Clean Up<br>OT - Other (describe) | PR - Part Replaced<br>RP - Repair Performed<br>RE - Repainted<br>SP - Security Provided<br>DP - Dirt Work Performed<br>SM - Site Cleaned-Up<br>OT - Other (describe) |

| Date | Print Order | Valve Tag No. | Serial Number | Location / Service | Service Performed | Condition of Valve/Site | Deficiency | Action Required | Action Taken |
|---|---|---|---|---|---|---|---|---|---|
| 11/01/01 | 111 | 42-37B | BR549 | #3 Pump Room | VI | NR | DP | PR | RE |
| 11/13/01 | 112 | 54-18C | XJ242 | Pump Station 14 | MO | RR | LK | RP | PR |
| | | | | | | | | | |
| | | | | | | | | | |
| | | | | | | | | | |
| | | | | | | | | | |
| | | | | | | | | | |
| | | | | | | | | | |
| | | | | | | | | | |
| | | | | | | | | | |
| | | | | | | | | | |
| | | | | | | | | | |
| | | | | | | | | | |
| | | | | | | | | | |

SAMPLE

REMARKS: __________

DIVISION: __________ STATION: __________ INSPECTED BY: __________

105F13.EPS

*Figure 13* ◆ Valve inspection report.

### 3.2.0 Exposed Pipe Reports

When exposed pipe is discovered, it must be inspected for evidence of external damage, corrosion, and the extent of damage or corrosion. Corrective action is taken as required. The inspection and action are recorded on a pipeline inspection report form, a form that asks for information on the reason for inspection, the pipe and material conditions, general corrosion, defects or damage, and defect, damage, or corrosion measurements (see *Figure 14*). A sketch of the damaged pipe is also required. The form is dated and signed by the inspector.

Classroom

**Refer trainees to Figure 14. Discuss the types of damage that may occur on exposed pipe.**

## PIPELINE INSPECTION REPORT

*Frequency: At Time of Inspection*

**PART A — LOCATION**

Reason For Inspection ____________

1. Location: Division ____________ County ____________ State ______ Date of Inspection ____________

A) Mainline
Alignment Sheet ________ Pipeline Size/Segment ________
Beg. M.P. ________ End M.P. ________
Beg. Sta. No. ________ End Sta. No. ________

Note: Station numbers shall be measured from a known landmark (Road Crossing, etc.)

B) Company Facility: Pump Station, Tank Farm, etc.
Location Name ____________
Physical Location of Pipe Inspected ____________
Milepost ____________

**PART B — MATERIAL/CONDITIONS**

2. Piping:
Pipe Size ________ In. O.D. Norm. W.T. ________ In. Actual W.T. ________ Grade ________

3. External Pipe Coating:

A) Coating
☐ Bare
☐ LAC 450
☐ Taped
☐ Insulated
☐ Other ________

B) Application
☐ Mill Coated
☐ Yard Coated
☐ Field Coated
☐ Other

C) Coating Thickness
Factory Applied ____ mile
Field Applied ____ mile

D) Condition of Coating
☐ Good ☐ Fair
☐ Poor

E) Bond
☐ Good
☐ Fair
☐ Poor

F) Cause of Coating Failure
☐ Damage
☐ Disbondment
☐ Defective Application
☐ Decomposition
☐ Other ________

4. Environment
☐ Clay ☐ Loam ☐ Sand ☐ Gumbo ☐ Caliche ☐ Rock ☐ Other ________
☐ Wet ☐ Dry ☐ Moist Depth of Cover (Inches) ________
Soil Resistivity Measurement ____________ (ohm-cm) Measurement Method ____________

5. Coating Repaired ☐ No ☐ Yes
Repair System ____________

6. Pipe-to-Soil Potential at Inspected Location (At Grade) ________ (volts) (At Pipe/Soil Interface) ________ (volts)

7. Were any Non-Destructive Tests Completed? ☐ No ☐ Yes Type ________

8. Evidence of: ☐ Corrosion ☐ Defect ☐ Damage ☐ Stress Corrosion Cracking (S.C.C.) ☐ Calcareous Deposits

9. Test Station Added ☐ No ☐ Yes I.D. No. ____________

NOTES:
1) If S.C.C. or Corrosion is Noted, Complete Below and Back of Report (External Only, See Note 4)
2) If Defect or Damage is Noted, Complete Back of Report
3) If Repairs are Required, AAPL Form 502 "REPAIR REPORT FOR LEAK DAMAGE, CORROSION OR TEST FAILURE" Must be Attached.
4) AAPL Form 508 "INTERNAL CORROSION INSPECTION REPORT" Should be Completed When Internal Corrosion Is Detected.

**PART C — EXTERNAL CORROSION**

10. General Corrosion Information:

A. DESCRIPTION
☐ Less than 10% of Norm. Wall Thickness, Complete Front of Report
☐ Greater than 10% of Norm. W.T., Complete Front and Back of Report
☐ Pitting ☐ General ☐ Localized
Max Depth ______ (In.) Longitudinal Length ______ (In.)

B. STATUS
☐ Active
☐ Inactive
☐ Unknown

C. CAUSE
☐ Galvanic
☐ Stray Current
☐ Unknown
☐ Other ______

11. Cathodic Protection: ☐ Impressed ☐ Galvanic

12. Pipe-to-Soil Potential:
a) Last Pipe-to-Soil Measurement at Nearest Point Each Side of Corroded Area
Upstream ________ (volts) Downstream ________ (volts)
b) Distance from Corroded Area to Measurement Points:
Upstream ________ (feet) Downstream ________ (feet)
c) Date of Measurements: Upstream ________ Downstream ________

13. pH of Soil Near Corroded Area ________ pH of Moisture Under Coating ________

14. Pipe Surface Temperature if Fluid is Heated ________ (deg F)

Remarks ____________

Distribution: (Original) Corr. Ctrl./Eng (Copy) Division Office (Copy) Other ________

________ Technician or Inspector ___/___/___ Date ________ Supervisor ___/___/___ Date

SAMPLE

105F14A.EPS

*Figure 14* ◆ Pipeline inspection report (Page 1 of 2).



Instructor's Notes:

## PIPELINE INSPECTION REPORT

*Frequency: At Time of Inspection*

**PART D — DEFECT, DAMAGE OR CORROSION INVESTIGATION**

Notes:
1) Tabulate all Dents, Gouges, Slivers, Corrosion, S.C.C., etc.
2) List Reference Point - Valve, Tee, Center Road etc., as shown on Alignment Sheet with appropriate Station Number, if available.
3) Indicate Location of Longitudinal Seam. If Pipe is Vertical Substitute Compass direction for o'clock Position. For Spiral Weld Pipe, Note as Such.
4) Defects Such as Dents, Slivers, Gouges, etc., shall be tabulated individually. Corrosion Cracks and/or Corrosion Pits Closely Grouped May be Tabulated as One Defect.
5) Defect Width and Length Will be Measured Along the Girth and Longitudinal Axis Respectively.

Reference Point: ☐ Upstream End (Station No.) ______________ Description ______________________

☐ Downstream End (Station No) ______________ Description ______________________

Upstream End — Downstream End

**SKETCH DEPICTING PIPE CUT AT BOTTOM AND LAID FLAT**
**NOTE: Indicate Girth Weld Location If Known**

O'Clock Position →

← Length of Inspected Area →



**PART E — DEFECT, DAMAGE OR CORROSION MEASUREMENTS AND LOCATION**

| No. | Type | Distance from Girth Weld | Distance from Long Seam | Width (Circumferencial) | Length (Longitudinal) | Maximum Depth | Number of Anomalies |
|---|---|---|---|---|---|---|---|
| 1. | | | | | | | |
| 2. | | | | | | | |
| 3. | | | | | | | |
| 4. | | | | | | | |
| 5. | | | | | | | |
| 6. | | | | | | | |
| 7. | | | | | | | |
| 8. | | | | | | | |
| 9. | | | | | | | |
| 10. | | | | | | | |
| 11. | | | | | | | |
| 12. | | | | | | | |
| 13. | | | | | | | |
| 14. | | | | | | | |

If Repair is Required, send to Division Manager for Approval.

Repair Required ☐ No ☐ Yes Type of Repair ______________________

Approved by ______________________ Title ______________________

Remarks: ______________________

105F14B.EPS

*Figure 14* ◆ Pipeline inspection report (Page 2 of 2).

**Classroom**

**Refer trainees to Figure 15. Review the steps followed in aerial patrol reports and follow-up reports.**

**Refer trainees to Figure 16. Review the tank inspection process.**

**Ask trainees if they have questions about the material covered so far.**

**Homework**

**Ask trainees to read Sections 3.6.0–3.10.0 for the next class session.**

**Materials**

**Ensure that you have everything required for teaching this session.**

**Classroom**

**Ask trainees to give examples of pipe changes. Discuss why pipe changes must be carefully planned.**

**Review the safety-related conditions that must be reported on active pipelines and on pipelines in service.**

### 3.3.0 Aerial Patrol Reports

As the aerial patrol flies the route of line, the pilot notes any abnormal conditions of the flight log, which is typed up on the return to base. This log is distributed to all division and operations offices. The aerial patrol report form or log contains the following information:

- Date
- Location
- Time
- Remarks
- Report number

A report number is assigned by the pilot to each abnormal sighting when it is reported. There is also a space on the form to indicate if the sighting is a new one or an old one.

### 3.4.0 Aerial Patrol Followups

The person(s) who investigate the reported abnormalities complete an ROW investigation report, a hard copy of which is given to the originating pilot to check against the log (see *Figure 15*). The original ROW investigation report goes to the division compliance records coordinator and then to division DOT files. The report details the reported information, the date and result of the investigation, and the action taken. The form is signed and dated by the inspector.

### 3.5.0 Tank Inspection Reports

Breakout tanks are inspected externally once a month for stability, physical condition, venting and safety valves, drains, containment structures, equipment, grounding, and alarms. A more detailed inspection by a qualified person is also required at least once per calendar year, not to exceed 15 months between inspections. A visual inspection is made, and any problems are recorded and corrected as necessary. Tanks are inspected externally by API 653-certified personnel every five years. They are inspected internally at one half the corrosion rate of the bottom or every 20 years, whichever is less.

Inspection results are recorded on the Inspection Report for Oil Storage Tanks, with internal inspection noted if performed (see *Figure 16*). The form asks for the location of the tank and then a description of the condition of the physical parts of the tank and all of the associated equipment. Repairs that are required and/or made must also be described. The form is dated and signed by the inspector.

### 3.6.0 Pipe Repair/Replacement Reports

Pipe changes involve replacing portions of pipe, installing fittings, moving pipe, and making permanent repairs to pipe. These changes may be required for relocation or rearrangement of the line, repairs, changes in pressure requirements, or for other reasons.

Pipe change events must be planned and controlled to avoid spills and interruption in operations. Emergency pipe changes are planned ahead as much as possible, based on events that could occur. All pipe changes must be explained and documented on a pipe report.

### 3.7.0 Safety-Related Condition Reports

Safety-related conditions must be reported if they occur on an active pipeline or facility that is:

- Within 220 yards of any building intended for human occupancy or outdoor place of assembly
- Within the right-of-way of an active railroad, paved road, street, or highway
- Offshore
- At locations where an oil spill could reasonably be expected to pollute any stream, river, lake, reservoir, or other body of water

Safety-related conditions involving pipelines in service that must be reported are:

- General corrosion that has reduced wall thickness to less than that of maximum operating pressure (*OSHA Part 195.55*)
- Unintended movement or abnormal loading of a pipeline
- Any material defect or physical damage impairing serviceability
- Any malfunction or operating error causing pressure to rise above MOP
- A leak constituting an emergency
- Any condition that could lead to a hazard, a reduction in pressure, or shutdown

All employees are expected to learn how to recognize safety-related conditions and report these to the manager. The manager will initiate action and file the report.



Instructor's Notes:

**RIGHT-OF-WAY INVESTIGATION REPORT** Report Date ______

*Frequency: As Required by Aerial Patrol Report*

Note: Right-of-Way personnel should be consulted before confronting enroachment situations for advice pertaining to right-of-way restrictions and agreements.

**PART A — REPORTED INFORMATION**

1. Right-of-Way Investigation Report Number: ______
2. Date Condition Reported: ______
3. Milepost Location Reported: ______
4. Descriptive Location Reported: ______
5. Problem Reported: ______

**PART B — INVESTIGATION INFORMATION**

1. Date Condition Investigated: ______
2. Actual Milepost Location: ______
3. Actual Location of Condition: ______
4. Actual Problem Found: ______

☐ Encroachment Advisory (AAPL Form 202) was given to encroaching party

Name of Encroaching Party: ______

**PART C — ACTION TAKEN**

**PART D — DISTRIBUTION INFORMATION**

(Original) Division Compliance Records Coordinator
(Copy) Forward to Pilot at the Arizona Division Office
(Copy) Right-of-Way

*Technician or Inspector* *Date* *Supervisor* *Date*

**Retention:** Two Years

SAMPLE

105F15.EPS

*Figure 15* ◆ ROW investigation report.

**Materials**

**Ensure that you have everything required for teaching this session.**

**INSPECTION REPORT FOR OIL STORAGE TANKS**

*(Frequency: Every Calendar Year, Not to Exceed 15 months)*

**PART A — LOCATION**

Division: ______________________ Station / M.P.: ______________________

County: ______________________ Tank No.: ______________________

State: ______________________

**PART B — GENERAL CONDITION**

Roof Surface ______________________
Roof Seal ______________________
Roof Drain ______________________
Roof Vents ______________________
All Pontoons ______________________
Gauge Hatches ______________________
Roof Legs ______________________
Auto Tank Gauge ______________________
Tank Exterior Walls ______________________
Tank Exterior Insulation & Straps ______________________
Rolling Ladder ______________________
Spiral Ladder and Platforms ______________________
Valves ______________________
Tank Grounding ______________________
Foundation ______________________
Mixers ______________________
Dikes and Tank Area Grounds ______________________
Dike Drains ______________________
Alarms ______________________
Fire Protection Equipment (As Provided) ______________________
Repairs Required ______________________
Repairs Made ______________________
Remarks ______________________

Notes:
1. If an internal tank inspection is performed type "Internal Inspection" at the top of this form.
2. If the inspection is performed by a third party attach all reports to this form.

Distribution: (Original) Division Office (Copy) Other ______________________

______________________ Technician or Inspector ___/___/___ Date ______________________ Supervisor ___/___/___ Date

SAMPLE

105F16.EPS

*Figure 16* ◆ Tank inspection report.

5.16 PIPELINE MAINTENANCE LEVEL ONE — TRAINEE MODULE 62105-02

Instructor's Notes:

### 3.8.0 Accident Reports

All pipeline accidents must be reported according to company procedure. The manager will initiate investigation into the matter and ensure that a record of accidents and safety-related conditions is maintained regardless of whether or not the accident is reportable to a government agency. The company's accident form will be used for this recordkeeping (see *Figure 17*).

The manager or regulatory compliance department will make reports to government agencies. Written accident reports to DOT are required to be completed for any accident on the pipeline that meets the following criteria:

- Explosion or fire not intentionally set by operator
- Loss of five or more barrels of liquid
- Escape to the atmosphere of more than five barrels a day of highly volatile liquids
- Death of any person
- Bodily harm to any person resulting in loss of consciousness, the need to be carried from the scene, the need for medical treatment, or disability preventing normal duties or activities
- Property damage exceeding $50,000

The report must be filed on DOT form 7000-1 or a facsimile (see *Figure 18*).

### 3.9.0 Work Orders

Some companies require work orders for all maintenance work to be done on the pipeline. Because any disruption of service on the line affects everyone on the entire length of the line, all such work must be carefully controlled and planned. Work orders and permits are a part of that control.

Sometimes work orders for routine preventive maintenance are created automatically by computer scheduling. Other work orders must be initiated by appropriate personnel as needed and distributed to maintenance crews who will use them to plan and perform the work. Completed work orders are returned to be entered into the recordkeeping system and filed to document what has been done. Work orders are also entered on work order logs as they are issued and completed.

### 3.10.0 Permits

Permits are required for a variety of jobs on the pipeline, including confined space work and hot tapping. These permits must be provided to maintenance by management. Permits are also entered on a log as they are issued and as the work is completed. *Figure 19* shows a work permit form.

## 4.0.0 ◆ DOCUMENT MANAGEMENT

It is important to keep documents, but with the number produced, they must be managed to make them usable. Document management includes the following factors:

- Retention rates
- Document control
- Liquid inspection intervals
- Gas inspection intervals

### 4.1.0 Retention Rates

Retention time for documents is set by company policy based on government regulations. The following sections describe what DOT says must be retained and for how long. Always follow company policy.

#### *4.1.1 Construction Records*

A complete record that shows the number of girth welds, pipe size and cover, crossing locations, and valve and test station locations must be kept for the life of the facility.

#### *4.1.2 Hydrostatic Testing Records*

A record must be made of each required hydrostatic test, and the record of the latest test must be retained as long as the tested facility is in use. This must include charts, data, names, date, time, pressure, medium, facility and apparatus description, explanations, and profile with test sites.



**Classroom**

**Refer trainees to Figures 17 and 18. Review the criteria for written accident reports to the U.S. Department of Transportation (DOT).**

**Homework**

**Ask trainees to read Sections 4.0.0–4.1.5 for the next class session.**

**Performance Testing**

**Have each trainee complete Performance Tasks 2 and 3 to your satisfaction. Fill out a Performance Profile Sheet for each task for each trainee.**

**Audiovisual**

**Show Transparency 4. Leave the transparency up during your discussion of Sections 4.0.0–4.1.5. Use the transparency again for your discussion of Sections 4.2.0–4.4.0 in the next class session.**

**Classroom**

**Review DOT requirements for record retention. Stress the importance of following company policy.**

**Emphasize the importance of keeping construction records for the life of the facility.**

**List required items in a hydrostatic testing record. Note the importance of keeping a record of the latest test as long as the facility is in use.**

**Classroom**

**Refer trainees to Figure 19. Ask a trainee to describe why work orders are important.**

**Describe the types of jobs requiring permits.**

**Teaching Tip**

**See the Teaching Tip at the end of this module.**

## LEAK/ACCIDENT REPORT

*Frequency: When ANY leak or accident occurs*

**PART A — TIME AND LOCATION OF ACCIDENT**

1. Date: *(Month, Day, Year)* ____________________
2. Hour: *(24 hour clock)* ____________________
3. Give County or City: ____________________
4. Did accident occur on Federal Land? ☐ Yes ☐ No
5. Specify location. (*If location is near buildings, other land marks such as highways, waterways, or railroads, attach a sketch or drawing showing relationship to these landmarks).*

____________________

____________________

____________________

**PART B — ORIGIN OF RELEASE OF LIQUID** *(Check all applicable items)*

1. Part of system involved:
   ☐ Mainline ☐ Tank Farm ☐ Pump Station
2. Item Involved:
   ☐ Pipe ☐ Valve ☐ Scraper Trap ☐ Pump ☐ Welded Fitting
   ☐ Girth Weld ☐ Tank ☐ Bolted Fitting ☐ Longitudinal Weld
   Other *(Specify)* ____________________
   ____________________
3. Year Item Installed ____________
4. Depth of Cover ____________ inches

**PART C — CAUSE OF ACCIDENT**

☐ Corrosion ☐ Failed Pipe ☐ Failed Weld ☐ Outside Force Damage
☐ Malfunction of Control Relief Equipment ☐ Incorrect Operation by Operator Personnel
☐ Other *(Specify)* ____________________
____________________



**PART D — DEATH OR INJURY**

1. Number of persons killed
   ____________ Operator Employees ____________ Non-Employees
2. Number of persons injured and requiring medical treatment other than on-site first aid.
   ____________ Operator Employees ____________ Non-Employees

**PART E — ESTIMATED TOTAL COST**

$ ____________

**PART F — AMOUNT OF OIL INVOLVED**

1. Estimated amount of oil spilled
   ________ barrels spilled ________ barrels recovered
2. Was there an explosion?
   ☐ Yes ☐ No
3. Was there a fire?
   ☐ Yes ☐ No
4. Was there a rupture?
   ☐ Yes ☐ No
5. * Area covered by spill
   ________ Feet X ________ Feet
6. * Depth of Soil Saturation
   ________ Inches
   * (Attach Sketch)
7. Was contaminated soil removed?
   ☐ Yes ☐ No

____________________ Technician or Inspector ___/___/___ Date ____________________ Supervisor ___/___/___ Date

Retention: Life of System

105F17A.EPS

*Figure 17* ◆ Accident report (Page 1 of 2).



Instructor's Notes:

## LEAK/ACCIDENT REPORT

*Frequency: When ANY leak or accident occurs*

INSTRUCTIONS: *Answer Sections G, H or I only if it applies to the particular accident being reported.*

**PART G — OCCURRED IN LINE PIPE**

1. Nominal diameter (inches) ____________
2. Wall Thickness (inches) ____________
3. SMYS (psi) ____________
4. Type joint: ☐ Welded ☐ Flanged ☐ Threaded ☐ Coupled ☐ Other
5. Pipe was ☐ Below Ground ☐ Above Ground
6. Maximum operating pressure ____________
7. Pressure at time and location of accident *(psig)* ____________
8. Had there been a pressure test on the system? ☐ Yes ☐ No
9. Duration of test *(hrs)* ____________
10. Maximum test pressure *(psig)* ____________
11. Date of last test ____________

**PART H — CAUSED BY CORROSION**

1. Location of Corrosion ☐ Internal ☐ External
2. Facility Coated? ☐ Yes ☐ No
3. Facility Under Cathodic Protection? ☐ Yes ☐ No
4. Type of Corrosion? ☐ Galvanic ☐ Other *(specify)* ____________

**PART I — CAUSED BY OUTSIDE FORCE**

1. ☐ Damage by operator or by AAPL Contractor
   ☐ Damage by Others
   ☐ Damage by Natural Forces
   ☐ Landslide
   ☐ Subsidence
   ☐ Washout
   ☐ Earthquake
   ☐ Mudslide
2. Was damage prevention program in effect?
   ☐ Yes ☐ No
3. If yes, was the program
   ☐ "one-call" ☐ Other ________
4. Did excavator call?
   ☐ Yes ☐ No
5. Was the pipeline location temporarily marked?
   ☐ Yes ☐ No

**PART J — ACCOUNT OF ACCIDENT**



Retention: Life of System

105F17B.EPS

*Figure 17* ◆ Accident report (Page 2 of 2).

NOTICE: This report is required by 49 CFR Part 195. Failure to report can result in a civil penalty not to exceed $25,000 for each violation for each day that such violation persists except that the maximum civil penalty shall not exceed $500,000 as provided in 49 USC 60122

Form Approved
OMB No. 2137-0047

U.S. Department of Transportation
Research and Special Programs Administration

**ACCIDENT REPORT – HAZARDOUS LIQUID PIPELINE SYSTEMS**

Report Date ______________

No. ______________
(DOT Use Only)

**INSTRUCTIONS**

***Important:*** *Please read the separate instructions for completing this form before you begin. They clarify the information requested and provide specific examples. If you do not have a copy of the instructions, you can obtain one from the Office Of Pipeline Safety Web Page at http://ops.dot.gov.*

**PART A – GENERAL REPORT INFORMATION** Check: ☐ **Original Report** ☐ **Supplemental Report** ☐ **Final Report**

1. a. Operator's OPS 5-digit Identification Number (if known) /__/__/__/__/__/
2. b. If Operator does not own the pipeline, enter Owner's OPS 5-digit Identification Number (if known) /__/__/__/__/__/
   c. Name of Operator ______________________________
   d. Operator street address ______________________________
   e. Operator address ______________________________
   City, County, State and Zip Code

**IMPORTANT: IF THE SPILL IS SMALL, THAT IS, THE AMOUNT IS AT LEAST 5 GALLONS BUT IS LESS THAN 5 BARRELS, COMPLETE THIS PAGE ONLY, UNLESS THE SPILL IS TO WATER AS DESCRIBED IN 49 CFR §195.52(A)(4) OR IS OTHERWISE REPORTABLE UNDER §195.50 AS REVISED IN CY 2001.**

2. Time and date of the accident
   /__/__/__/__/ hr. /__/__/ month /__/__/ day /__/__/ year
3. Location of accident
   *(If offshore, do not complete a through d. See Part C.1)*
   a. Latitude: ________ Longitude: ________
   *(if not available, see instructions for how to provide specific location)*
   b. ______________________________
   City, and County or Parish
   c. ______________________________
   State and Zip Code
   d. Mile post/valve station O or survey station no. O
   (whichever gives more accurate location)
   ______________________________
4. Telephone report
   /__/__/__/__/__/__/ NRC Report Number /__/__/ month /__/__/ day /__/__/ year

5. Losses *(Estimated)*

**Public/Community Losses reimbursed by operator:**

Public/private property damage $__________
Cost of emergency response phase $__________
Cost of environmental remediation $__________
Other Costs $__________
(describe) ______________________________

**Operator Losses:**

Value of product lost $__________
Value of operator property damage $__________
Other Costs $__________
(describe) ______________________________
**Total Costs** $__________

6. Commodity Spilled O Yes O No
(If Yes, complete Parts a through c where applicable)
   a. Name of commodity spilled ______________________________
   b. Classification of commodity spilled:
   O HVLs /other flammable or toxic fluid which is a gas at ambient conditions
   O $CO_2$ or other non-flammable, non-toxic fluid which is a gas at ambient conditions
   O Gasoline, diesel, fuel oil or other petroleum product which is a liquid at ambient conditions
   O Crude oil

**c. Estimated amount of commodity involved :**
O **Barrels**
O **Gallons (check only if spill is less than one barrel)**

**Amounts:**
**Spilled :** __________
**Recovered:** __________

**CAUSES FOR SMALL SPILLS ONLY (5 gallons to under 5 barrels) :** ***(For large spills [5 barrels or greater] see Part H)***

O **Corrosion** O **Natural Forces** O **Excavation Damage** O **Other Outside Force Damage**
O **Material and/or Weld Failures** O **Equipment** O **Incorrect Operation** O **Other**

**PART B – PREPARER AND AUTHORIZED SIGNATURE**

______________________________ (type or print) Preparer's Name and Title
______________________________ Area Code and Telephone Number
______________________________ Preparer's E-mail Address
______________________________ Area Code and Facsimile Number
______________________________ Authorized Signature
______________________________ (type or print) Name and Title
__________ Date
______________________________ Area Code and Telephone Number

Form RSPA F 7000-1 ( 01-2001 ) *Page 1 of 4*

*Figure 18* ◆ DOT accident report form (page 1 of 4).

105F18A.PDF



Instructor's Notes:

**PART C – ORIGIN OF THE ACCIDENT (Check all that apply)**

1. Additional location information
   a. Line segment name or ID ____________
   b. Accident on Federal land other than Outer Continental Shelf O Yes O No
   c. Is pipeline interstate? O Yes O No

Offshore: O Yes O No *(complete d if offshore)*

d. Area ____________ Block # ____________

State / / / or Outer Continental Shelf ☐

2. Location of system involved (*check all that apply*)
   - ☐ Operator's Property
   - ☐ Pipeline Right of Way
   - ☐ High Consequence Area (HCA)?
     Describe HCA____________

3. Part of system involved in accident
   - O Above Ground Storage Tank
   - O Cavern or other below ground storage facility
   - O Pump/meter station; terminal/tank farm piping and equipment, including sumps
   - O Other *Specify:* ____________
   - O Onshore **pipeline**, including valve sites
   - O Offshore **pipeline**, including platforms

*If failure occurred on **Pipeline**, complete items a - g:*

a. Type of leak or rupture
   - OLeak: O Pinhole O Connection Failure *(complete sec. H5)*
     O Puncture, diameter *(inches)* ____________
   - ORupture: O Circumferential – Separation
     O Longitudinal – Tear/Crack, length *(inches)* ____________
     Propagation Length, total, both sides *(feet)* ____________
   - ON/A
   - OOther ____________

b. Type of block valve used for isolation of immediate section:
   Upstream: ☐ Manual ☐ Automatic ☐ Remote Control ☐ Check Valve
   Downstream: ☐ Manual ☐ Automatic ☐ Remote Control ☐ Check Valve

c. Length of segment isolated ______ ft

d. Distance between valves ______ ft

e. Is segment configured for internal inspection tools? OYes O No

f. Had there been an in-line inspection device run at the point of failure? O Yes O No O Don't Know
   O Not Possible due to physical constraints in the system

g. If *Yes*, type of device run (*check all that apply*)

| | |
|---|---|
| ☐ High Resolution Magnetic Flux tool | Year run: ______ |
| ☐ Low Resolution Magnetic Flux tool | Year run: ______ |
| ☐ UT tool | Year run: ______ |
| ☐ Geometry tool | Year run: ______ |
| ☐ Caliper tool | Year run: ______ |
| ☐ Crack tool | Year run: ______ |
| ☐ Hard Spot tool | Year run: ______ |
| ☐ Other tool | Year run: ______ |

4. Failure occurred on

| | | |
|---|---|---|
| O Body of Pipe | O Pipe Seam | O Scraper Trap |
| O Pump | O Sump | O Joint |
| O Component | O Valve | O Metering Facility |
| O Repair Sleeve | O Welded Fitting | O Bolted Fitting |
| O Girth Weld | | |

Other *(specify)* ____________

Year the component that failed was installed: / / / / /

5. Maximum operating pressure *(MOP)*
   a. Estimated pressure at point and time of accident: ______ *PSIG*
   b. MOP at time of accident: ______ *PSIG*
   c. Did an overpressurization occur relating to the accident? OYes O No

**PART D – MATERIAL SPECIFICATION**

1. Nominal pipe size *(NPS)* / / / / / / in.
2. Wall thickness / / / / / / in.
3. Specification ____________ SMYS / / / / / / / /
4. Seam type ____________
5. Valve type ____________
6. Manufactured by ____________ in year / / / / /

**PART E – ENVIRONMENT**

1. Area of accident
   - O Under pavement
   - O Underground
   - O Inside/under building
   - O In open ditch
   - O Above ground
   - O Under water
   - O Other ____________
2. Depth of cover: ____________ inches

**PART F – CONSEQUENCES**

1. **Consequences** *(check and complete all that apply)*

a.

| | Fatalities | Injuries |
|---|---|---|
| Number of operator employees: | | ______ |
| Contractor employees working for operator: | | ______ |
| General public: | | ______ |
| **Totals:** | | ______ |

b. Was pipeline/segment shutdown due to leak? OYes O No
   If Yes, how long? ______ days ______ hours ______ minutes

c. Product ignited OYes O No  d. Explosion OYes O No

e. ☐ Evacuation *(general public only)* / / / / / people
   Reason for Evacuation:
   - O Precautionary by company
   - O Evacuation required or initiated by public official

f. Elapsed time until area was made safe: / / / hr. / / / min.

2. **Environmental Impact**
   a. Wildlife Impact: Fish/aquatic O Yes O No; Birds O Yes O No; Terrestrial O Yes O No
   b. Soil Contamination O Yes O No
      If Yes, estimated number of cubic yards: ____________
   c. Long term impact assessment performed: O Yes O No
   d. Anticipated remediation O Yes O No
      If Yes, check all that apply: ☐ Surface water ☐ Groundwater ☐ Soil ☐ Vegetation ☐ Wildlife
   e. Water Contamination: O Yes O No *(If Yes, provide the following)*
      Amount in water ______ barrels
      Ocean/Seawater O No O Yes
      Surface O No O Yes
      Groundwater O No O Yes
      Drinking water O No O Yes *(If Yes, check below.)*
      O Private well O Public water intake

Form RSPA F 7000-1 ( 01-2001 ) Page 2 of 4

105F18B.PDF

*Figure 18* ◈ DOT accident report form (page 2 of 4).

**PART G – LEAK DETECTION INFORMATION**

1. Computer based leak detection capability in place? O Yes O No

2. Was the release initially detected by? (check one):
O CPM/SCADA-based system with leak detection
O Static shut-in test or other pressure or leak test
O Local operating personnel, procedures or equipment
O Remote operating personnel, including controllers
O Air patrol or ground surveillance
O A third party O Other (specify) ________

3. Estimated leak duration days ____ hours ____

**PART H – APPARENT CAUSE**

***Important:*** *There are 25 numbered causes in this Part H. Check the box corresponding to the primary cause of the accident. Check one circle in each of the supplemental categories corresponding to the cause you indicate. See the instructions for guidance.*

**H1 – CORROSION**

1. ☐ External Corrosion

2. ☐ Internal Corrosion

(Complete items a – e where applicable.)

a. Pipe Coating: O Bare O Coated

b. Visual Examination: O Localized Pitting O General Corrosion O Other ________

c. Cause of Corrosion: O Galvanic O Atmospheric O Stray Current O Microbiological O Cathodic Protection Disrupted O Stress Corrosion Cracking O Selective Seam Corrosion O Other ________

d. Was corroded part of pipeline considered to be under cathodic protection prior to discovering accident?
O No O Yes, Year Protection Started: /__/__/__/__/

e. Was pipe previously damaged in the area of corrosion?
O No O Yes ⇒ Estimated time prior to accident: /__/__/ years /__/__/ months Unknown ☐

**H2 – NATURAL FORCES**

3. ☐ Earth Movement ⇒ O Earthquake O Subsidence O Landslide O Other ________
4. ☐ Lightning
5. ☐ Heavy Rains/Floods ⇒ O Washouts O Flotation O Mudslide O Scouring O Other ________
6. ☐ Temperature ⇒ O Thermal stress O Frost heave O Frozen components O Other ________
7. ☐ High Winds

**H3 – EXCAVATION DAMAGE**

8. ☐ Operator Excavation Damage (including their contractors/Not Third Party)
9. ☐ Third Party *(complete a-f)*
   a. Excavator group
   O General Public O Government O Excavator other than Operator/subcontractor
   b. Type: O Road Work O Pipeline O Water O Electric O Sewer O Phone/Cable
   O Landowner-not farming related O Farming O Railroad
   O Other liquid or gas transmission pipeline operator or their contractor
   O Nautical Operations O Other ________
   c. Excavation was: O Open Trench O Sub-strata (boring, directional drilling, etc…)
   d. Excavation was an ongoing activity (Month or longer) O Yes O No If Yes, Date of last contact /__/__/__/
   e. Did operator get prior notification of excavation activity?
   O Yes; Date received: /__/__/ mo. /__/__/ day /__/__/__/__/ yr. O No
   Notification received from: O One Call System O Excavator O Contractor O Landowner
   f. Was pipeline marked as result of location request for excavation? O No O Yes *(If Yes, check applicable items i - iv)*
   i. Temporary markings: O Flags O Stakes O Paint
   ii. Permanent markings: O
   iii. Marks were (check one) : O Accurate O Not Accurate
   iv. Were marks made within required time? O Yes O No

**H4 – OTHER OUTSIDE FORCE DAMAGE**

10. ☐ Fire/Explosion as primary cause of failure ⇒ Fire/Explosion cause: O Man made O Natural
11. ☐ Car, truck or other vehicle not relating to excavation activity damaging pipe
12. ☐ Rupture of Previously Damaged Pipe
13. ☐ Vandalism

Form RSPA F 7000-1 ( 01-2001 ) *Page 3 of 4*

105F18C.PDF

*Figure 18* ◆ DOT accident report form (page 3 of 4).

5.22 PIPELINE MAINTENANCE LEVEL ONE — TRAINEE MODULE 62105-02

Instructor's Notes:

**H5 – MATERIAL AND/OR WELD FAILURES**

**Material**

14. ☐ Body of Pipe ⇒ O Dent O Gouge O Bend O Arc Burn O Other ____________
15. ☐ Component ⇒ O Valve O Fitting O Vessel O Extruded Outlet O Other ____________
16. ☐ Joint ⇒ O Gasket O O-Ring O Threads O Other ____________

**Weld**

17. ☐ Butt ⇒ O Pipe O Fabrication O Other ____________
18. ☐ Fillet ⇒ O Branch O Hot Tap O Fitting O Repair Sleeve O Other ____________
19. ☐ Pipe Seam ⇒ O LF ERW O DSAW O Seamless O Flash Weld
O HF ERW O SAW O Spiral O Other ____________

*Complete a-g if you indicate **any** cause in part H5.*

a. Type of failure:
O Construction Defect ⇒ O Poor Workmanship O Procedure not followed O Poor Construction Procedures
O Material Defect

b. Was failure due to pipe damage sustained in transportation to the construction or fabrication site? O Yes O No

c. Was part which leaked pressure tested before accident occurred? O Yes, *complete d-g* O No

d. Date of test: /__/__/__/__/ yr. /__/__/ mo. /__/__/ day

e. Test medium: O Water O Inert Gas O Other ____________________

f. Time held at test pressure: /__/__/ hr.

g. Estimated test pressure at point of accident: ____________________ *PSIG*

**H6 – EQUIPMENT**

20. ☐ Malfunction of Control/Relief Equipment ⇒ O Control valve O Instrumentation O SCADA O Communications
O Block valve O Relief valve O Power failure O Other ____________
21. ☐ Threads Stripped, Broken Pipe Coupling ⇒ O Nipples O Valve Threads O Dresser Couplings O Other ____________
22. ☐ Seal Failure ⇒ O Gasket O O-Ring O Seal/Pump Packing O Other ____________

**H7 – INCORRECT OPERATION**

23. ☐ Incorrect Operation

a. Type: O Inadequate Procedures O Inadequate Safety Practices O Failure to Follow Procedures
O Other ____________________

b. Number of employees involved who failed a post-accident test: drug test: /__/__/__/ alcohol test /__/__/__/

**H8 – OTHER**

24. ☐ Miscellaneous, *describe:* ____________________
25. ☐ Unknown
O Investigation Complete O Still Under Investigation *(submit a supplemental report when investigation is complete)*

**PART I – NARRATIVE DESCRIPTION OF FACTORS CONTRIBUTING TO THE EVENT** *(Attach additional sheets as necessary)*

Form RSPA F 7000-1 (01-2001) *Page 4 of 4*

105F18D.PDF

*Figure 18* ◈ DOT accident report form (page 4 of 4).

PIPELINE DOCUMENTATION — TRAINEE MODULE 62105-02 **5.23**

**Pipeline Work Permit**

No:

| Effective From: Time: Date: | Effective To: Time: Date: |
|---|---|
| Work Location: | |
| Specific Work To Be Done: | |

| | | |
|---|---|---|
| 1. OCC notification required? | ☐ Yes | ☐ No |
| 2. Alyeska representative required on site when work is being performed? | ☐ Yes | ☐ No |
| 3. P&CM required on site when work is being performed? | ☐ Yes | ☐ No |
| 4. All required government permits in hand (Title 16,C.O.E.)? | ☐ Yes | ☐ No |
| 5. All required Company Work Permits in hand? | ☐ Yes | ☐ No |
| 6. Type of work being performed has potential to adversely affect the buried Fiber Optics Cable? **Note:** If answer is "Yes" contact GCI, 1-888-560-2022 for cable locate. | ☐ Yes | ☐ No |
| 7. Responsible Work Supervisor informed of the need to avoid damage to TAPS Survey Monuments/Markers, and that any damage (pre-existing or otherwise) should be immediately reported? | ☐ Yes | ☐ No |
| 8. Mining and/or materials required from Company Operations Material Sites (OMS)? **NOTE:** Mineral material used for any work will require tracking and reporting on a daily basis to the local Maintenance Coordinator utilizing the "Daily Mineral Material Hauling Voucher," Company Form #10164. | ☐ Yes | ☐ No |

Specifications/Procedures To Be Utilized:

Special Instructions, Re-notification Milestones, etc.:

| Responsible Work Supervisor: | Date: | Maintenance Coordinator Signature: | Date: |
|---|---|---|---|
| Job Completed: Time: Date: | Initial: | Permit Cancelled: Time: Date: | Initial: |

SAMPLE

3459 (03/01)

105F19.EPS

*Figure 19* ◆ Work permit form.



Instructor's Notes:

*4.1.3 Operation and Maintenance Maps and Records*

Each operator shall maintain current maps and records of its pipeline systems, including location and identification of facilities, all crossings, maximum operating pressures, and specifications of all pipe, until these are revised.

*4.1.4 Daily Operating Records*

Each operator shall maintain for at least three years daily operating records that indicate discharge pressure at each pump station and any emergency or abnormal operation.

*4.1.5 Miscellaneous Records*

There are many types of miscellaneous records. These records vary in how long they must be retained. The following shows types of records and how long they must be retained:

- Retained for the life of the system, facility, pipe, or part:
  - Pipe repair description
  - Pressure testing
  - Weld qualification tests
  - Welder qualification tests
  - Non-destructive testing of welds
- Retained for three years or until the next inspection:
  - Liaison with fire, police and other official
  - Review of personnel training
  - Supervisor knowledge records
- Retained for two years or until the next inspection, whichever is greater:
  - ROW inspections
  - Cathodic protection testing reports
  - Rectifier inspections
  - Inspections of exposed buried pipe
  - Investigation of corrosive effects
  - Internal corrosion monitoring tests
  - Inspections of internal pipe surface
  - Station valve inspections
  - Main-line valve inspections
  - Overpressure safety device inspections
  - Fire-fighting equipment inspections and tests
  - Breakout tank inspections
  - Continuing public education programs
  - Review of training effectiveness
  - Procedure review record
- Retained until the next inspection or test:
  - Navigable waterway inspection
  - Corrosion testing of bare pipe
  - Records of each inspection and test

### 4.2.0 Document Control

All control documents generated are assigned numbers according to the company numbering system. Documents are distributed within the company according to specific distribution lists. Documents published in the company are numbered and signed out to authorized persons. Revisions to these documents are recorded on revision sheets with descriptions, dates, and names and distributed to all document holders. Documents sent to DOT and other regulatory agencies are logged as sent. Logs and copies are kept on file.

### 4.3.0 Liquid Inspection Intervals

Inspection intervals are regulated by DOT Inspection Requirements, *Part 195, Subpart F.* This requirement sets the following intervals:

- Not exceeding fifteen months, but at least once per calendar year:
  - Operations and maintenance manual reviews
  - Public education material reviews
  - Procedure reviews
  - Training reviews of personnel
  - Training program revisions
  - Potential surveys of cathodic protection adequacy
  - Cathodic protection test leads inspections
  - Aboveground pipe/component coating inspections
  - Station valve inspections
  - Overpressure safety device inspections (non-HVL)
  - Breakout tanks (in service)
- Not exceeding seven and a half months, but at least twice per calendar year:
  - Corrosion inhibitor monitor examinations
  - Mainline valve inspections
  - Halon or inert gas system inspections
  - Overpressure safety device inspections (HVL)
- Not exceeding two and a half months, but at least six times per calendar year:
  - Rectifier inspections and readings



**Classroom**

**Note that operation and maintenance records are kept until revised.**

**Note that daily operating records must be kept for three years.**

**Review other types of records. Ask trainees why some records are kept for a facility's life and others are kept for limited periods.**

**Ask trainees if they have questions about the material covered so far.**

**Homework**

**Ask trainees to read Sections 4.2.0–4.4.0 and complete the Review Questions for the next class session.**

**Materials**

**Ensure that you have everything required for testing during this session.**

**Classroom**

**Outline document control procedures. If appropriate, share your company's document control system.**

**Review the DOT requirements on liquid inspection intervals.**

**Classroom**

**Review the conditions that dictate gas inspection intervals. Ask trainees if they have any questions on Sections 4.0.0–4.4.0.**

**Go over the Review Questions. Encourage trainees to ask questions. Have trainees prepare for the Module Examination.**

**Examination**

**Administer the Module Examination. Be sure to record the results of the Exam on Craft Training Report Form 200 and submit the results to the Training Program Sponsor.**

**Performance Testing**

**Ensure that all Performance Tasks have been completed and Performance Profile Sheets are filled out. Be sure to record the results of the testing on Craft Training Report 200 and submit the results to the Training Program Sponsor.**

- Not exceeding three weeks, but at least 26 times per calendar year:
  - Right-of-way and river crossing surface inspections
  - River crossing alignment and depth inspections
  - Bare pipe corrosion tests
- At regular intervals:
  - Fire extinguisher inspections (monthly, yearly)

### 4.4.0 Gas Inspection Intervals

Pipeline operators must have an inspection program to check for indications of leaks, construction activity, or any other factors affecting safety and operation. The frequency of inspections is determined by the size of the line, the operating pressures, the class location, terrain, weather, and other relevant factors.

### Summary

The creation and maintenance of maps and records is critical to the operation of the pipeline, as well as being required by regulatory agencies. Maps and drawings must be available to locate the pipeline, to perform maintenance, and to react to problems when they occur. Records are used to ensure that required work is done and to record all events and problems for operations planning.

From the initial design process and through construction, operations, and maintenance activities, maps are used to define the pipeline's route, its environment, and its characteristics. A variety of maps are used, including the original topographical and political maps that helped to determine the pipeline's characteristics and route, the surveyor's maps that were used to lay out the right-of-way, and the strip maps that combined different views used to correctly identify and locate the pipeline.

Piping and instrumentation diagrams (P&IDs) show the equipment and control systems that make up the pipeline and define the flow through the pipeline systems. Blueprints show the actual construction dimensions and layout of systems and facilities. As-built drawings are revised as changes are made to ensure that drawings are up-to-date with what really exists on the pipeline.

Many documents are required for recordkeeping, and they must be retained for specified lengths of time. This, and the need for maintenance of current maps and drawings, requires a systematic method of managing documents. Documents required cover the entire range of work done on and around the pipeline, from inspections to repairs, to daily operations activities. Maintenance personnel must provide detailed information about the pipeline, equipment, and surroundings any time field visits are made. Management is responsible for providing the necessary permits, work orders, and other forms required before work can be done. Safety and accident reports are required when certain incidents occur.

Many reports and forms are produced by computer or logged into the computer system for tracking. The computer is an important tool in the document management process, but the manual provision of correct, up-to-date information is essential for the document to be useful.



Instructor's Notes:

**Review Questions**

1. Topographical maps are also called _____ maps.
   a. earth
   b. graphical
   c. sea level
   d. topo

2. The GPS system uses _____ satellites in six orbits.
   a. twelve
   b. sixteen
   c. twenty-four
   d. thirty-six

3. P&IDs show the following *except* _____.
   a. natural features
   b. valve location
   c. pipe size
   d. flow direction

4. Valve inspection reports are separated into main-line valve reports and _____ valve reports.
   a. control
   b. station
   c. system
   d. secondary line

5. Aerial patrol pilots assign a _____ number to each abnormal sighting.
   a. condition
   b. station
   c. location
   d. report

6. Written accident reports are required to be sent to _____.
   a. OSHA
   b. DOT
   c. ANSI
   d. ISA

7. Confined space work requires a _____.
   a. hard hat
   b. breathing apparatus
   c. permit
   d. lockout

8. Construction records must be kept for _____.
   a. 5 years
   b. 10 years
   c. 15 years
   d. the life of the facility

9. Daily operating records must be kept for at least _____ years.
   a. 3
   b. 5
   c. 10
   d. 15

10. Breakout tanks must be inspected at least every _____ months, but at least _____ time(s) per year.
    a. 2½; six
    b. 4½; three
    c. 7½; two
    d. 15; one

GLOSSARY

# Trade Terms Introduced in This Module

**Alignment sheets:** Maps and drawings that identify the pipeline and its route in relation to direction, physical features, and political features.

**Blueprints:** Construction and engineering drawings showing the actual layout and measurements of the pipeline and facilities.

**Good working order:** All valves are required by *49 CFR Part 195 Paragraph 195.420(a)* to be in good working order, which can be interpreted to mean that a valve's performance, based on in-service criteria, meets all necessary functions of the valve.

**Piping and instrumentation diagrams (P&IDs):** Diagrammatic layouts of the pipe and equipment to show the flow through the line, using standard piping and instrumentation symbols.

**Right-of-way (ROW):** Area specifically set aside to contain the pipeline and provide access for maintenance and inspection.

**Station number:** The distance from the starting point on a survey map, shown as xxx + 00.



Instructor's Notes:

# MODULE 62105-02 — TEACHING TIPS

The following are suggested activities or instructional methods to help you teach the material in this AIG.

## General

When you call on someone to answer a question, the rest of the class relaxes or even tunes out because they expect that the question and answer will take place only between you and the trainee you called on. Instead, use this technique to involve more trainees in answering questions and to keep them on their toes.

1. Ask trainees to define a term or explain a concept.
2. After one trainee has answered, ask a trainee seated nearby if the answer is right. Then ask whether a trainee in the back of the room agrees.
3. Ask trainees to explain why they think an answer is right or wrong.
4. Use the session to clear up incorrect ideas and encourage trainees to learn from their mistakes.

## Section 2.1.2 *Topographic Maps*

This exercise will give trainees hands-on practice reading topographic maps. You will need a full-size topographic map of a local area that may be familiar to trainees. Review the map before the class session and note the topographical features. You will also need masking tape or pushpins. Trainees will need pencils and paper to take notes. Allow 45 to 60 minutes for this exercise.

1. Mount the map on a wall or bulletin board using the masking tape or pushpins.
2. Ask a trainee to identify a feature that can be found on the map (for example, the steepest part of a mountain). Ask another trainee to verify that the correct feature has been identified and to explain why the answer is correct (for example, lines closer together indicate steeper terrain).
3. Continue the exercise with other features and other trainees.
4. Ask trainees if they have any questions about reading topographic maps.

## Section 2.1.4 *Strip Maps*

This exercise will give trainees hands-on practice working with strip maps. You will need sample strip maps, including aerial, plan, elevation, and side views of one area with a pipeline superimposed on all views. You will also need a highlighter and masking tape or pushpins. Trainees will need pencils and paper to take notes. Allow 45 to 60 minutes for this exercise.

1. Mount the maps on a wall or bulletin board using the masking tape or pushpins. Have trainees gather around the maps.
2. Point out each type of map and use the highlighter to show trainees how each map shows a different view of the same area. Depending on the training and experience level of your trainees, you may wish to conduct a brief review of the different views from which a site may be portrayed. Use a box or a book to give trainees a simple, visual example of plan, elevation, and side views.
3. Describe how workers in the pipeline industry use the maps.
4. Ask trainees if they have questions.

## Section 2.3.0 *P&IDs*

This exercise will familiarize trainees with the standard P&ID piping symbols. You will need a sample P&ID, the P&ID piping symbols in the Trainee Guide (*Figures 8* and *9*), and masking tape or pushpins. Trainees will need pencils and paper to take notes. Allow 30 minutes for this exercise.

1. Mount the sample P&ID on a wall or bulletin board using the masking tape or pushpins.
2. Ask trainees to look at *Figures 8* and *9* in their Trainee Guides.
3. Call on a trainee to find and identify a piping symbol on the P&ID. Ask another trainee if the answer is correct and, if not, to correct it.
4. Repeat steps 3 and 4 with other trainees.
5. Ask trainees if they have any questions.

## Section 2.4.0 *Electrical Drawings*

This exercise will familiarize trainees with the standard symbols used on electrical blueprints. You will need a sample electrical blueprint, the standard electrical symbols in the Trainee Guide (*Figures 10* and *11*), and masking tape or pushpins. Trainees will need pencils and paper to take notes. Allow 30 minutes for this exercise.

1. Mount the sample electrical blueprint on a wall or bulletin board using the masking tape or pushpins. Ask trainees to look at *Figures 10* and *11* in their Trainee Guides.
2. Call on a trainee to find and identify an electrical symbol on the electrical blueprint. Ask another trainee if the answer is correct and, if not, to correct it.
3. Repeat steps 3 and 4 with other trainees.
4. Ask trainees if they have any questions.

## Section 4.1.0 *Retention Rates*

The administrative requirements for recordkeeping can make for rather dry material in the classroom. You can liven up this section and get trainees to appreciate the importance of what may be a dull but vital task by incorporating war stories—problems that occurred because records were not properly retained or managed. If you don't have a war story handy, spend some time before the class meets interviewing colleagues for their experiences or for help in drafting a list of problems resulting from poor record retention or management. Allow at least 45 minutes for the discussion.

## MODULE 62105-02 — ANSWERS TO REVIEW QUESTIONS

| Answer | Section |
|---|---|
| **1.** d | 2.1.2 |
| **2.** c | 2.2.0 |
| **3.** a | 2.3.0 |
| **4.** b | 3.1.0 |
| **5.** d | 3.3.0 |
| **6.** b | 3.8.0 |
| **7.** c | 3.10.0 |
| **8.** d | 4.1.1 |
| **9.** a | 4.1.4 |
| **10.** d | 4.3.0 |

## CONTREN™ LEARNING SERIES — USER UPDATES

The NCCER makes every effort to keep these textbooks up-to-date and free of technical errors. We appreciate your help in this process. If you have an idea for improving this textbook, or if you find an error, a typographical mistake, or an inaccuracy in NCCER's Contren™ textbooks, please write us, using this form or a photocopy. Be sure to include the exact module number, page number, a detailed description, and the correction, if applicable. Your input will be brought to the attention of the Technical Review Committee. Thank you for your assistance.

*Instructors* – If you found that additional materials were necessary in order to teach this module effectively, please let us know so that we may include them in the Equipment/Materials list in the Instructor's Guide.


**Write:** Curriculum Revision and Development Department
National Center for Construction Education and Research
P.O. Box 141104, Gainesville, FL 32614-1104

**Fax:** 352-334-0932

**E-mail:** curriculum@nccer.org


Craft ______ Module Name ______

Copyright Date ______ Module Number ______ Page Number(s) ______

Description ______

______

______

______

(Optional) Correction ______

______

______

(Optional) Your Name and Address ______

______

______

Module 62106-02

# *Preventing Pipeline Damage*

## *(CT 14 and 28)*

## MODULE OVERVIEW

This module explains how to prevent pipeline damage so that people, property, and the environment are protected. Detection and prevention of third-party and in-house damage are discussed along with other prevention measures.

## PREREQUISITES

Please refer to the Course Map in the Trainee Module. Prior to training with this module, it is recommended that the trainee shall have successfully completed the following:

Core Curriculum; Pipeline Core, Modules 66101-02 and 66102-02; Pipeline Maintenance, Level One, Modules 62103-02

## OBJECTIVES

Upon completion of this module, the trainee will be able to:

1. Explain the One-Call system.
2. Perform tasks to prevent third-party and in-house damage (CT 28).
3. Explain public awareness activities.
4. Place and maintain permanent line markers (CT 14).

## PERFORMANCE TASKS

Under the supervision of the instructor, the trainee should be able to:

1. Perform tasks or identify and describe tasks to prevent third-party and in-house damage per company procedure and/or using Performance Verification 28 as a guide.
2. Locate or simulate locating a line per company procedure and/or using Performance Verification 14.1 as a guide.
3. Install or simulate installing a line marker per company procedure and/or using Performance Verification 14.2 as a guide.
4. Inspect and maintain or simulate inspecting and maintaining a line marker per company procedure and/or using Performance Verification 14.3 as a guide.
5. Inspect and maintain or simulate inspecting and maintaining an aerial line marker per company procedure and/or using Performance Verification 14.4 as a guide.

## NCCER STANDARDIZED CRAFT TRAINING PROGRAM

The National Center for Construction Education and Research (NCCER) provides a standardized national program of accredited craft training. Key features of the program include instructor certification, competency-based training, and performance testing. The program provides trainees, instructors, and companies with a standard form of recognition through a National Craft Training Registry. The program is described in full in the *Guidelines for Accreditation*, published by the NCCER. For more information on standardized craft training, contact the NCCER by writing us at P.O. Box 141104, Gainesville, FL 32614-1104; calling 352-334-0911; or e-mailing info@nccer.org. More information may be found at our Web site, www.nccer.org.

## HOW TO USE THIS ANNOTATED INSTRUCTOR'S GUIDE

Each page presents two sections of information. The larger section displays each page exactly as it appears in the Trainee Module. The narrow column ties suggested trainee and instructor actions to each page and provides icons to call your attention to material, safety, audiovisual, or testing requirements. The bottom of each page includes space for your notes.

**Teaching Tip**

If you see the Teaching Tip icon, that means there is a teaching tip associated with this section. Also refer to any suggested teaching tips at the end of the module.

## SAFETY CONSIDERATIONS

Ensure that the trainees are equipped with appropriate personal protective equipment. Emphasize the importance of following all safety precautions and procedures when working with power tools.

## PREPARATION

Before teaching this module, you should review the Module Outline, Objectives, Performance Tasks, and the Materials and Equipment List. Be sure to allow ample time to prepare your own training or lesson plan and gather all required equipment and materials.

## MATERIALS AND EQUIPMENT LIST

Transparencies
Felt-tip markers/chalk
Posterboard
Blank acetate sheets
Transparency pens
Pencils and scratch paper
Tape or tacks
Strip maps (in addition to the trainee book)
Copies of the Performance Verification for Covered Tasks in 14 and 28†
Copies of the Quick Quiz*
Module Examinations**
Performance Profile Sheets**
Overhead projector and screen
Whiteboard/chalkboard
Appropriate personal protective equipment
Probe
Line locator

* Located at the end of this Annotated Instructor's Guide.
** Located in the Test Booklet packaged with this Annotated Instructor's Guide.
† Available for download at www.nccer.org.

## TEACHING TIME FOR THIS MODULE

An outline for use in developing your lesson plan is presented below. Note that each Roman numeral in the outline equates to one session of instruction. Each session has a suggested time period of 2½ hours. This includes 10 minutes at the beginning of each session for administrative tasks and one 10-minute break during the session. Approximately 15 hours are suggested to cover *Preventing Pipeline Damage*. You will need to adjust the time required for hands-on activity and testing based on your class size and resources.

| Topic | Planned Time |
|---|---|
| **Session I. One-Call System** | |
| A. One-Call System | ________ |
| 1. Laboratory | ________ |
| B. Marking Protocol | ________ |
| 1. Temporary Marking | ________ |
| 2. Permanent Marking | ________ |
| C. Emergency Locating | ________ |

**Session II. Preventing Third-Party Damage**

A. Maintaining the Pipeline Right-of-Way ____________

B. Maintaining Security ____________

1. Laboratory ____________

C. Performance Testing ____________

**Session III. Locating Underground Pipelines**

A. Permanent Line Markers ____________

B. Installing Line Markers ____________

1. Laboratory ____________

C. Performance Testing ____________

D. Inspecting and Maintaining Line Markers ____________

1. Laboratory ____________

E Performance Testing ____________

**Session IV. Inspecting and Maintaining Pipelines**

A. Patrolling the Pipeline ____________

B. Performance Testing ____________

C. Performance Testing ____________

**Session V. Preventing In-House Damage**

A. Excavation Procedures ____________

B. Lowering Pipelines ____________

C. Mapping Maintenance ____________

**Session VI. Public Awareness**

A. Public Awareness ____________

B. Excavators ____________

C. Gas Pipeline Emergency Procedures ____________

D. Summary ____________

1. Summarize module ____________

2. Answer questions ____________

E. Module Examination ____________

1. Trainees must score 70% or higher to receive recognition from the NCCER.
2. Record the testing results on Craft Training Report Form 200 and submit the results to the Training Program Sponsor.

F. Performance Testing ____________

1. Trainees must perform each task to the satisfaction of the instructor to receive recognition from the NCCER.
2. Record the testing results on Craft Training Report Form 200 and submit the results to the Training Program Sponsor.

Module 62106-02

# *Preventing Pipeline Damage*

## *(CT 14 and 28)*

Instructor's Notes:

Instructor's Notes:

# Introduction to the Pipeline Operator Qualification Rule, API, and NCCER

*It could be said that petroleum products keep America running. Petroleum products like oil and natural gas are the principal sources of energy in the United States and are used in many manufacturing and chemical processes. Petroleum products are primarily transported by pipeline. There are more than a million miles of liquid and natural gas pipeline in this country, ranging in diameter from two inches to four feet. These pipelines bring crude oil and natural gas from production fields to processing plants and refineries—and ultimately to the consumer.*

## Pipeline Operator Qualification

Petroleum products are considered hazardous, so pipeline personnel must take extra precautions and follow specific procedures when working with them. To help ensure the safe handling and transportation of petroleum products, the United States Department of Transportation (DOT) mandates safe practices as federal regulations.

One such regulation, the Operator Qualification (OQ) rule, requires pipeline operators to develop a written qualification plan to evaluate their personnel and their contract personnel (see *49 CFR Part 192 Subpart N—Gas,* and *49 CFR Part 195 Subpart G—Liquid*). The evaluation hinges on the operator's ability to perform certain covered tasks and to be able to recognize and respond to abnormal operating conditions that may be encountered while performing these covered tasks.

## Covered Tasks

A pipeline performance task is deemed *covered* if the activity passes the following four-part test:

1. Is it performed on a pipeline facility?
2. Is it an operations or maintenance task?
3. Is it performed as a requirement of *49 CFR Part 195 (Liquid)* or *49 CFR Part 192 (Gas)*?
4. Does it affect the operation or integrity of the pipeline facility?

If the answer to all four questions is *yes*, then the activity is considered *covered* and must figure into the evaluation process of pipeline personnel.

## American Petroleum Institute

The OQ rule and its related covered tasks provided the catalyst to develop a unified pipeline training and assessment program—so that pipeline operators and owners have ready access to the assessment and training tools they need. This effort was the direct result of the vision and leadership provided by The American Petroleum Institute (API) and API members.

Efforts to unify the industry are nothing new to API. Indeed, the impetus for forming the Institute in 1919 was the need to standardize engineering specifications for drilling and production equipment. Today, the most pressing issues revolve around public perceptions and government policies toward the industry. Speaking with one voice on these issues has become as essential as having interchangeable parts in the field. API's members determine the petroleum industry's positions on public policy as well as the standards that govern its day-to-day operations worldwide.

## Consortium on Operator Qualification

API assembled a focus group whose primary purpose was to identify covered tasks and provide compliance recommendations to pipeline operators.

The Consortium on Operator Qualification (COOQ) became the guiding force for the identification and development of covered tasks for the liquid pipeline industry. Their work product, API publication 1161, *Guidance Document for the Qualification of Liquid Pipeline Personnel,* formed the basis for the resulting Pipeliner Training and Assessment Program.

The covered task list created in API publication 1161 has now evolved to include gas covered tasks and is maintained by NCCER. The most current covered task list can be downloaded at www.nccer.org.

## National Center for Construction Education and Research

With the guidance and support of API, its members, and the Consortium on Operator Qualification, the National Center for Construction Education and Research (NCCER) was able to design and implement the Pipeliner Training and Assessment Program.

NCCER is a not-for-profit 501(c)(3) education foundation established in 1995 by the world's largest and most progressive construction and maintenance contractors and national contractor associations. NCCER was founded to address the

severe workforce shortage facing the construction and maintenance industries and to develop a standardized training process and curricula. Today, NCCER is supported by hundreds of leading construction and maintenance companies, manufacturers, and national associations and developed more than 35 construction and maintenance titles.

NCCER also maintains a National Registry, which contains the records of trainees and instructors who have received NCCER training. All completion information for every trainee is sent to the NCCER and filed within the Registry. The Registry can then confirm training and skills for workers as they move from company to company, state to state, or even within their own company.

### *Standardized Curriculum (NCCER's Contren™ Learning Series)*

NCCER curricula are developed using the expertise of industry subject matter experts (SMEs). The use of SMEs as an integral part of every new development or revision ensures that the curriculum is current, industry-driven, and indicative of industry practices. The development and/or revision process is facilitated from start to finish by NCCER project managers, who work closely with the SMEs and a select team of technical writer(s) on the development and presentation of the curriculum.

### *Pipeline: Five Training Paths*

The pipeline curriculum focuses on essential skills and competencies within five skill areas:

- Corrosion Control
- Electrical and Instrumentation
- Maintenance
- Mechanical
- Operations (Field and Control Center)

SMEs from the pipeline industry have determined the skills and competencies needed based on API's covered tasks and the consensus of the group/industry. Specific company information is generally not incorporated into the curriculum unless the SMEs determine that the information is the industry standard.

### *Training Modules and Covered Tasks*

The smallest unit of instruction in the curriculum is a *module*. Each curriculum is composed of modules (the equivalent of stand-alone chapters) that are logically sequenced into complete *levels* of instruction. Every module is task-oriented and requires the successful completion of a written (knowledge-based) examination and a performance (hands-on) test. Modules may be individually selected and/or combined for efficient task training. Modules in the pipeline curricula, to the extent possible, will align with API-developed covered tasks. While not all modules will relate to covered tasks, all covered tasks will relate to at least one module.

### *Benefits of Training to the Pipeline Marketplace*

Applying NCCER's standardized training process allows pipeline owners and contractors in the United States to establish the benchmark for acceptable skill and competency levels within the pipeline industry. This process will assist training organizations in developing and/or improving standardized pipeline training programs that effectively match their resources and operating conditions while responding to regulatory demands and to the needs of the industry. The process will also recognize training organizations that meet the minimum standards of competence, warranting regulatory, public, and professional confidence. This in turn will foster national unity within the pipeline industry for the mutual benefit of individual technicians, owners, contractors, and ultimately pipeline-users and society.

## Additional Resources

To review the Pipeliner Training and Assessment Program Implementation and Resource Guide and the latest version of pipeline covered tasks, visit NCCER's Web site at www.nccer.org, or contact:

**National Center for Construction Education and Research**
PO Box 141104
Gainesville, FL 32614-1104
(888) 622-3720
Web site: www.nccer.org

For more information on API, please contact:

**American Petroleum Institute**
1220 L Street, Northwest
Washington, D.C. 20005
(202) 682-8000
Web site: www.api.org

To order *API Pub 1161*, contact Global Engineering Documents by calling 1-800-854-7149 or visiting their Web site at www.global.his.com.

Additional sources of information on the pipeline operator qualification regulation include:

**DOT – Office of Pipeline Safety**
www.ops.dot.gov

**Transportation Safety Institute**
www.tsi.dot.gov



Instructor's Notes:

## COURSE MAP

This course map shows all of the modules in the first level of the Pipeline Maintenance curriculum. The suggested training order begins at the bottom and proceeds up. Skill levels increase as you advance on the course map. The local Training Program Sponsor may adjust the training order.

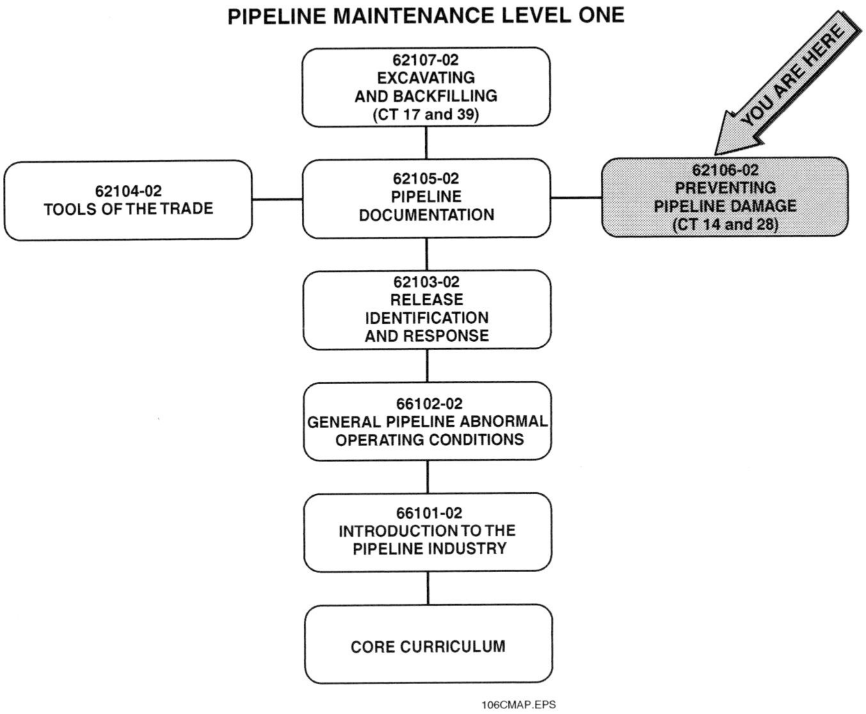






**Homework**

**Assign reading of Sections 1.0.0–2.2.0 of this module.**

Instructor's Notes:

## MODULE 62106-02 CONTENTS

Instructor's Notes:

MODULE 62106-02

# Preventing Pipeline Damage (CT 14 and 28)

### Objectives

When you have completed this module, you will be able to do the following:

1. Explain the One-Call system.
2. Perform tasks to prevent third-party and in-house damage (CT 28).
3. Explain public awareness activities.
4. Place and maintain permanent line markers (CT 14).

### Prerequisites

Before you begin this module, it is recommended that you successfully complete the following: Core Curriculum; Pipeline Core, Modules 66101 and 66102; Pipeline Maintenance Level One, Module 62103.

### Required Trainee Materials

1. Pencil and paper
2. Appropriate personal protective equipment

## 1.0.0 ◆ INTRODUCTION

Preventing pipeline damage is critical for the protection of people, property, and the environment.

The two leading causes of pipeline releases are third-party and in-house damage. Detection and prevention of both of these are discussed in this module along with other prevention measures. This module is divided into the following sections:

- One-Call system
- Preventing third-party damage
- Preventing in-house damage
- Public awareness

## 2.0.0 ◆ ONE-CALL SYSTEM

Before beginning an excavation, residents or companies performing the excavation must notify the local utility companies to have the underground line locations identified. Notification and utility location can be handled either by utility operators individually or by a collaborative multicompany effort. Some operators may elect to set up their own system for handling calls from excavators and for locating and marking underground lines. However, the most common method used is an operator-sponsored centralized program known as a One-Call system. The One-Call system is defined as a communication system established by two or more utilities, governmental agencies, or other operators of underground facilities to provide one telephone number for excavating contractors and the general public to call to notify of their intent to use equipment for excavating, demolition, or other work. It requires that the participating members identify and give the location of their facilities.

The One-Call system is generally set up state-by-state. Some states have statewide One-Call systems in place. Other states have multiple systems, but no area is without coverage. There are a few states that have areas that are not covered.

Pipeline company offices have communication systems connected to the One-Call system. A dedicated phone line and printer may be set up to receive excavation notices (see *Figure 1*). Notification may also be received via fax, email, or the Internet. When notices are received, the company researches the area to confirm it is the location of the pipe. If the pipe is not threatened, the excavation will be okayed. If the pipe may be threatened, the company will arrange to provide temporary markings and oversee the excavation.



**Materials**

**Ensure that you have everything necessary to teach the course. Check the Materials and Equipment List at the front of this module.**

**Classroom**

**Explain that the terms shown in bold are defined in the Glossary at the back of this module.**

**Audiovisual**

**Show Transparency 1, Course Objectives.**

**Show Transparency 2, Performance Tasks.**

**Teaching Tip**

**See the general Teaching Tip at the end of this module.**

**Classroom**

**Discuss how One-Call systems work. Ensure that all trainees understand the reasons for contacting the local One-Call system.**

## Laboratory

**Refer to Figure 1. Pair trainees and ask one to simulate calling in an excavation notice. Have the other trainee fill in the call investigation report. Ask trainees how a company responds if its pipe is threatened.**

CALL INVESTIGATION REPORT

| | |
|---|---|
| COMPANY NAME: | DATE: |
| CALLER'S NAME: | TIME: |
| CONTACT TELEPHONE NUMBERS/NAMES: | |
| NAME: | NUMBER: |
| NAME: | NUMBER: |
| AREA/LOCATION OF REPORT: | |
| DESCRIPTION: | |
| NOTIFIED NAME: | DATE: |
| CONTROLLER'S NAME: | TIME: |

SAMPLE

106F01.EPS

*Figure 1* ◆ Call investigation report form.



Instructor's Notes:

If the call is an emergency, the response must be made as soon as possible. The One-Call system records all communications on the One-Call Log.

Response to the One-Call system is addressed in the sections that follow:

- Marking protocol
- Emergency locating

## 2.1.0 Marking Protocol

Marking protocol includes the use of visible markers and color codes. Permanent signs are placed near the pipeline to inform the public that a line is nearby, what type of line it is, the name of the operator, and a number to call in case of a leak, another emergency condition, or for reporting any abnormal condition. *Figure 2* shows a sample marker sign.

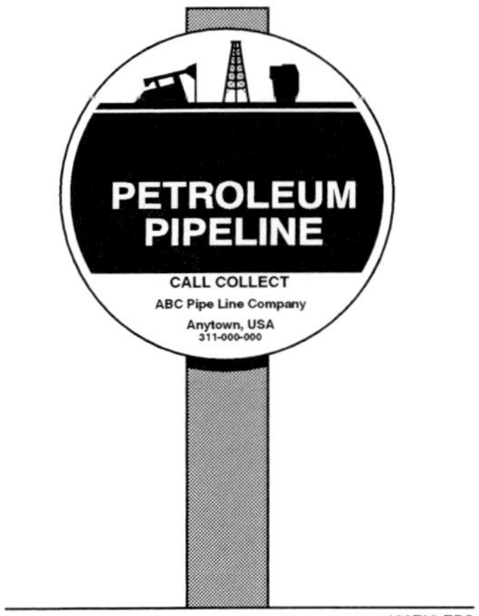


106F02.EPS

*Figure 2* ◆ Sample marker sign.

### *2.1.1 Temporary Marking*

For temporary markings, flags, stakes, or paint will be used to mark the line location. The International Color Code for marking underground facilities is used for marking.

### *2.1.2 Permanent Marking*

Permanent **line markers** are required at road crossings, railroad crossings, and water crossings, and spaced at intervals along the line. These signs inform the public that a pipeline is nearby. The signs may not actually be on top of the line, the line may not follow a straight path from sign to sign, and the signs do not provide information on the depth or number of pipelines.

All lines need to be marked accurately and continuously. Both temporary and permanent markings are needed. All markings should follow the color code and temporary markings must be made to last throughout any proposed work. The color code is shown in *Table 1*.

**Table 1** Color Code

| Marking Color | Utility Lines |
|---|---|
| Red | Electric |
| Yellow | Natural Gas/Oil/Steam/ Hazardous Materials |
| Orange | Communication/CATV |
| Blue | Water |
| Purple | Reclaimed Water |
| Green | Sewer |
| Pink | Temporary Survey Markings |
| White | Proposed Excavation |

106T01.EPS

## 2.2.0 Emergency Locating

In the case of emergencies or problem situations, operators are informed of the problem and asked to provide immediate assistance in locating their underground facilities. If necessary, authorized persons may have to begin emergency excavation or demolition of facilities to protect life, health, or property before the line can be marked. They are expected to exercise reasonable precautions to protect the underground facilities.

## 3.0.0 ◆ PREVENTING THIRD-PARTY DAMAGE

Preventing third-party damage to pipelines involves making sure that the location of the pipeline is clearly marked, that excavators and the public are aware of the pipeline, that the line is inspected regularly, and that activities near the line are investigated and monitored. It also means that all pipeline facilities must be secure against willful damage. Public education is covered in another section. Otherwise, preventing third-party damage to the pipeline involves:

- Maintaining the pipeline right-of-way
- Maintaining security
- Locating underground pipelines
- Installing and maintaining markers
- Patrolling the pipeline



**Classroom**

**Refer to Figure 2. Ask trainees to explain what types of information are conveyed on permanent pipeline signs. Review the International Color Code for marking.**

**Ask trainees for examples of temporary and permanent marking.**

**Refer to Table 1. Pair trainees and have them quiz each other on which marking colors are used for which types of utility lines.**

**Teaching Tip**

**See the Teaching Tip at the end of this module.**

**Homework**

**Assign reading of Sections 3.0.0–3.2.2 for the next class session.**

**Materials**

**Ensure that you have everything required for demonstrations and testing during this session.**

**Classroom**

**Discuss the five actions involved in preventing third-party damage to pipelines. Ask trainees if they have questions.**

**Demonstration**

**Using chalk and chalkboard or marker and posterboard, draw samples of right-of-way (ROW) width during construction and after construction. Emphasize the safety reasons for maintaining the pipeline ROW.**

**Classroom**

**Review how the ROW is patrolled and what observers should look for.**

**Audiovisual**

**Show Transparency 3. Ask trainees to name eight things that help maintain security. Review items that trainees missed.**

**Classroom**

**Refer trainees to Figure 3. Discuss the steps for providing pipeline security.**

**Laboratory**

**Using pencil and paper, have trainees practice the nine steps to provide pipeline security. Give them a hypothetical security violation for each step and ask them to detect and record it.**

### 3.1.0 Maintaining the Pipeline Right-of-Way

The pipeline right-of-way (ROW) is the area along the pipeline which has been legally permitted to be used by the pipeline company to construct, operate, and maintain the pipeline. The ROW extends the entire length of the line. During construction, it is wide enough to allow for construction equipment and cathodic protection equipment installed on the line. In some cases, space for access roads is included in the ROW. Rights-of-way can be 50 to 100 feet wide during construction. Once construction is completed, the ROW typically becomes much smaller. It may be reduced to just a few feet on either side of the pipeline, or even just the width of the pipeline itself. Before beginning any work, you must be sure you have verified the ROW for the section of pipeline you are maintaining. After the pipeline is built, the ROW is returned as close to its original condition as possible, including restoring fences across it, so that the use of the adjoining land is not obstructed. The ROW itself is not fenced in, except for aboveground equipment and facilities.

The ROW is patrolled regularly by plane, vehicle, or on foot to observe conditions, construction activity, encroachments, markers, soil condition, uncovered pipe, facilities condition and security, leaks, and any other indications of problems.

### 3.2.0 Maintaining Security

In addition to accidental damage, the pipeline system is also subject to willful damage by outside parties. All facilities and aboveground equipment must be protected against unauthorized access or vandalism. Each pumping station, breakout tank area, and other exposed facility, such as a **main-line valve**, must be kept secure at all times.

Maintaining security can involve any, or all, of the following:

- Fences
- Gates
- Permits
- Locks
- Crew/employee awareness
- Signs
- Cameras
- Guards

Stations must be completely fenced and the fences well maintained. Main-line valves and other exposed equipment may also be fenced or protected by guardrails (see *Figure 3*). Gates are provided at entry points to facilities and are to be locked at all times or controlled by security guards. All main-line valves are chain-locked in the open position, as are all block valves in the relief piping. Locks are installed on all gates and facility doors, limiting access to authorized persons.

In addition, many larger facilities have security systems and other **intrusion devices** that have **remote monitoring** capabilities and allow the control center to watch multiple facilities simultaneously.

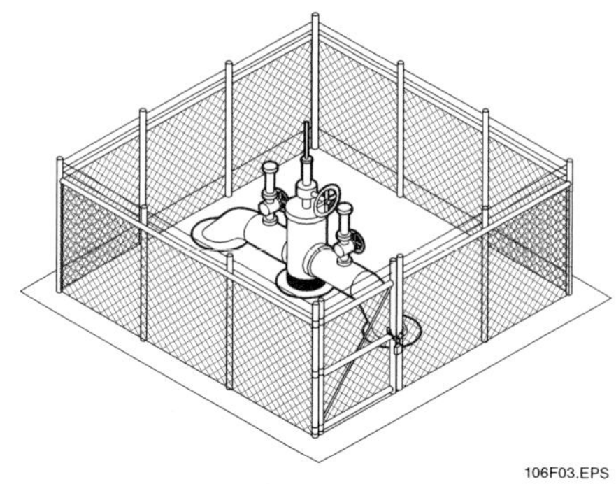


*Figure 3* ◆ Fenced main-line valve.

Crews in the field are responsible for maintaining the line security. All employees are informed of security provisions and of the importance of security for everyone. Although no two companies' security procedures are the same, the following are basic steps you should take to provide security for pipeline facilities you maintain.

*Step 1* Notify appropriate company personnel before entering remotely monitored facilities if the security system cannot be deactivated at the facility.

*Step 2* Inspect the facility fencing, gates, or other entry barriers for integrity and possible damage.

*Step 3* Inspect all locks for possible damage and ensure that locks are in good working order.

*Step 4* Inspect signs for proper placement, visibility to the public, and to ensure that the information on the sign is legible and correct.

*Step 5* Inspect local monitoring devices, which may include motion sensors, alarms, and cameras, for operability, damage, or required maintenance.



Instructor's Notes:

*Step 6* Test intrusion devices to ensure operability. Remotely monitored intrusion devices must be tested periodically.

*Step 7* Perform repairs to facility security equipment as appropriate and if you are qualified, or prepare a work order for the appropriate personnel.

*Step 8* Reactivate facility security system or notice appropriate personnel upon leaving remotely monitored facilities so they can reactivate the system.

*Step 9* Verify closure to prevent a call-out.

### *3.2.1 Securing Facilities*

Company signs are displayed at all facilities, such as pump stations and breakout tanks. The signs are located around the facility so that anyone approaching from any direction accessible to the public is able to see the signs. Aboveground facilities are surrounded by fences with locked or guarded gates that should be kept in good repair. Aboveground main-line valves should be locked to prevent outside interference.

### *3.2.2 Encroachment Prevention*

Encroachment is activity on the company's right-of-way by outside parties. A particular encroachment may be authorized or not, so each situation requires specific handling. Unauthorized activities on the pipeline right-of-way that may affect the pipeline are often spotted by patrols doing line inspection and reported by them to management. The encroaching party is advised to cease activity until the company can give consent. The company then investigates the activity and communicates with the encroaching party the final resolution. The company's right-of-way department usually handles these investigations.

## 3.3.0 Locating Underground Pipelines

Individuals involved in locating underground pipelines need to be proficient in reading maps and alignment sheets. Maps and alignment sheets are created to furnish pertinent, permanent information about the location of various pipeline installations. Prior to dispatching on a pipe location assignment, use the most current version of your company's maps and alignment sheets to identify the location of the target pipeline and determine if there are any special access requirements that must be met. If there is a question about the accuracy of the strip maps and alignment sheets being used, obtain a surveyor's map or blueprint to pinpoint the exact location of the pipeline right-of-way. *Figure 4* shows a pipeline strip map.

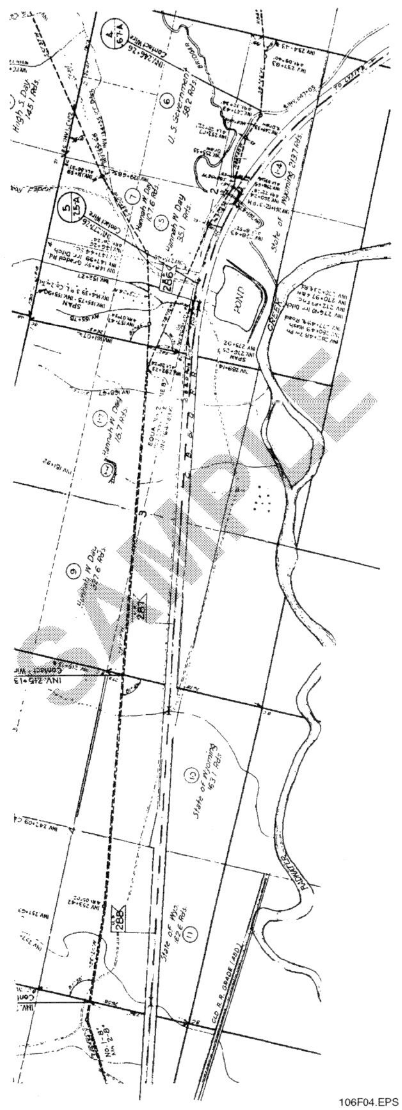

106F04.EPS

*Figure 4* ◆ Pipeline strip map.



**Classroom**

**Ask a trainee to explain a main-line valve. Ask another trainee to define the various threats to exposed facilities, such as main-line valves.**

**Performance Testing**

**Have each trainee demonstrate Performance Task 1 to your satisfaction. Fill out a Performance Profile Sheet for each trainee.**

**Homework**

**Assign reading of Sections 3.3.0–3.4.2 for the next class session.**

**Materials**

**Ensure that you have everything required for demonstrations and testing during this session.**

**Demonstration**

**Refer to Figure 4. Show trainees how to locate the target pipeline and determine if there are special access restrictions. Explain why access restrictions strengthen pipeline safety.**

**Laboratory**

**Hang sample strip maps throughout the room. Have trainees pair up, with one locating the target pipeline and the other determining if there are special access restrictions. Have trainees check their partner's answers.**

**Safety**

**Stress that maps and markers may only partially delineate the location of pipelines.**

**Performance Testing**

**Have each trainee demonstrate Performance Task 2 to your satisfaction. Fill out a Performance Profile Sheet for each trainee.**

**Safety**

**Stress the importance of maintaining permanent line markers. Discuss the consequences of not maintaining markers.**

**Homework**

**Assign reading of Sections 3.5.0–3.5.4 for the next class session.**

**Materials**

**Ensure that you have everything required for demonstrations and testing during this session.**

**Classroom**

**Review and discuss the four steps used when installing permanent line markers. Ensure that trainees understand the proper use of line markers.**

**Discuss where permanent line markers need to be installed. Stress the importance of knowing the location through line of sight.**

> **NOTE**
> Remember that, unless the lines have been recently, accurately surveyed and had the GPS coordinates established, the maps, alignment sheets, and as-built drawings may NOT be 100% accurate. Originally, the drawings may have only been "close", and since then the lines may have been abandoned, moved, or new ones built. If in doubt, always pinpoint and mark the location of the line.

Alignment sheets show the location and size of each pipe installed, all crossings of public roads, railroads, rivers and streams, as well as buried utilities and pipelines operated by other companies. Locations should be accurate as shown on the alignment sheet. *Figure 5* shows a pipeline alignment sheet.

### 3.4.0 Permanent Line Markers

To prevent or keep to a minimum any third-party damage, the U.S. Department of Transportation requires the use of permanent line markers to indicate the location of underground pipelines. These markers are located at all road crossings, railroad crossings, and at intervals along the line sufficient to allow the linewalker to maintain a clear line-of-sight view of the pipeline ROW. Specifications are described in *DOT 192* and *195*. Line markers should display the company name and location of the operations office at the top of the sign. The signs display the word "WARNING," identify the product in the line, the operator, and the toll-free number where the company may be contacted in the event of an emergency. Markers must be maintained and are routinely inspected for damage, displacement, vandalism, and illegible messages. In cases where the markers are damaged or stolen, they must be replaced.

Markers contain no indication of depth or of the number of lines in the area. These marker signs are critical to the safety of the general public. It is an offense punishable by fines up to $5,000 and/or a year of imprisonment for defacing or destroying a pipeline marker.

> **NOTE**
> It is important to remember that the markers indicate location only; they do not contain specific information regarding the depth of cover or the number of lines in the area.

Some exceptions for buried pipelines where markers are not required are:

- Mains and transmission lines located offshore, or at crossings of or under waterways and other bodies of water
- Mains in Class 3 or Class 4 locations where a damage prevention program is in effect
- Transmission lines in Class 3 or 4 locations where placement of a line marker is impractical
- Pipelines aboveground

#### *3.4.1 Installing Line Markers*

Line markers only indicate the approximate location of the buried pipe. They are typically not placed directly over the line, but are placed and maintained as close as possible over each buried main and transmission line and at each public road and railroad crossing. Markers must be securely installed to provide a long-term indication of the presence of the pipe. In order to install pipeline markers, you might follow the following steps:

*Step 1* Determine the proper marker location and orientation of the sign. Signs should be located directly over the pipeline if possible, but offset orientation is permissible. Signs should be installed:
- At public road crossings
- At railroad crossings
- At above ground locations

*Step 2* Install a sufficient number of markers so that the location of the pipeline is accurately known. (i.e. line of sight)

*Step 3* Safely and securely install the marker or the support for the sign. All signs need to have a solid foundation.

*Step 4* Properly attach the correct sign for the location as necessary. Each sign needs to have correct product identification and information.

#### *3.4.2 Inspecting and Maintaining Line Markers*

It is very important that, once the correct markers have been placed, they are routinely inspected, and maintained. If signs are damaged or missing, they must be repaired or replaced, as needed. The following steps are one procedure that could be used to inspect and maintain line markers:

Step 1. Identify the section of the pipeline to be inspected. This is necessary to ensure inspection of the proper line.



Instructor's Notes:

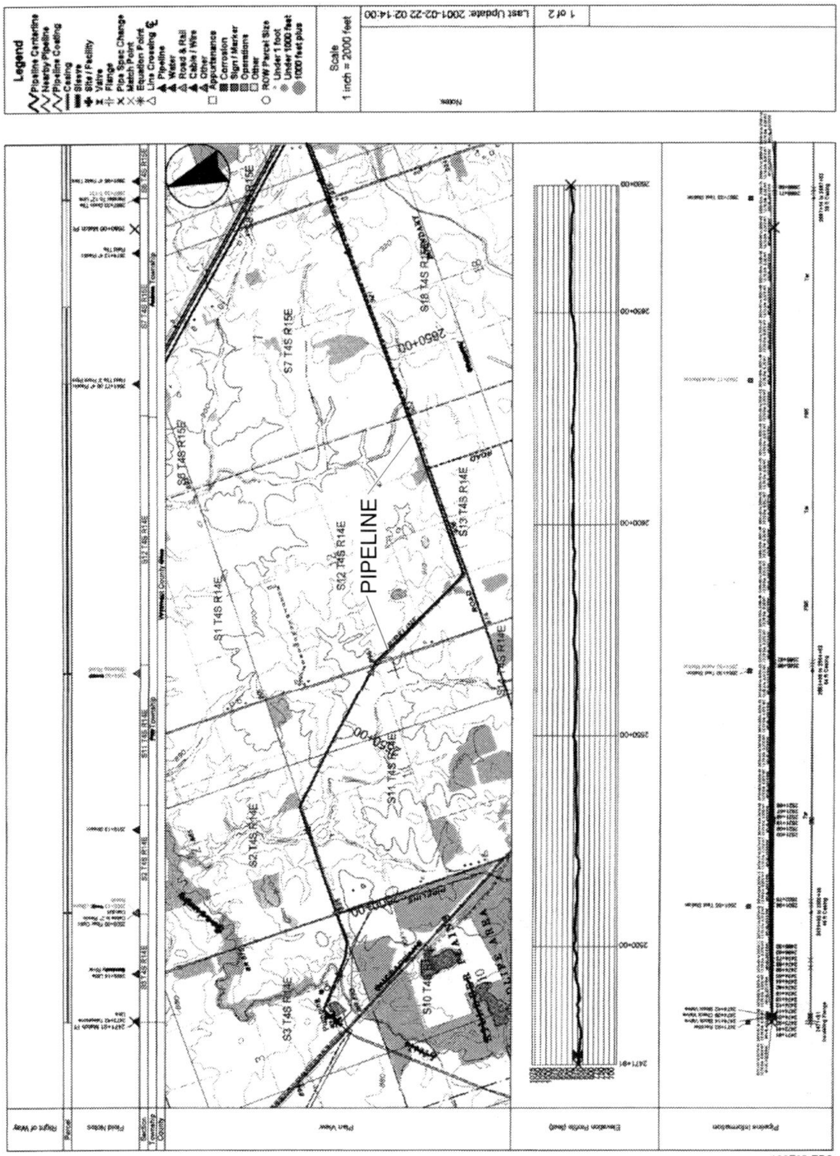


106F05.EPS

*Figure 5* ◆ Pipeline alignment sheet.



### Laboratory

**Hang a large pipeline alignment sheet on the wall. Show trainees how to find line crossings, nearby pipelines, and public roads. Stress how the ability to identify these elements improves safety.**

### Classroom

**Refer to Figure 5. Ask trainees to point out public roads and pipelines operated by other companies.**

### Safety

**Discuss how permanent line markers protect the public. Ask a trainee to explain the penalty for defacing or destroying a line marker.**

### Performance Testing

**Have each trainee demonstrate Performance Task 3 to your satisfaction. Fill out a Performance Profile Sheet for each trainee.**

**Classroom**

**Refer trainees to Table 2. Ask the minimum number of patrols per year for a class 1 or 2 gas pipeline crossing a railroad.**

**Demonstration**

**Have a trainee demonstrate how to do a vehicle/foot patrol and write log notes. Correct any errors. Ask trainees if they have questions.**

**Performance Testing**

**Have each trainee demonstrate Performance Task 4 to your satisfaction. Fill out a Performance Profile Sheet for each trainee.**

Step 2. Locate the most recent version of maps, records, and alignment drawings, and use them to accurately locate the section of pipeline to be inspected.

Step 3. Determine the access requirements for the section to be inspected or maintained. Some ROWs are restricted.

Step 4. Once on the ROW, inspect the signs for correct information, visibility, and orientation. Replace if marker is missing, damaged, or the incorrect information is shown. Also ensure that a sufficient number of signs are present, and that they are placed in the required locations.

Frequently, pipelines have **aerial markers** that also need to be inspected or maintained. They are located in the same way as ground line markers, and inspected for the same information. The only difference is that you must also ensure that the markers are visible from the air.

### 3.5.0 Patrolling the Pipeline

Pipeline operators must have a patrol program in place to check for indications of leaks, construction activity, or any other conditions affecting safety and operation. The frequency of patrols is determined by the size of the line, the operating pressures, the product in the pipeline, terrain, weather, and other relevant factors. Maximum regulated intervals between patrols for gas pipelines are shown in *Table 2*. The company's policies and procedures will determine the actual patrol interval for a given pipeline. Methods of patrolling include walking, driving, flying, or other appropriate means of traversing the ROW.

**Table 2** Maximum Intervals Between Patrols

| Class Location of Line | Maximum Interval Between Patrols | |
|---|---|---|
| | At Highway and Railroad Crossings | At All Other Places |
| 1 and 2 | 7½ months, but at least twice each calendar year | 15 months, but at least once each calendar year |
| 3 | 4½ months, but at least four times each calendar year | 7½ months, but at least twice each calendar year |
| 4 | 4½ months, but at least four times each calendar year | 4½ months, but at least four times each calendar year |

106T02.EPS

#### *3.5.1 Vehicle/Foot Patrol*

Experienced maintenance crews often obtain reliable and useful information about specific situations from on-site surveys. They observe existing conditions and record details observed during every trip to, or along, the pipeline ROW.

Linewalker is the name given the ground patrol members who may walk or use a vehicle to inspect the line. Although the aerial patrol has taken over much of the line inspection, linewalkers are still needed in certain situations, such as densely populated areas where flying at low height may not be allowed. Covering the ROW at ground level is still required when emergencies occur during darkness or when bad weather prevents aerial patrol.

Where foot patrols are routinely used, inspections are done every two to three weeks. Vehicle and foot patrols are effective methods for closely observing the ROW conditions, construction activity at or near the line, encroachments, line marker condition, soil erosion, exposed pipe, condition of pipeline facilities, breaches of security, leaks, and any other condition that could place the pipeline, public or the environment in danger. Patrols should also carefully observe all markers to ensure that the required information is on the marker and in a readable condition. Steps should be taken to replace a marker if any information is incorrect or missing.

#### *3.5.2 Water Crossing Inspections*

Federal regulations require that all navigable water crossings be inspected at least every five years. To ensure the integrity of the pipeline, water crossings may require inspections more frequently to determine the impact of flooding, upstream construction, or other activities. Inspections will determine any changes in the horizontal alignment of the pipe and the river bottom profile to verify that the pipe has adequate cover.

These inspections are done with appropriate surveying equipment and techniques, and the results are compared to existing alignment drawings that show the crossings, stationing of known landmarks, and as-built river profile. The as-built river profile will also include the pipeline location, river bottom profile of previous surveys, the water levels of previous surveys, and the description and location of two bench marks on each side of the river as applicable. *Figure 6* shows an as-built river profile.



Instructor's Notes:

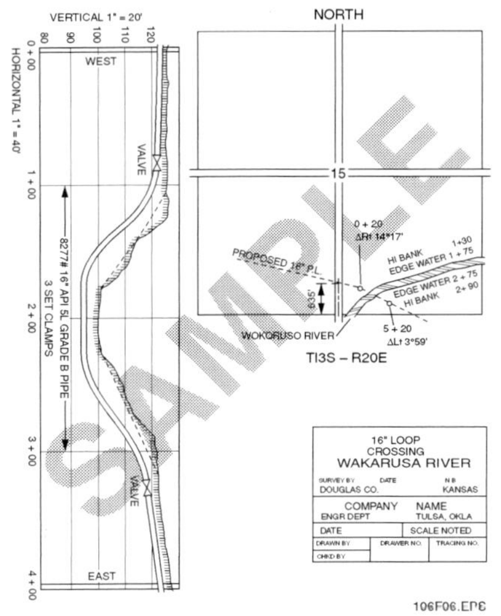


*Figure 6* ◆ As-built river profile.

### 3.5.3 Aerial Patrols

The most common form of information gathering is done by aerial patrol on a weekly or biweekly schedule. During these patrols, the inspectors watch for and report developments or changes on, and adjacent to, the ROW that may result in the need for maintenance or investigation. Prior to departing for an aerial survey, obtain the most current version of your company's maps and alignment sheets to identify the location of the target pipeline to be surveyed. If there is a question about the accuracy of the strip maps and alignment sheets being used, obtain a surveyor's map or blueprint to pinpoint the exact location of the pipeline ROW. Identify the section of line to be surveyed. If there is a restriction against low-level flying over the pipe ROW, notify your immediate supervisor to determine how the task will be completed.

Frequently, the first abnormal sign the observer sees is a spot of wilted vegetation, indicating the possibility of a leak. Often, however, changes are gradual and may take several years to reach stages that require maintenance. Soil erosion, stream bed changes, weathering of markers and signs, and growth of bushes and trees are examples. Construction work near the ROW is noted and reported. Local pipeline personnel investigate such activities and make arrangements to prevent pipeline damage.

Finding leaks, reporting their locations, and estimating their size and nature are the prime responsibilities of air patrol observers. Special flights are made to investigate leaks indicated by drops in pressure, meter readings, and other data. Radio communication between the plane and repair crew headquarters allow for quick response in case of emergencies.

Aerial patrols operate under visual flight rules conditions during daylight hours. Aerial patrols are conducted at a safe altitude above the terrain and obstacles. Patrol altitude is usually less than 1,000 feet above ground level. Special permission is required for regular flights at these low altitudes. The small planes and helicopters used are capable of flying at slow speeds to allow for better observation. Permanently colored, high-visibility markers are used to enable the pilot and observer to easily track the pipeline ROW. Any threatening condition observed will be reported immediately to the pipeline operations control center. Lost or damaged markers and markers with missing or incorrect information must be reported so they can be repaired or replaced.

### 3.5.4 Patrol Follow-Up Reports

The patrol assigns a ROW investigation report number to each abnormal condition observed on the ROW. The pilot or observer may radio any findings to the control center, and notes the specific details in a log. If a follow-up is deemed necessary, personnel are dispatched to investigate the reported condition and provide further detailed information in their report back to operations. These reports are signed by investigators, verifying that conditions have been investigated. Most pipeline company procedures call for the originating patrol pilot to obtain a hard copy of this report to compare with the patrol log to confirm that all sightings have been acted upon (*Figure 7*).

## 4.0.0 ◆ PREVENTING IN-HOUSE DAMAGE

During normal daily operations, various methods are used to prevent or minimize damage to the pipeline from outside forces. On occasion, however, the damage is caused by the company's own operations. Pipeline damage comes in many forms. Damaging the coating on the pipe or denting the pipe during handling can lead to corrosion problems. Heavy equipment running over pipe can damage it if the pipe is not adequately covered. Contraction and expansion of the pipe can cause ruptures. Inaccurate location of piping during excavations can cause it to be damaged.



**Classroom**

**Ask trainees to explain how aerial patrols track threatening conditions on a pipeline. Give examples from your own experience.**

**Performance Testing**

**Have each trainee demonstrate Performance Task 5 to your satisfaction. Fill out a Performance Profile Sheet for each trainee.**

**Classroom**

**Refer to Figure 7. Ask trainees to call out the four items that need follow-up.**

**Homework**

**Assign reading of Sections 4.0.0–4.3.0 for the next session. Answer any questions from trainees.**

**Materials**

**Ensure that you have everything required for teaching this session.**

ALL AMERICAN PIPELINE COMPANY
AAPL Form 204
Rev. 12/97

**PIPELINE AERIAL PATROL REPORT**

*Sample of information format distributed through the computer network*

Date: ____________________

| LOCATION | TIME | OLD / NEW | REMARKS | REPORT NO. |
|---|---|---|---|---|
| CADIZ | 05:53 | | STATION OK. ROW OK. | |
| LUDLOW | 06:15 | | STATION OK. 1) ½ MILE EAST OF MP 145, DIGGING 50' NORTH OF LINE. TOOK PICTURE, REPORTED TO BARSTOW. | 060391-1 |
| 12 GAUGE | 06:49 | | STATION OK, ROW OK. | 060391-2 |
| MOJAVE | 07:10 | | STATION OK. 1) ½ MILE EAST MP 36, NEW FENCE CROSSING ROW. REPORTED TO UNIT #205. | |
| TEJON | 07:32 | | STATION OK. ROW OK. | |
| EMIDIO | 07:43 | | STATION OK. ROW OK. | |
| PENTLAND | 07:50 | | STATION OK. ROW OK. | |
| 16" | | | 1) DIGGING IN 3 PLACES NEAR LINE BETWEEN MP 38 & MP 39. REPORTED TO UNIT #237. | 060391-3 |
| BELRIDGE | 08:11 | | STATION OK. ROW OK. DEADHEAD TO 30" LINE. | |
| MP 49 | 09:02 | | FOG. DEADHEAD TO LA PAZ. | |
| LA PAZ | 13:16 | | STATION OK 1) MP 379, NUMBER ON GROUND. | |
| GILA | 14:08 | | STATION OK. ROW OK. | |
| COOLIDGE | 14:44 | | STATION OK. ROW OK. | |
| TOM MIX | 14:57 | | STATION OK. ROW OK. END PATROL. | |
| DATE: 06/05/91 | | | | |
| TOM MIX | 05:15 | | STATION OK. ROW OK. | |
| HOT SPRINGS | 05:43 | | STATION OK. ROW OK. | |
| LORDSBURG | 06:30 | | STATION OK. ROW OK. | |
| ANTHONY | 07:24 | | STATION OK. ROW OK. | |

SAMPLE

106F07.EPS

*Figure 7* ◆ Pilot's log.



Instructor's Notes:

Some ways of preventing in-house damage are the following:

- Proper excavation procedures
- Lowering pipelines
- Mapping maintenance

### 4.1.0 Excavation Procedures

When a portion of the pipeline must be excavated for maintenance and repair or for operational requirements, it is necessary to accurately locate the pipeline and determine its depth. **Line locators** and **probes** are used to find the precise location before excavation. It is critical that the exact location be precisely known before any excavation begins. Ground cover is removed carefully to avoid damaging the pipe, and access is further widened by digging around the pipe.

The covering material removed during excavation is saved for reuse. When the repair is finished, the ditch is backfilled as required by your company's policies and procedures. Appropriate backfill material is required to protect the pipe in rocky areas. The backfill is carefully added around the pipe to avoid coating damage and is then compacted with heavy equipment. The ground surface is restored to avoid erosion from rain water and may be further stabilized by covering with rock or other materials. *Figure 8* shows how different materials are used to correctly bury a pipe.

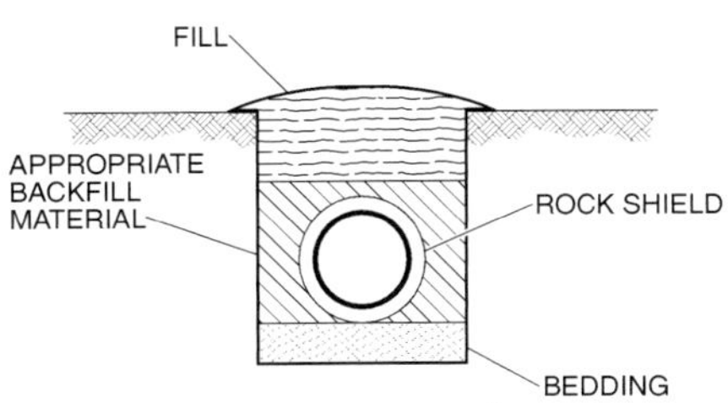


*Figure 8* ◈ How different materials are used to correctly bury a pipe.

### 4.2.0 Lowering Pipelines

The lowering of pipe into a ditch requires precision and skill on the part of the equipment operator. In some cases, two or more side boom tractors act in unison to lift larger pipe sections, using belts or slings of sufficient width to avoid pressure damage to the pipe coatings. To prevent buckling or other damage, the tractors are spaced so that the weight of the unsupported pipe is evenly divided. *Figure 9* shows a side boom tractor lowering a section of pipe into position.

First, the pipe is suspended above the ditch in such a position that it will not scrape the sides when lowered. The pipe is then lowered to a few inches above the bottom of the ditch. When more than one tractor is used to perform the job, one end is carefully lowered until the rear tractor lines become slack; then the sling can be unlatched on one side and withdrawn without damaging the coating. The other end is lowered the few remaining inches; the belt is unlatched and worked carefully out from under the pipe.

Although pipelines are capable of withstanding tension and compression forces, substantial differences in temperature must be considered during construction. Pipe laid in hot weather will contract in length during cold weather; pipe laid in cold weather will expand in length during hot weather or when filled with hot oil. Providing slack in the line or allowing for pipe movement can help ease these temperature-related changes. When screw-threaded pipe ends and collars are used, the pipe can be laid alternately along one side of the ditch and then the other side. This provides slack in the line and avoids tension as the piping expands. Mechanically coupled pipelines should always be laid so that they are compressed longitudinally. Mechanically coupled pipes should not be subjected to high stresses from tension because the couplings have less longitudinal strength than the pipe does.

*Figure 9* ◈ Side boom tractor lowering a section of pipe into position.



**Classroom**

**Emphasize the importance of preventing in-house damage. Ask trainees for examples.**

**Teaching Tip**

**See the Teaching Tips at the end of this module.**

**Classroom**

**Refer to Figure 8. Explain how the different types of materials shown protect a buried pipe.**

**Review weather and temperature stress on pipelines. Ask trainees for examples.**

**Homework**

**Assign reading of Sections 5.0.0–6.00 and completion of Review Questions for the next session. Answer any questions from trainees.**

**Materials**

**Ensure that you have everything required for testing during this session.**

**Classroom**

**Review the methods for creating positive public relationships. Ask trainees what a company might include in its public education materials. Ask trainees if they have questions.**

### 4.3.0 Mapping Maintenance

When pipelines are constructed, as-built maps are drawn up to indicate the pipeline location and depth and to identify facilities and equipment along the line. These maps use local landmarks, ground terrain and features, and roads as well as longitude and latitude readings to provide accurate location of the pipe. Any change in the pipeline or equipment must be documented on the appropriate maps and drawings so that personnel are kept up to date on changes. This will help avoid damage resulting from an operator's lack of understanding of the types and locations of all pipeline systems and equipment.

## 5.0.0 ◆ PUBLIC AWARENESS

A positive public relationship is important to all pipeline companies as a means of protecting people, property, the environment, having good community relations, and contributing to everyone's safety. Methods for developing a positive public relationship include:

- Establishing and maintaining public awareness
- Establishing and maintaining the awareness of local excavators
- Establishing and maintaining emergency response contacts

### 5.1.0 Public

As a part of a typical pipeline company public awareness program, pipeline safety information is periodically sent to all known tenants, lessees, and property owners living on property traversed by the pipeline. Also included are other residents living near the ROW for the pipeline system. The purpose of this program is to make known the location of the pipeline route, to assist the public in recognizing and reporting an emergency, and to establish a working relationship with local organizations. The program must be conducted in English and in other languages understood by a significant number of the non-English speaking population in the area.

Establishing and maintaining public awareness involves:

- Public education
- Public meetings
- Meetings between pipeline companies
- Documentation

#### *5.1.1 Public Education*

Public education should be an on-going, continuous program and not limited to a specific time of year. The materials used for this purpose are reviewed for accuracy and effectiveness before being sent out to the public. Information may include materials published by API, information about One-Call systems, or other materials. This material will include information on the pipeline and its location, what safety measures have been taken, how to prevent damage or accidents, how to recognize leaks or problems, what to do in case of a leak or problem, and relevant telephone numbers. Up-to-date material is published and distributed about every two years to ensure that information is current and that current residents have been contacted.

#### *5.1.2 Public Meetings*

Public meetings are another method of distributing information, particularly in an area where an incident may have occurred. This gives everyone a chance to ask questions and receive immediate answers and to voice personal concerns. It also gives the pipeline company an opportunity to learn information about the area that may not be available otherwise. As a public relations activity, a meeting allows the public to deal face-to-face with the people involved.

#### *5.1.3 Meetings Between Pipeline Companies*

Pipeline companies located near each other in a given area can often deal more effectively with public issues as a group. They can meet to discuss and plan consistent, combined approaches and programs to further public education and answer public concerns.

#### *5.1.4 Documentation*

Continuing public education records are maintained by the area offices and may include:

- Name, address, and telephone number of each recipient
- Examples of all materials used in the program
- Emergency response contact information
- A copy of all correspondence and meeting notes



Instructor's Notes:

### 5.2.0 Excavators

Education and information similar to the above is also distributed to contractors who may be involved in excavation activities. A current list of all area contractors who normally engage in excavation activities is maintained by the pipeline operators and in the One-Call system files. Establishing and maintaining excavator awareness involves:

- Focus group meetings
- Face-to-face meetings
- Crew/employee awareness

#### *5.2.1 Focus Group Meetings*

Focus group meetings among pipeline operators are used as forums to identify and implement ways of working with local contractors to maintain awareness of pipeline safety. Different companies may have found effective ways of doing this and can share their expertise with others during these meetings. This also helps to present a unified approach to contractors.

#### *5.2.2 Face-to-Face Meetings*

Meeting face-to-face with local contractors gives the pipeline companies the same types of benefits found in public meetings. It also gives both parties the chance to discuss future plans and decide on methods of preparation.

#### *5.2.3 Crew/Employee Awareness*

Each contractor must ensure that their crews and employees are well informed on pipeline safety. Materials provided to contractors should provide that information for the people who will be in the field doing the work.

### 5.3.0 Emergency Response Contacts

It is a DOT requirement that pipeline companies have procedures for establishing and maintaining liaison with area fire, police, and other potential emergency responders to learn the responsibilities and resources of each organization who may respond to a pipeline emergency. While learning about the responders' resources and responsibilities, the pipeline company must familiarize the agencies with the pipeline system and emergency response planning. The emergency responders in each community through which the pipeline runs will be included in this activity; such communications are often the responsibility of the right-of-way department of the company. Establishing and maintaining responder contacts involves supplying printed materials and making personal contact.

#### *5.3.1 Printed Materials*

The same material provided to the general public is also provided to these agencies. In addition, public officials may receive Material Safety Data Sheets for all products shipped and an information letter or brochure with emergency phone numbers. Maps may be included that show the pipeline route, valve and station locations, and immediate response locations that may present an eminent hazard to the public or environment.

#### *5.3.2 Personal Contact*

An initial visit to each fire department, law enforcement agency, and local emergency organization is conducted to establish contacts. These contacts are maintained by mail, telephone, or personal contact on a recurring basis and, at a minimum, every two years.

Company representatives must be able to answer detailed questions about the company's operation and applicable emergency procedures. The representative will also be responsible for collecting and updating company records regarding any change in agency support abilities. All contacts and information are recorded on company-specific forms which are kept on file.

## 6.0.0 ◆ GAS PIPELINE EMERGENCY PROCEDURES

Pipeline operators must have written procedures to minimize the hazard resulting from a gas pipeline emergency. The procedures must cover the following:

- Receiving, identifying, and classifying notices of events that require immediate response
- Establishing and maintaining communication with fire, police, and other pubic officials
- Prompt and effective response to a notice of each type of emergency
- The availability of personnel, equipment, tools, and materials, required at the scene of an emergency
- Actions directed toward protecting people first and then property
- Emergency shutdown and pressure reduction in any section of the pipeline system necessary to minimize hazards to life or property



**Teaching Tip**

**See the Teaching Tip at the end of this module.**

**Classroom**

**Ensure that trainees learn the emergency response contact procedures. Stress a company representative's role in being able to answer detailed questions.**

**Safety**

**Stress the importance of ensuring that established emergency response procedures are in place. Discuss some of the consequences of not having proper procedures in place.**

**Classroom**

**Have trainees state the 10 emergency procedures for a gas pipeline. Stress the importance of personal contacts.**

**Go over the Review Questions. Encourage trainees to ask questions. Have trainees prepare for the Module Examination.**

- Making any actual or potential hazard to life or property safe
- Notifying appropriate fire, police, and other public officials of gas pipeline emergencies and coordinating with them
- Safely restoring any service outage
- Train the appropriate operating personnel to assure that they are knowledgeable of the emergency procedures

### Summary


Efforts to prevent or minimize pipeline damage are critical to protection of people, property and the environment. Pipelines can be damaged in a variety of ways, but the most prevalent source of damage is third-party damage.

Marking the pipeline is the first step in protecting it. Permanent marking is required for all road, railroad, and water crossings and at regular intervals along the line. Signs are also required at all facilities. The purpose of the markers is to ensure that anyone in the area can see that a pipeline or pipeline facility is present. The signs identify the type of line, the company responsible, and the number to call for observed problems or when planning to excavate.

The One-Call system is set up to provide the public and any excavating organization one number to call to find out what pipelines or other buried structures exist in the area. It is the responsibility of anyone planning to dig to call this number to find out what is located there and to notify the owners of the structures. The One-Call number is on marker signs and in company publications issued to the public and officials. Calling any of the known local utilities in the phone book also provides access to the One-Call system.

Routine patrols of the pipeline ROW are performed to keep an eye on anything that is going on along the pipeline route. Aerial patrols and foot/vehicle patrols check for any signs of encroachment by outside parties, signs of threatening activities or vandalism, and signs of problems along the line. Any evidence of these signs are reported and investigated immediately. Security at pipeline surface facilities includes fencing, locks, guards, and surveillance cameras.

When maintenance requiring excavation is being done on the company's own pipeline, the same precautions apply. Preventing damage during excavations requires careful locating of the pipeline and other structures. Maps and pipeline locators are used to do this, and company representatives should be present from any company owning any structures in the area.

For all activities, it is important that maps and drawings of the pipeline be strictly maintained to accurately show all parts of the system as they actually exist.

Public awareness is also important to these activities. Owners of land along the ROW, local companies involved in excavations, and local officials must be provided with the information they need about the pipeline, how it affects them, and what to do in case of emergency or when they plan to do something that might affect the pipeline. The pipeline company provides this information through publications, meetings, and other contacts. The pipeline company must also keep records of whom to call for local emergency response.




Instructor's Notes:

**Review Questions**

1. The most common pipeline location system is called the _____ system.
   a. emergency
   b. alert
   c. pipe location
   d. One-Call

2. If an emergency One-Call is received the company must respond _____.
   a. as soon as possible
   b. within an hour
   c. within six hours
   d. within two days

3. Temporary marking uses the _____ color code.
   a. company
   b. state
   c. international
   d. pipeline

4. The area along the pipeline that has been legally permitted to the pipeline company is the _____.
   a. pipe spread
   b. line marker
   c. right-of-way
   d. marker area

5. Specifications for line markers are described in *DOT 192* and _____.
   a. local code
   b. *OSHA CFR 29*
   c. *DOT 195*
   d. *API 653*

6. The ground patrol personnel that inspect the pipeline are called _____.
   a. patrollers
   b. inspectors
   c. linewalkers
   d. pilots

7. All navigable water crossings must be inspected at least every _____ years.
   a. two
   b. three
   c. four
   d. five

8. Aerial patrols are performed on a weekly or _____ basis.
   a. bi-weekly
   b. monthly
   c. quarterly
   d. yearly

9. Pipe laid in hot weather will contract during _____.
   a. startup
   b. cold weather
   c. product flow
   d. rainy weather

10. Emergency response personal contacts must be made at least every _____ years.
    a. two
    b. three
    c. four
    d. five



## Examination

**Administer the Module Examination. Record the results of the exam on Craft Training Report Form 200 and submit the results to the Training Program Sponsor.**

## Performance Testing

**Ensure that all Performance Tasks have been completed to your satisfaction. Fill out a Performance Profile Sheet for each task for each trainee. Be sure to record the results of the testing on Craft Training Report Form 200 and submit the results to the Training Program Sponsor.**

GLOSSARY

# *Trade Terms Introduced in This Module*

***Main line valve:*** A device for controlling main-line fluid flow.

***One-Call system:*** An organization that excavators may contact (required in some states) to determine the presence of underground substructures within the area of their proposed excavation activities, and to advise operators of such substructures of the pending activity that may impact their facility.

***Intrusion device:*** Cameras, motion sensors, gate alarms, or other equipment capable of detecting entry into a facility.

***Remote monitoring:*** Monitoring at a location other than the facility being inspected.

***Line locator:*** An electronic device used to define the lateral lines of the pipeline.

***Line marker:*** A highly visible device used to communicate the location of the pipeline.

***Probe:*** A tee-handled steel rod used to probe for the location of the pipeline.

***Aerial marker:*** A sign that is specifically designed to be visible from a flying aircraft.



Instructor's Notes:

# MODULE 62106-02 — TEACHING TIPS

The following are suggested activities or instructional methods to help you teach the material in this AIG.

## General

When you call on someone to answer a question, the rest of the class relaxes or even tunes out because they expect that the question and answer will take place only between you and the trainee you called on. Instead, use this technique to involve more trainees in answering questions and to keep them on their toes.

1. Ask trainees to define a term or explain a concept.
2. After one trainee has answered, ask a trainee seated nearby if the answer is right. Then ask whether a trainee in the back of the room agrees.
3. Ask trainees to explain why they think an answer is right or wrong.
4. Use the session to clear up incorrect ideas and encourage trainees to learn from their mistakes.

## Section 2.1.0 *Marking Protocol*

This Quick Quiz will reinforce the trainees' understanding of how pipelines are marked. You will need photocopies of the quiz provided on the following page. Trainees will need pencils. If you allow trainees to use the Trainee Guide, decrease the amount of time you give them to complete the quiz.

1. Make as many photocopies as you will need of the quiz—one for each trainee.
2. Give trainees 5 to 10 minutes to complete the quiz.
3. Go over the answers to the quiz.
4. Ask trainees whether they have questions about how to mark pipelines.

### *Answers to Quick Quiz*

1. d. communication/CATV lines
2. a. direct top of the pipeline
3. c. yellow
4. b. the depth of the pipeline

# Quick Quiz *Marking Pipelines*

Circle the correct answer.

1. The International Color Code for marking mandates using orange markings for _____.
   a. electric lines
   b. gas lines
   c. water lines
   d. communication/CATV lines

2. Permanent line markers are required on all of the following *except* _____.
   a. direct top of the pipeline
   b. road crossings
   c. railroad crossings
   d. intervals along the pipeline

3. The International Color Code signal for natural gas lines is _____.
   a. white
   b. red
   c. yellow
   d. green

4. A permanent marker sign contains all of the following information *except* _____.
   a. the type of pipeline
   b. the depth of the pipeline
   c. the name of the operator
   d. the number to call in an emergency

## Sections 4.1.0–4.2.0

### *Excavating and Lowering Pipelines*

This exercise will help trainees learn the proper safety and skill techniques for use in excavating and lowering pipelines. You will need copies of *Figure 9* from the Trainee Guide. Trainees will need paper and pencil to take notes for later discussion. Allow 30 minutes for this exercise.

1. Refer to *Figure 9.* Ask trainees to identify three proper safety and skill techniques shown in the drawing. The following are the three proper techniques:
   - Using a belt or sling of sufficient width to avoid pressure damage
   - Ensuring that the weight of the unsupported pipe is evenly divided
   - Suspending the pipe so that it will not scrape the sides when lowered
2. Discuss proper safety and skill techniques in the field. Provide examples from your own experience.
3. Answer any questions trainees may have on safety and skill techniques for use in excavating and lowering pipelines.

## Section 4.1.0

### *Excavation Procedures*

This exercise will give trainees hands-on experience using the probe and line locator. Arrange a field trip to a nearby pipeline. Invite a member of the operations personnel to speak to the class while touring the site. You will need a probe, line locator, and appropriate personal protective equipment. Trainees will need paper and pencil to take notes for later discussion. If a site visit is not possible, demonstrate how to use probes and line locators in the classroom.

1. Ensure that trainees are wearing appropriate personal protective equipment.
2. Ensure that trainees observe the procedures carefully and take notes.
3. Demonstrate how to use a probe.
4. If time permits and if it is appropriate, have trainees practice using the probe.
5. Demonstrate how to use the line locator.
6. If time permits and if it is appropriate, have trainees practice using the line locator.

## Section 5.2.0

### *Excavators*

This exercise will give trainees an opportunity to learn how pipeline companies work with contractors and excavators. Invite a pipeline company representative to speak to the class. Trainees will need a pencil and paper to take notes for later discussion. Allow at least an hour for the discussion and questions.

1. Tell the trainees that a guest speaker will be discussing how pipeline companies work with contractors and excavators.
2. Have trainees brainstorm questions for the guest speaker before the speaker arrives. Questions could address topics such as common challenges faced in selecting contractors and excavators, proper business relations with contractors and excavators, the approval process for excavations, and system design considerations.
3. Introduce the guest speaker. Remember to include the speaker's name, title, company, and relevant credentials. Allow 45 minutes for the speaker and 15 minutes for questions.
4. Following the presentation, prompt trainees to ask the guest speaker their questions.

## MODULE 62106-02 — ANSWERS TO REVIEW QUESTIONS

| Answer | Section |
|---|---|
| **1.** d | 2.0.0 |
| **2.** a | 2.0.0 |
| **3.** c | 2.1.1 |
| **4.** c | 3.1.0 |
| **5.** c | 3.4.0 |
| **6.** c | 3.5.1 |
| **7.** d | 3.5.2 |
| **8.** a | 3.5.3 |
| **9.** b | 4.2.0 |
| **10.** a | 5.3.2 |

## CONTREN™ LEARNING SERIES — USER UPDATES

The NCCER makes every effort to keep these textbooks up-to-date and free of technical errors. We appreciate your help in this process. If you have an idea for improving this textbook, or if you find an error, a typographical mistake, or an inaccuracy in NCCER's Contren™ textbooks, please write us, using this form or a photocopy. Be sure to include the exact module number, page number, a detailed description, and the correction, if applicable. Your input will be brought to the attention of the Technical Review Committee. Thank you for your assistance.

*Instructors* – If you found that additional materials were necessary in order to teach this module effectively, please let us know so that we may include them in the Equipment/Materials list in the Instructor's Guide.


**Write:** Curriculum Revision and Development Department
National Center for Construction Education and Research
P.O. Box 141104, Gainesville, FL 32614-1104

**Fax:** 352-334-0932

**E-mail:** curriculum@nccer.org


Craft ______________________ Module Name ______________________

Copyright Date ____________ Module Number ____________ Page Number(s) ____________

Description

______________________________________________

______________________________________________

______________________________________________

______________________________________________

(Optional) Correction

______________________________________________

______________________________________________

______________________________________________

(Optional) Your Name and Address

______________________________________________

______________________________________________

______________________________________________

Module 62107-02

# *Excavating and Backfilling*
## *(CT 17 and 39)*

## MODULE OVERVIEW

This module introduces trainees to the procedures involved in excavating and backfilling a pipeline trench. Covered topics include OSHA requirements, locating pipelines, marking pipelines, permits, exposing pipeline, damage inspection, reporting obvious defects, water inspections, and backfilling.

## PREREQUISITES

Please refer to the Course Map in the Trainee Module. Prior to training with this module, it is recommended that the trainee shall have successfully completed the following:

Core Curriculum; Pipeline Core, Modules 66101-02 and 66102-02; Pipeline Maintenance, Level One, Modules 62103-02 through 62106-02

## OBJECTIVES

Upon completion of this module, the trainee will be able to:

1. Explain OSHA requirements for excavating.
2. Explain locating pipelines (CT 17).
3. Explain permits.
4. Explain exposing pipeline.
5. Explain damage inspection.
6. Explain reporting obvious defects.
7. Explain water inspection.
8. Explain backfilling (CT 39).

## PERFORMANCE TASKS

Under the supervision of the instructor, the trainee should be able to:

1. Locate or simulate locating a line per company procedure and/or using Performance Verification 17.1 as a guide.
2. Install or simulate installing a line marker per company procedure and/or using Performance Verification 17.2 as a guide.
3. Inspect and maintain or simulate inspecting and maintaining a line marker per company procedure and/or using Performance Verification 17.3 as a guide.
4. Explain backfilling a trench following maintenance per company procedure and/or using Performance Verification 39 as a guide.
5. Using a company's manuals or procedure books, identify and explain what trainees should do if they encounter any damage or defects while performing any of the above tasks.

## NCCER STANDARDIZED CRAFT TRAINING PROGRAM

The National Center for Construction Education and Research (NCCER) provides a standardized national program of accredited craft training. Key features of the program include instructor certification, competency-based training, and performance testing. The program provides trainees, instructors, and companies with a standard form of recognition through a National Craft Training Registry. The program is described in full in the *Guidelines for Accreditation*, published by the NCCER. For more information on standardized craft training, contact the NCCER by writing us at P.O. Box 141104, Gainesville, FL 32614-1104; calling 352-334-0911; or e-mailing info@nccer.org. More information may be found at our Web site, www.nccer.org.

## HOW TO USE THIS ANNOTATED INSTRUCTOR'S GUIDE

Each page presents two sections of information. The larger section displays each page exactly as it appears in the Trainee Module. The narrow column ties suggested trainee and instructor actions to each page and provides icons to call your attention to material, safety, audiovisual, or testing requirements. The bottom of each page includes space for your notes.

Teaching Tip

If you see the Teaching Tip icon, that means there is a teaching tip associated with this section. Also refer to any suggested teaching tips at the end of the module.

## SAFETY CONSIDERATIONS

Ensure that the trainees are equipped with appropriate personal protective equipment. Emphasize the importance of following all safety precautions and procedures when working with power tools.

## PREPARATION

Before teaching this module, you should review the Module Outline, Objectives, Performance Tasks, and the Materials and Equipment List. Be sure to allow ample time to prepare your own training or lesson plan and gather all required equipment and materials.

## MATERIALS AND EQUIPMENT LIST

Transparencies and blank transparencies

Markers/chalk

Pencils and scratch paper

Empty 3' × 3' × 3' cardboard box

Tape measure

Stick locator

Sample permanent pipeline marker

Electromagnetic (battery-operated) locator

Split box locator

Induction set

Copies of the Quick Quiz*

Module Examinations**

Performance Profile Sheets**

Copies of the Performance Verification for Covered Tasks 17 and 39†

Overhead projector and screen

Whiteboard/chalkboard

Appropriate personal protective equipment

The instructor will determine the number of the following items needed based on class size:
- Plastic buckets
- Two-liter water bottles
- Small bags of various soil types (sand, potting soil, clay, fill dirt, and so on)
- Garden trowels or large spoons

* Located at the end of this Annotated Instructor's Guide.

** Located in the Test Booklet packaged with this Annotated Instructor's Guide.

† Available for download at www.nccer.org.

## TEACHING TIME FOR THIS MODULE

An outline for use in developing your lesson plan is presented below. Note that each Roman numeral in the outline equates to one session of instruction. Each session has a suggested time period of 2½ hours. This includes 10 minutes at the beginning of each session for administrative tasks and one 10-minute break during the session. Approximately 17½ hours are suggested to cover *Excavating and Backfilling (CT 17 and 39)*. You will need to adjust the time required for hands-on activity and testing based on your class size and resources.

| Topic | Planned Time |
|---|---|
| **Session I. OSHA Requirements and Locating Pipelines, Part One** | |
| A. OSHA Requirements | ______ |
| B. Locating Pipelines | ______ |
| 1. One-Call System | ______ |
| 2. Alignment Maps | ______ |
| 3. Installing Temporary Markers | ______ |
| 4. Determining Pipeline and Cable Depth | ______ |
| 5. Mechanical Pipe Locators | ______ |
| a. Using Probe Rods | ______ |
| b. Hand Digging and Pot-Holing | ______ |
| c. Using Vacuum-Powered Tools | ______ |
| d. Using a Backhoe | ______ |
| **Session II. Locating Pipelines, Part Two** | |
| A. Electronic Pipe Locators | ______ |
| 1. Stick Locators | ______ |
| a. Laboratory | ______ |
| 2. Electromagnetic (Battery-Operated) Locators | ______ |
| a. Laboratory | ______ |
| 3. Transmitter Signal Effectiveness | ______ |
| 4. Finding Foreign Lines in a Right-of-Way—Parallel or Crossing | ______ |
| a. Laboratory | ______ |
| 5. Using an Induction Set | ______ |
| a. Laboratory | ______ |
| 6. Using Inductive Couplers/Clamps | ______ |
| a. Laboratory | ______ |
| 7. Using Direct Connection | ______ |
| 8. Other Locating Techniques | ______ |
| 9. Underwater Locating | ______ |
| B. Other Methods | ______ |
| **Session III. Marking Pipelines; Permits; and Exposing Pipeline, Part One** | |
| A. Marking Pipelines | |
| 1. Permanent Markers | ______ |
| 2. Temporary Markers | ______ |
| B. Permits | ______ |
| C. Exposing Pipeline | |
| 1. Excavating | ______ |
| a. Excavating and Shoring | ______ |
| b. Soil Properties | ______ |
| c. Sloping Requirements | ______ |

d. Support Systems ______
e. Dewatering Soil ______
D. Pressure Reduction ______

**Session IV. Exposing Pipeline, Part Two**

A. Danger Awareness
1. Cave-Ins ______
2. Flooding ______
3. Gases ______
4. Electricity ______
5. Falling Objects ______
6. Fire or Explosion ______
B. Pipe Supports/Plugs ______
1. Supports ______
2. Plugs ______

**Session V. Damage Inspection, Reporting Obvious Defects, and Water Inspections**

A. Damage Inspection
1. Dents and Gouges ______
2. Corrosion ______
3. Coating Damage ______
4. Pipeline Stability ______
B. Reporting Obvious Defects ______
C. Water Inspections ______

**Session VI. Backfilling, Part One**

A. Backfilling
1. Preparing Fill ______
2. Removing and Positioning Skid Piles and Sandbags ______
3. Placing Bedding ______
a. Stabilization ______
b. Bedding ______
c. Initial Backfill ______

**Session VII. Backfilling, Part Two**

A. Performing Soil Compaction ______
B. Steps for Backfilling ______
C. Summary
1. Summarize module ______
2. Answer questions ______
D. Module Examination ______
1. Trainees must score 70% or higher to receive recognition from the NCCER.
2. Record the testing results on Craft Training Report Form 200 and submit the results to the Training Program Sponsor.
E. Performance Testing ______
1. Trainees must perform each task to the satisfaction of the instructor to receive recognition from the NCCER.
2. Record the testing results on Craft Training Report Form 200 and submit the results to the Training Program Sponsor.

Module 62107-02

# *Excavating and Backfilling*
## (*CT 17, 32, and 39*)

Instructor's Notes:

Instructor's Notes:

# Introduction to the Pipeline Operator Qualification Rule, API, and NCCER

*It could be said that petroleum products keep America running. Petroleum products like oil and natural gas are the principal sources of energy in the United States and are used in many manufacturing and chemical processes. Petroleum products are primarily transported by pipeline. There are more than a million miles of liquid and natural gas pipeline in this country, ranging in diameter from two inches to four feet. These pipelines bring crude oil and natural gas from production fields to processing plants and refineries—and ultimately to the consumer.*

### *Pipeline Operator Qualification*

Petroleum products are considered hazardous, so pipeline personnel must take extra precautions and follow specific procedures when working with them. To help ensure the safe handling and transportation of petroleum products, the United States Department of Transportation (DOT) mandates safe practices as federal regulations.

One such regulation, the Operator Qualification (OQ) rule, requires pipeline operators to develop a written qualification plan to evaluate their personnel and their contract personnel (see *49 CFR Part 192 Subpart N—Gas*, and *49 CFR Part 195 Subpart G—Liquid*). The evaluation hinges on the operator's ability to perform certain covered tasks and to be able to recognize and respond to abnormal operating conditions that may be encountered while performing these covered tasks.

### *Covered Tasks*

A pipeline performance task is deemed *covered* if the activity passes the following four-part test:

1. Is it performed on a pipeline facility?
2. Is it an operations or maintenance task?
3. Is it performed as a requirement of *49 CFR Part 195 (Liquid)* or *49 CFR Part 192 (Gas)*?
4. Does it affect the operation or integrity of the pipeline facility?

If the answer to all four questions is *yes*, then the activity is considered *covered* and must figure into the evaluation process of pipeline personnel.

### *American Petroleum Institute*

The OQ rule and its related covered tasks provided the catalyst to develop a unified pipeline training and assessment program—so that pipeline operators and owners have ready access to the assessment and training tools they need. This effort was the direct result of the vision and leadership provided by The American Petroleum Institute (API) and API members.

Efforts to unify the industry are nothing new to API. Indeed, the impetus for forming the Institute in 1919 was the need to standardize engineering specifications for drilling and production equipment. Today, the most pressing issues revolve around public perceptions and government policies toward the industry. Speaking with one voice on these issues has become as essential as having interchangeable parts in the field. API's members determine the petroleum industry's positions on public policy as well as the standards that govern its day-to-day operations worldwide.

### *Consortium on Operator Qualification*

API assembled a focus group whose primary purpose was to identify covered tasks and provide compliance recommendations to pipeline operators.

The Consortium on Operator Qualification (COOQ) became the guiding force for the identification and development of covered tasks for the liquid pipeline industry. Their work product, API publication 1161, *Guidance Document for the Qualification of Liquid Pipeline Personnel*, formed the basis for the resulting Pipeliner Training and Assessment Program.

The covered task list created in API publication 1161 has now evolved to include gas covered tasks and is maintained by NCCER. The most current covered task list can be downloaded at www.nccer.org.

### *National Center for Construction Education and Research*

With the guidance and support of API, its members, and the Consortium on Operator Qualification, the National Center for Construction Education and Research (NCCER) was able to design and implement the Pipeliner Training and Assessment Program.

NCCER is a not-for-profit 501(c)(3) education foundation established in 1995 by the world's largest and most progressive construction and maintenance contractors and national contractor associations. NCCER was founded to address the

severe workforce shortage facing the construction and maintenance industries and to develop a standardized training process and curricula. Today, NCCER is supported by hundreds of leading construction and maintenance companies, manufacturers, and national associations and developed more than 35 construction and maintenance titles.

NCCER also maintains a National Registry, which contains the records of trainees and instructors who have received NCCER training. All completion information for every trainee is sent to the NCCER and filed within the Registry. The Registry can then confirm training and skills for workers as they move from company to company, state to state, or even within their own company.

### *Standardized Curriculum (NCCER's Contren™ Learning Series)*

NCCER curricula are developed using the expertise of industry subject matter experts (SMEs). The use of SMEs as an integral part of every new development or revision ensures that the curriculum is current, industry-driven, and indicative of industry practices. The development and/or revision process is facilitated from start to finish by NCCER project managers, who work closely with the SMEs and a select team of technical writer(s) on the development and presentation of the curriculum.

### *Pipeline: Five Training Paths*

The pipeline curriculum focuses on essential skills and competencies within five skill areas:

- Corrosion Control
- Electrical and Instrumentation
- Maintenance
- Mechanical
- Operations (Field and Control Center)

SMEs from the pipeline industry have determined the skills and competencies needed based on API's covered tasks and the consensus of the group/industry. Specific company information is generally not incorporated into the curriculum unless the SMEs determine that the information is the industry standard.

### *Training Modules and Covered Tasks*

The smallest unit of instruction in the curriculum is a *module*. Each curriculum is composed of modules (the equivalent of stand-alone chapters) that are logically sequenced into complete *levels* of instruction. Every module is task-oriented and requires the successful completion of a written (knowledge-based) examination and a performance (hands-on) test. Modules may be individually selected and/or combined for efficient task training. Modules in the pipeline curricula, to the extent possible, will align with API-developed covered tasks. While not all modules will relate to covered tasks, all covered tasks will relate to at least one module.

### *Benefits of Training to the Pipeline Marketplace*

Applying NCCER's standardized training process allows pipeline owners and contractors in the United States to establish the benchmark for acceptable skill and competency levels within the pipeline industry. This process will assist training organizations in developing and/or improving standardized pipeline training programs that effectively match their resources and operating conditions while responding to regulatory demands and to the needs of the industry. The process will also recognize training organizations that meet the minimum standards of competence, warranting regulatory, public, and professional confidence. This in turn will foster national unity within the pipeline industry for the mutual benefit of individual technicians, owners, contractors, and ultimately pipeline-users and society.

## Additional Resources

To review the Pipeliner Training and Assessment Program Implementation and Resource Guide and the latest version of pipeline covered tasks, visit NCCER's Web site at www.nccer.org, or contact:

**National Center for Construction Education and Research**
PO Box 141104
Gainesville, FL 32614-1104
(888) 622-3720
Web site: www.nccer.org

For more information on API, please contact:

**American Petroleum Institute**
1220 L Street, Northwest
Washington, D.C. 20005
(202) 682-8000
Web site: www.api.org

To order *API Pub 1161*, contact Global Engineering Documents by calling 1-800-854-7149 or visiting their Web site at www.global.his.com.

Additional sources of information on the pipeline operator qualification regulation include:

**DOT – Office of Pipeline Safety**
www.ops.dot.gov

**Transportation Safety Institute**
www.tsi.dot.gov



Instructor's Notes:

## COURSE MAP

This course map shows all of the modules in the first level of the Pipeline Maintenance curriculum. The suggested training order begins at the bottom and proceeds up. Skill levels increase as you advance on the course map. The local Training Program Sponsor may adjust the training order.

### PIPELINE MAINTENANCE LEVEL ONE

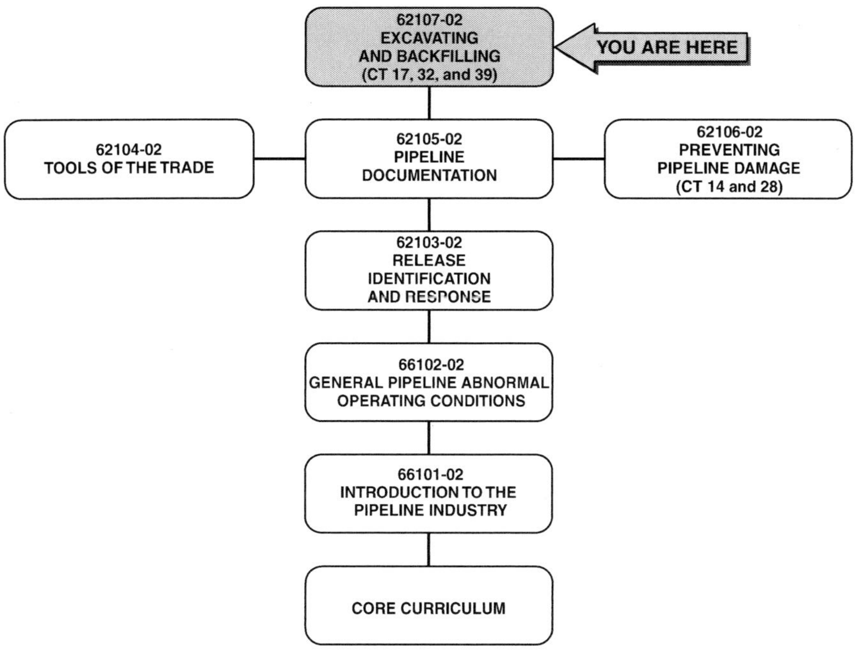


107CMAP.EPS





**Homework**

**Assign reading of Module 62107-02.**

Instructor's Notes:

## MODULE 62107-02 CONTENTS

Instructor's Notes:

## Figures



## Tables

Instructor's Notes:

MODULE 62107-02

# Excavating and Backfilling (CT 17, 32, and 39)

### Objectives

When you have completed this module, you will be able to do the following:

1. Explain OSHA requirements for excavating.
2. Explain locating pipelines (CT 17).
3. Explain permits.
4. Explain exposing pipeline.
5. Explain damage inspection.
6. Explain reporting obvious defects.
7. Explain water inspection.
8. Explain backfilling (CT 39).

### Prerequisites

Before you begin this module, it is recommended that you successfully complete the following: Core Curriculum; Pipeline Core, Modules 66101 and 66102; Pipeline Maintenance Level One, Modules 62103 through 62106.

### Required Trainee Materials

1. Pencil and paper
2. Appropriate personal protective equipment

## 1.0.0 ◆ INTRODUCTION

Excavating and **backfilling** are necessary when sections of the pipeline must be exposed. For example, you may need to repair or replace damaged or leaking pipe. It may be necessary to inspect the existing pipe. Both excavation and backfilling should be done in a manner that does not damage the pipe.

Backfilling provides protection for the pipeline. Current federal and local specifications must be followed when replacing backfill. Careful planning is necessary for both excavation and backfilling. Excavating and backfilling are explained in the following sections:

- OSHA requirements
- Locating pipelines
- Marking pipelines
- Permits
- Exposing pipeline
- Damage inspection
- Reporting obvious defects
- Water inspections
- Backfilling

## 2.0.0 ◆ OSHA REQUIREMENTS

OSHA has strict and specific requirements that apply to any open excavation work, including trenches. The purpose of these requirements is to ensure the safety of excavation workers, because accidents during excavation are among the leading causes of injuries and deaths in the construction industry. These requirements are explained in detail in *29 CFR Part 1926.650-652*. The following is a summary of the general requirements:

- All surface encumbrances that are located so as to create a hazard to employees shall be removed or supported, as necessary, to safeguard employees.
- The estimated location of utility installations, which may be encountered during excavation, shall be determined prior to opening an excavation.
- Structural ramps that are used solely by employees as a means of access and egress from excavations or ramps for equipment shall be designed by a **competent person** and built according to design.



**Materials**

**Ensure that you have everything required to teach the course. Check the Materials and Equipment List at the front of this module.**

**Classroom**

**Explain that the trade terms shown in bold are defined in the Glossary at the end of the module.**

**Audiovisual**

**Show Transparency 1, Course Objectives.**

**Show Transparency 2, Performance Tasks.**

**Teaching Tip**

**See the general Teaching Tip at the end of this module.**

**Classroom**

**Ask trainees to define backfilling and competent person.**

**Teaching Tip**

**See the Teaching Tip at the end of this module.**

Classroom

**Stress the importance of precisely locating pipelines.**

**Refer trainees to Figures 1 and 2. Review how the One-Call system works.**

- Employees exposed to public vehicular traffic shall be provided with, and shall wear, warning vests or other suitable garments made of or marked with reflectorized or high-visibility material.
- No employee shall be permitted underneath loads handled by lifting or digging equipment. Employees shall be required to stand away from any vehicle being loaded or unloaded to avoid being struck by any spillage or falling materials.
- When mobile equipment is operated adjacent to an excavation, or when such equipment is required to approach the edge of an excavation, and the operator does not have a clear and direct view of the edge of the excavation, a warning system shall be utilized such as barricades, hand or mechanical signals, or stop logs.
- Tests and controls shall be utilized to prevent exposure to harmful levels of atmospheric contaminants and to assure acceptable atmospheric conditions in the excavation.
- Employees shall not work in excavations in which there is accumulated water, or in excavations in which water is accumulating, unless adequate precautions have been taken to protect employees against the hazards posed by water accumulation.
- Where the stability of adjoining buildings, walls, or other structures is endangered by excavation operations, support systems such as shoring, bracing, or underpinning shall be provided to ensure the stability of such structures for the protection of employees.
- Adequate protection shall be provided to protect employees from loose rock or soil that could pose a hazard by falling or rolling from an excavation face.
- Daily inspections of excavations, the adjacent areas, and protective systems shall be made by a competent person for evidence of a situation that could result in possible cave-ins, indications of failure of protective systems, hazardous atmospheres, or other hazardous conditions. Evidence of such conditions shall require the removal of employees from the area until such conditions have been corrected.
- Where employees or equipment are required or permitted to cross over excavations, walkways or bridges with standard guardrails shall be provided. Adequate barrier physical protection shall be provided at all remotely located excavations.

In addition, the code lists specific design and construction requirements for the protective systems to be provided. Some of these are discussed in this module.

## 3.0.0 ◆ LOCATING PIPELINES

It is critical to precisely locate the pipeline before any excavation begins. Not only will digging in the wrong place be a waste of time and expense, but it may also inadvertently damage the pipe. Location of the company's pipeline should be clearly shown on company alignment maps, to be used along with visible markers.

However, your company may not be the only pipeline or utility with lines in the area, so any plan to excavate must be preceded by alerting the One-Call system. When this is done, any company with buried lines in the area of planned excavation will be alerted and asked to provide temporary markings and information such as depths and line descriptions. An emergency number is shown on permanent marker signs in the area (*Figure 1*).

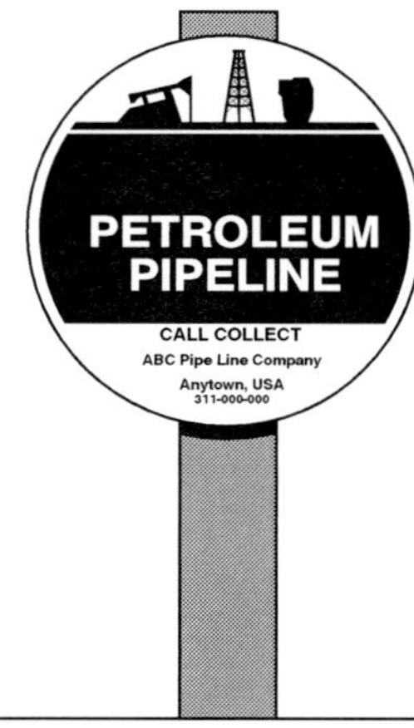


107F01.EPS

*Figure 1* ◆ Company emergency number on a sign.

### 3.1.0 One-Call System

Before beginning an excavation, you are required to notify local utilities and have the underground line locations identified. Notification and utility location can be handled either by utility operators individually or by a collaborative multicompany effort. However, the most common method used is an operator-sponsored centralized program known as a One-Call system. The One-Call system is defined as a communication system established by two or more utilities, governmental agencies, or other operators of underground facilities to pro-



Instructor's Notes:

vide one telephone number for excavating contractors and the general public to call for notification of their intent to use equipment for excavating, demolition, or other work. It requires that the participating members identify and give the location of their facilities.

The One-Call system is generally set up state-by-state. Some states provide a state-wide One-Call system. Other states have multiple systems, but no area is without coverage. A few states have areas that are not covered.

Pipeline company offices have communication systems connected to the One-Call system. A dedicated phone line and printer may be set up to receive excavation notices. When notices are received, the company researches the area to confirm it is the location of the pipe. If the pipe is not threatened, the excavation will be okayed. If the pipe may be threatened, the company will arrange to provide temporary markings and oversee the excavation.

If the call is an emergency One-Call, the company must respond appropriately within 2 hours. The One Call system records all communications on the One-Call log. *Figure 2* shows a call investigation report form.

### 3.2.0 Alignment Maps

Alignment maps are used to help locate the pipeline (or other line) with relation to surface features and by depth. Alignment maps can be of several different types or combination of types of maps. These maps will have the pipeline superimposed on the plan view with indications of lengths and angles.

Individuals who locate underground pipelines and cables must be proficient in reading maps and alignment sheets. Maps and alignment sheets are created to furnish pertinent, permanent information about the location of various pipeline and cable installations.

If there is a question about the accuracy of the maps or alignment sheets being used, a surveyor's map or blueprint should be obtained to pinpoint the exact location of the pipeline or cable right-of-way (ROW). In instances where no other information is available, it may be necessary to obtain the system installation as-built specifications. *Figure 3* shows a pipeline strip map.

> **NOTE**
> Remember that unless the lines have been recently and accurately surveyed and have had the Global Positioning System (GPS) coordinates established, the maps, alignment sheets, and as-built drawings may not be 100% accurate. Originally, the drawings may have been close, but since then, the lines may have been abandoned, moved, or new ones built. If in doubt, always pinpoint and mark.

### 3.3.0 Installing Temporary Markers

The standard *49 CFR 195.410* requires that each pipeline operator must accurately install and maintain line markers over buried pipelines. Upon receiving a One-Call notification, qualified individuals are dispatched to the area to investigate the proposed work site. Only trained and qualified individuals should be assigned inspection and repair tasks on or near pipelines.

The pipeline company's first responsibility is to physically locate the pipe. Temporary white markers may be used to mark the proposed excavation or construction location. Once the line is physically located, the work crew must replace any missing markers, repair damaged markers, and replace any incorrectly installed markers.

If the pipeline company determines that work should be allowed and a permit is issued for the work to proceed, small yellow flags or stakes are inserted in the ground along the pipeline ROW. The exact location will be plainly seen by anyone who will be performing excavation or blasting. Temporary pipeline markers should be evenly spaced along the location of the line or cable.

### 3.4.0 Determining Pipeline and Cable Depth

Depth of cover for pipe and cable is usually specified on engineering blueprints and elevations or as-built charts and typically varies according to the location and use of the area over the installed system. While some installations may follow set standards, logging, construction and other uses may be approved on a case-by-case basis. Before excavation, the depth of cover must be verified by a qualified individual. The actual cover of a pipeline or cable may have changed after installation due to flooding or other influences. No excavation can be allowed until the pipeline has been definitely located.

Unless specifically exempted, all oil and gas new construction pipelines must be buried below the level of cultivation for protection. However, your company may have deeper depth-of-cover standards than those indicated in *CFR 195.248*. *Tables 1* and *2* show examples of company depth requirements.



**Classroom**

**Ask a trainee to describe how alignment maps are used. Discuss what is done when there is a question about their accuracy.**

**Discuss the use of temporary white markers and yellow flags.**

**Refer trainees to Tables 1 and 2. Ask a trainee to state why pipeline or cable cover can change. Stress the importance of following company standards for pipeline cover.**

**CALL INVESTIGATION REPORT**

| | |
|---|---|
| COMPANY NAME: | DATE: |
| CALLER'S NAME: | TIME: |
| CONTACT TELEPHONE NUMBERS/NAMES: | |
| NAME: | NUMBER: |
| NAME: | NUMBER: |
| AREA/LOCATION OF REPORT: | |
| DESCRIPTION: | |
| NOTIFIED NAME: | DATE: |
| CONTROLLER'S NAME: | TIME: |



107F02.EPS

*Figure 2* ◆ Call investigation report form.



Instructor's Notes:

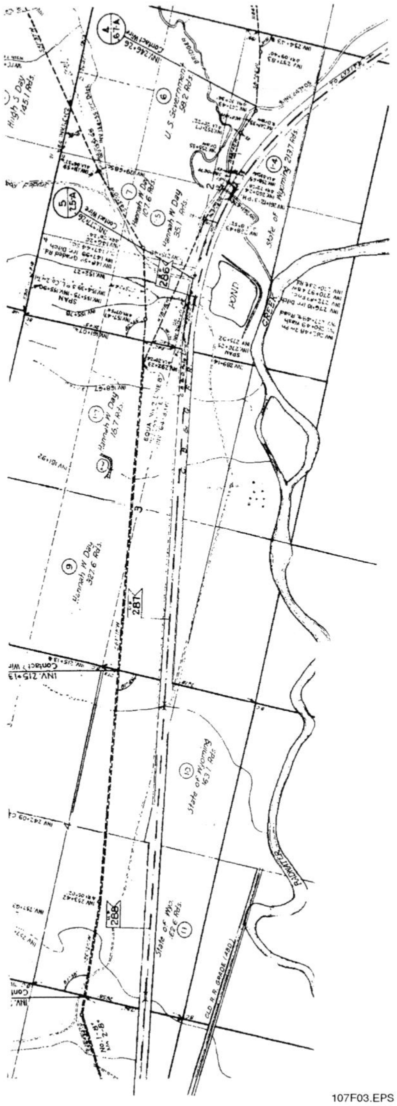

107F03.EPS

*Figure 3* ◈ Pipeline strip map.

**Table 1** New Construction Liquid Pipeline Depth Requirements

| Location | Cover (inches) | Notes |
|---|---|---|
| Industrial, commercial, and residential areas | 36 | |
| Inland water of at least 100 ft width high water to high water | 48 | Water crossings |
| Drainage ditches at public roads and railroads | 36 | Bar-Ditch |
| Deepwater port safety zone | 48 | |
| Gulf of Mexico and offshore water less than 12 ft at low tide | 36 | |
| All other areas | 30 | Railroads, highways, other |
| Underwater | up to 96 | |

107T01.EPS

**Table 2** New Construction Gas Pipeline Depth Requirements

| Location | Normal Soil (inches) | Consolidated Rock (inches) |
|---|---|---|
| Class 1 locations | 30 | 18 |
| Class 2, 3, and 4 locations | 36 | 24 |
| Drainage ditches at public roads and railroads | 36 | 24 |

107T02.EPS

**NOTE**

Many companies require more cover than these standards. Check your company's depth-of-cover requirements.

## 3.5.0 Mechanical Pipe Locators

Trained individuals use certain methods and tools for physically locating underground pipelines and cables. The physical location of pipes and cables can be accomplished by **probing**, **pot-holing** by hand, and by using vacuum units. Electronic locators are also used to locate the pipe, but only some indicate depth. Power equipment, such a backhoe, is not normally used until the exact location of the system is established.

**Classroom**

**Refer trainees to Figure 4. Describe how the probe rod is used.**

**Safety**

**Stress the importance of using a probe made of nonconductive material to prevent electrocution.**

**Classroom**

**Ask a trainee to explain pot-holing.**

**Safety**

**Stress the importance of proceeding slowly and using tools made of nonconductive material to prevent electrocution.**

**Classroom**

**Ask a trainee to explain how a water-vac unit works. Ask another trainee to explain how an air-vac unit works. Discuss the advantages of using vacuum units over other tools.**

**Safety**

**Stress the importance of wearing appropriate personal protective equipment when using air-vac units.**

### *3.5.1 Using Probe Rods*

A probe bar is a steel or fiberglass rod used by some companies to determine the location and depth of pipelines and utility cables. It has a smooth, rounded end and generally has a "T" handle attached (see *Figure 4*). To locate a pipe, the rod is pushed into the ground until contact is made with the pipe. Probe rods must be used carefully to avoid breaking the pipeline coating and should only be used by trained and qualified persons.

> **WARNING!**
> Proceed slowly and carefully when probing. An electrocution hazard from electrical lines may be present. To offset this risk, use only tools made of a nonconductive material (such as fiberglass) or tools equipped with insulated handles.

> **NOTE**
> Some companies do not allow probing at all. Check your company's policies.

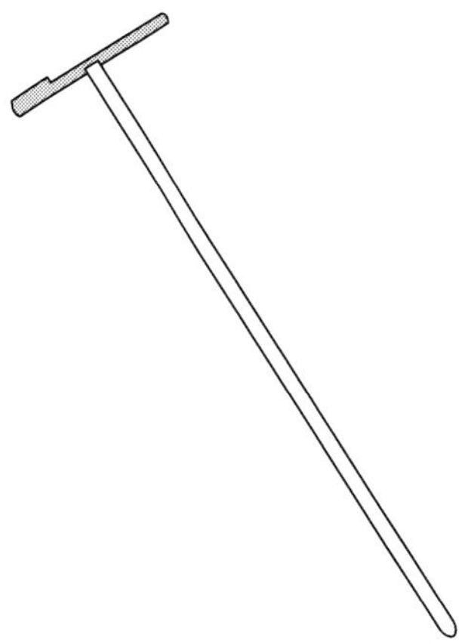


*Figure 4* ◆ Typical probe rod.

### *3.5.2 Hand Digging and Pot-Holing*

Pot-holing is a method of physically locating the pipeline by hand digging small holes. Pot-holing may be more hazardous than probing because tools with sharp edges could potentially damage the pipeline or other utility lines. Caution should be used. When pot-holing, individuals should proceed with caution until the pipeline is exposed. Any tool used should have insulated handles.

> **WARNING!**
> When hand digging or pot-holing, use extreme caution and proceed slowly because electrical lines could be present. It is critical that you only use tools made of nonconductive materials or tools equipped with electrically insulated handles to offset the risk of electrocution.

### *3.5.3 Using Vacuum-Powered Tools*

A water-vac unit (see *Figure 5*) is an effective tool that hydraulically bores and air vacuums out a hole over a pipeline. It opens a clean, smooth hole with no damage to the pipeline or cable. Water-vac units minimize disruption of other work on the site.

Air-vac units operate on the same basic principle as water-vac equipment. Air-vac vacuum trucks bore holes with compressed air rather than water and vacuum the debris from the hole over the pipeline.

> **WARNING!**
> It is imperative that you wear appropriate personal protective equipment (PPE) for eye protection because air-vac units tend to propel more dirt and rock particles into the air than a water-vac unit.

*Figure 5* ◆ Water-vac truck.



*Instructor's Notes:*

Both types of vacuum units are designed to reduce the risk associated with using a pick, shovel, posthole digger, or other sharp metal tools and probes to expose the pipeline.

### *3.5.4 Using a Backhoe*

A backhoe is a large, powerful piece of construction equipment that can be safely used to quickly expose underground pipe and cables by moving large amounts of cover. The size and power of the backhoe can significantly reduce the time needed to perform the job.

> **CAUTION**
> Exercise care while digging around known pipeline and cable locations. Digging too deeply or off center from the target area could cause damage to pipes and cables before the backhoe operator is able to reverse the direction of the equipment. The initial excavation often destroys the temporary line markings such as flags, stakes, and paint. Therefore, to help the backhoe operator, the line should be periodically relocated in the hole as the hole deepens and the pipeline is exposed.

Before resorting to heavy equipment to expose a pipeline or utility service, make sure that other critical preliminary steps have been completed. These could include any or all of the following:

- Consulting available strip maps and alignment sheets
- Verifying that locating equipment items are operational
- Physically locating the line
- Verifying that the location and depth are accurate
- Ensuring that appropriate supervisors are informed
- Using alternate methods (hand digging or potholing) to perform the work, if possible
- Proceeding with caution

## 3.6.0 Electronic Pipe Locators

Electronic locators are a class of lightweight, portable detectors used to locate a variety of underground utility lines. More advanced models can locate a specific type of utility, even in congested utility areas. They are accurate, but the physical pipe or cable location may need to be verified with a physical location tool before excavation is allowed to proceed. *Figure 6* shows an electronic pipe and cable locator in use.

107F06.EPS

*Figure 6* ◆ Electronic pipe and cable locator in use.

Locating an underground utility line with an electronic locator unit typically requires that you follow the steps outlined below:

*Step 1* Once in the field, review maps, alignment sheets, and as-builts.

*Step 2* Look for indications of foreign lines such as valves, valve box covers, manhole covers, pedestals, junction boxes, or utility poles with downward cables for visible clues of what might be underground.

*Step 3* Use the passive modes in the receiver, if available, to sweep the area for indications of signals on lines (50/60 Hz-power, radio, cathodic protection signal, CATV).

*Step 4* Determine if induction, connection, or two-person sweep method is to be used.

*Step 5* Always connect the transmitter when possible, because the signal will be cleaner and will travel further.

*Step 6* Set the transmitter controls.

*Step 7* Use the locator receiver to locate and mark the target pipes or cables.

*Step 8* Verify the identification of the target.

*Step 9* Determine the depth of the target.

*Step 10* Notify concerned supervisors, planners, and work crews of your finding.



**Audiovisual**

**Show Transparency 3. Stress the steps to be taken before using a backhoe.**

**Safety**

**Stress the importance of relocating the pipeline during backhoe operations.**

**Homework**

**Have trainees review Sections 3.6.0–3.7.0 for the next class session.**

**Materials**

**Ensure that you have everything required for demonstrations and testing during this session.**

**Classroom**

**Explain when to use a physical location tool to verify electronic pipe locator readings. Stress the importance of following the manufacturer's operating and maintenance instructions.**

**Classroom**

**Refer trainees to Figure 7 or bring a stick locator to class. Discuss the advantages and disadvantages of the stick locator.**

**Safety**

**Stress the importance of verifying that the line is dead before proceeding.**

**Demonstration**

**If you bring a stick locator to class, demonstrate its use for trainees. Remind trainees to always follow the manufacturer's instructions.**

**Laboratory**

**If you bring a stick locator to class, have trainees practice using it under your supervision. Ask trainees to observe and point out the locator's features.**

> **NOTE**
> These are generic steps for operating a typical set of equipment to locate a buried utility cable. They do not include enough information to use a specific item of equipment. Detailed information must be obtained from the original equipment manufacturer's (OEM) operating and maintenance instruction manual.

### *3.6.1 Stick Locators*

This class of lightweight, portable detectors, also referred to as metal detectors or ferromagnetic locators, are quick and easy to use in the field to locate a variety of underground items, including:

- Pipe
- Deadman
- Anchor
- Tank
- Survey Pins

> **CAUTION**
> Do not use near steel-toed boots.

Stick locators (see *Figure 7*) typically consist of two encased sensors in a tube or shaft. These antenna sense the magnetic field from the iron metal and produce increased noise over the object or line. Some models also have a meter to help in noisy environments, a 50/60 Hz mode to indicate live electric current flow, and/or an erase mode to eliminate detection of iron debris (nuts, bolts and welding rods).

> **WARNING!**
> These detectors will locate both dead and live lines. Always assume the line is live until you confirm otherwise.

> **NOTE**
> It can be a slight disadvantage that these locators only detect iron. They are not sensitive to tracer tape, tracer wire or cables, electric wiring, telephone lines, or control system wiring.

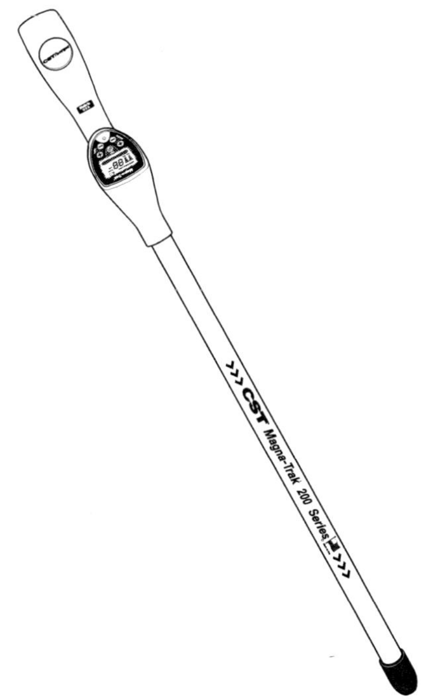


107F07.EPS

*Figure 7* ◆ Typical stick locator.

To effectively use a stick locator, walk across the area where the item you are seeking should be located, sweeping the locator back and forth above the ground in front of you. Move with the locator in a 360° circle (approximately 6 feet out) around the spot to determine if the ferro/magnetic indication is a pipeline or some smaller metal item, such as a deadman, anchor, barrel, or piece of iron trash.

Hold the locator upright while walking in a smaller circle to narrow the search and pinpoint the item. Use the erase mode, if available, and repeat the process to confirm the finding.

> **NOTE**
> These generic directions for operating a typical magnetic locator do not include enough information to use a specific item of equipment. Detailed operating instructions can be obtained from the original equipment manufacturer's (OEM) instruction manuals.



Instructor's Notes:

### 3.6.2 Electromagnetic (Battery-Operated) Locators

Electromagnetic locators have become the standard tools for locating and tracing buried pipe and utility cables. They range from simple units able to detect the presence of pipes and cables to sophisticated equipment able to identify systems, depth, and system faults in the most complex situations. Locators contain two major components: a transmitter and a receiver.

Transmitters apply an easily identifiable signal to the line with which you are working. The technique employed will vary with the site and situation. The user will gain a better understanding of which technique works best in a given situation through field experience. Techniques include direct connection, clamping, two-person sweep, and induction.

When the signal is detected, the receiver provides a signal to the user, indicating the location of the pipe or cable being sought. Following this signal enables the user to trace, pinpoint, and identify the target line or locate other, unidentified lines in the target area. Receivers can also detect the presence of passive signals. Examples of passive signals include:

- 60 Hz AC electrical power
- Cathodic protection signals
- Radio broadcast signals of 17-28 Hz
- Cable TV signals at 31 kHz

Transmitters can operate over a wide frequency range from 4 Hz to 500 kHz. Transmitter signal power ranges from a very low 0.1 watt to 5 watts in a portable transmitter to very strong signals of 150 watts output power.

Different models of electronic locators' transmitters are capable of transmitting signals in a variety of frequency ranges as shown in the *Table 3*.

**Table 3** Active Frequencies

| Connect Only[1] | Connect or Clamp Around | Induction[2] |
|---|---|---|
| 4 Hz/8 Hz | 8 kHz | 83 kHz |
| 98 Hz | 9.8 kHz | 200 kHz |
| 440 Hz | 22 kHz | 400 kHz |
| 512 Hz | 29 kHz | 500 kHz |
| 577 Hz | 33 kHz | |
| 815 Hz | 65 kHz | |
| 1000 Hz | | |

1. Best for wire or diagnostics; needs more power on pipes
2. Best on tracer tape and tracer wire

107T03.EPS

Note that some signals are better for connecting, some for signal clamping, while others are best suited for induction.

The more powerful the signal, the greater the distance over which it can be detected and the deeper the line can be and still be located. When pinpointing deeper lines (+10 feet), however, it can be more difficult to get accurate depth readings because of the fading signal and interference.

This equipment operates on the principle of induction. A coil antenna will produce an AC voltage signal proportional to the strength of the AC field or AC signal passing through it. Receiver antennas or aerials usually take this shape and are typically fitted with ferrite rods through the center of the coil to make them more effective at amplifying the small signals they pick up.

**CAUTION**

In the induction mode, the transmitter can put a signal on any buried line, so the operator must ensure the signal is being applied to the company's pipeline.

Locator operating modes are discussed below:

- *Null mode* – The Null mode gives a quiet response or no signal when you are directly over the target line, with an increased response on either side of the line. The numeric display makes it easier to interpret the response in noisy areas. Many units provide a visual indication by including right/left arrows that switch directly over the line.
- *Peak - Single Horizontal* – This setting provides a wider, maximum signal or noise over the line and tends to be more sensitive.
- *Peak - Dual Horizontal* – This setting provides a narrow, more precise pinpoint and gets rid of airborne interference if electronics/software are designed correctly. It provides a maximum or high noise/signal strength on top of the line.

Two problems can occur when locating with a Null antenna. First, varying amounts of airborne interference signals exist in the field along the line route. These signals range from VLF to electric line current (60 Hz) to radio frequencies. These interference signals can dramatically distort the locating signal that runs down the line from the transmitter. This distortion can cause the instrument to inaccurately locate up to 50 inches of the top center of the pipe or cable.

For example, overhead power lines, especially high power transmission lines, can cause the locator to hum, to have no left to right null arrow



**Classroom**

**Refer trainees to Table 3. Describe the use of electromagnetic locators. Review the locator operating nodes. Ask if trainees have any questions.**

**Demonstration**

**If possible, bring an electromagnetic locator to class and demonstrate its use for trainees.**

**Laboratory**

**If you bring an electromagnetic locator to class, have trainees practice using it under your supervision. Ask trainees to observe and explain the locator's features.**

Audiovisual

**Show Transparency 4. Discuss the variables affecting transmitter signal effectiveness.**

change and/or no clear null point, or no signal at the locale.

Another problem occurs if two lines run parallel and are relatively close together (approximately 8 feet or less depending on depth of cover) and/or bonded together. In this situation, the transmitted signal can be on both lines. When this occurs, the signals from both lines can merge or blend together, causing major distortion. In this situation, if a null vertical antenna is used to locate these lines, the receiver often inaccurately indicates that there is only one line present, not two. Under these conditions, the line will appear to be located between the actual location of the two lines.

If the null vertical antenna receiver has left/right arrows, the arrow switchover can occur in between the lines. In some cases, one arrow will pop up and the other never does. In neither case will you get an accurate locate result.

To accurately locate under these interference conditions, the receiver must be capable of operating in a dual horizontal peak antenna mode and have appropriate electronics and software. Changing to this mode of operation will reject the interference and accurately locate target lines. In this mode, lower transmission frequencies (below 10 kHz) and lower Maud gain settings improve locator performance.

Depth readings taken with an electronic locator are most accurate when taken in the active transmitter mode. Operating steps for this unit are listed below:

*Step 1* Apply transmitter signal.

*Step 2* Turn unit ON.

*Step 3* Select frequency.

*Step 4* Select peak and line modes.

*Step 5* Adjust gain as necessary.

*Step 6* Keeping locator unit in a vertical position, sweep the area.

*Step 7* Mark point of strongest signal.

*Step 8* At the mark, rotate locator 90 degrees until minimum signal (null) is achieved.

*Step 9* Rest the receiver on the ground. Depress the depth key. The display will show the target depth.

When measuring conductor depth by triangulation, begin by determining the center line of the conductor. There are two ways to estimate depth. Once the center line is known, position the receiver (split box in this example) directly over and parallel to the line. Holding the receiver as close to the ground as possible, tilt the receiver at a 45-degree angle.

**NOTE**
Perform this step carefully because any deviation from the 45-degree angle will affect the depth analysis.

With the receiver at 45 degrees and facing the center line of the target conductor, back away and mark the point where the minimum signal is present. Complete the same process on the opposite side of the target line. Measure the distance from the receiver to the center line. Subtract the height of the receiver above the ground (approx. 5 inches). This value will be the distance of the target line below ground level.

A second method of triangulation, with no math or 45-degree angles, is the 100.0-70.0-100.0 method using a dual horizontal peak antenna. Put the receiver on the top center line and adjust the sensitivity to exactly 100.0 (or to 99.8 if it is a digital display that does not go to 100.0). Then move perpendicular to the right of the line until the sensitivity is exactly 70.0. Mark the spot. Move the receiver back to the center line and make sure it is still reading 100.0 (99.8). Move left of the line in the same manner until the reading is 70.0. Mark the spot. The distance between the two spots is the depth to the center of the pipe/signal.

**NOTE**
All locators sense depth to the center of the signal, roughly the center of the pipe. To get cover depth, subtract ½ of the pipe diameter from the depth measured between the spots to get actual cover depth.

### *3.6.3 Transmitter Signal Effectiveness*

The effectiveness of locator transmitted signals and, therefore, the quality and distance of the signal varies according to a number of signal characteristics. These variables include the following:

- *Conductor metal* – Transmitted signals will travel a greater distance if the metal is a good conductor. A list of conducting materials that must often be located or traced include gold, silver, copper, aluminum, steel, cast iron, and ductile iron. Gold is the best conductor, and ductile iron is the least effective conductor.
- *Line size* – The physical size of the line to which the signal is applied also determines the distance the signal travels along that line. The smaller the



Instructor's Notes:

line, the further the signal runs. Smaller lines offer less surface area to ground and less resistance to current flow than do larger lines.

- *Line continuity* – The better the continuity of a line, the further a signal is able to travel along the line. More efficient lines are those that contain fewer splices, fewer welded connections, and a limited number of mechanical or physical couplings tying smaller sections of line together.
- *Coating integrity and quality* – The quality and integrity of line coating effects how well the line carries a signal. The better the coating/sheath, the further the signal runs.
- *Signal frequency* – Lower frequency signals travel better than higher frequencies, assuming the signal will travel down the carrier line. Higher frequency signals tend to travel shorter distances and often experience a greater loss of signal strength to other lines, fences, and rebar in nearby structures.
- *Soil type* – The more organic matter, the further the signal runs, up to a point. Beyond that point, too much organic matter decreases the effective travel distance. Also, signal distance is shorter in sand.
- *Soil moisture* – The more water in the soil, the further the signal travels. As with soil types, this is true only up to a point. Very dry soil or very wet, swampy conditions increase signal loss, reduce effectiveness, and shorten the distance a signal will travel.
- *Transmitter output* – As the transmitter output signal strength increases, the signal travels further and can be detected at a greater distance from the input location. Signal travel is even more effective with proper impedance matching.
- *Signal interference* – As interference from other signals and other lines decreases, the further the transmitter signal runs and can be detected. Dual antennas increase the unit's capability to pick up weak signals.
- *Input connection* – Signals run farther if input through a direct connection, which results in 100% of the transmitter signal strength being input onto the line. In comparison, clamping results in only 60-90% of the signal from the transmitter entering the line. Induction, the least effective, only transfers between 10-50% of the signal from the transmitter to the line.

**CAUTION**
Be sure to secure the transmitters against theft.

### 3.6.4 *Finding Foreign Lines in a Right-of-Way – Parallel or Crossing*

There are three ways to detect foreign lines in the ROW, whether they are parallel or crossing. Some companies require the use of two or even all three methods.

The first method involves the use of one of the stick metal detectors. Sweep the area wider than you will be excavating. Remember that these locators locate metal pipes and objects, not "hot" electric cables, telephone lines, or fiber optic cables. Follow the general directions in this document plus the manufacturer's instructions and training.

**CAUTION**
Make sure you keep the locator away from steel toe boots.

The second method uses a split box locator. When using a split box locator to find lost or unknown metal pipes or cables in the inductive mode, the locator transmitter emits a signal that induces an electromagnetic field around the object you are seeking that can be detected by the receiver. The locator is carried on a long handle provided for that purpose when initially locating the pipe or cable.

As you cross over the target, the locator receiver provides the user with both sound and meter indications that you are standing over a conductor. Beyond the target, both the sound and the meter reading indications decrease. To establish the path of the target, move 5–10 yards up and down the line and take additional readings. Drawing a line along these three points will show you the path of the conductor.

The third method is the two-person sweep. Start 25–30 feet apart with both transmitter and receiver tuned to the higher frequency (+60 kHz). Walk parallel to each other. When the individual with the receiver gets an indication of a signal, the individual with the transmitter should place it on the ground. The receiver is then used to pinpoint the target line. Mark the spot. The transmitter is then moved back and forth to locate the strongest signal point and that spot is marked.

At this point, both the transmitter and receiver are over the target pipe or cable. Continue this process until the entire area is swept in this direction. The area is then swept in a direction that is perpendicular to the first sweep. To ensure that no lines are overlooked, two additional sweeps oriented diagonally to the first two must be completed.



**Audiovisual**

**Show Transparency 5. Discuss the three ways to detect foreign lines in a right-of-way.**

**Demonstration**

**If possible, bring a split-box locator to class and show trainees how to use it.**

**Laboratory**

**If you bring a split-box locator to class, have trainees practice the two-person sweep method under your supervision.**

**Audiovisual**

**Show Transparency 6. Discuss the advantages and disadvantages of using an induction set.**

**Demonstration**

**Bring an induction set to class. Demonstrate its use for trainees.**

**Laboratory**

**Under your supervision, have trainees practice using the induction set.**

**Classroom**

**Explain the use of inductive couplers/clamps. Describe the advantages and disadvantages of this approach.**

**Demonstration**

**Note that most locators have induction couplers or clamps. Demonstrate their use for trainees.**

**Laboratory**

**Under your supervision, have trainees practice working with the couplers or clamps.**

**Audiovisual**

**Show Transparency 7. Discuss use of direct connection and review methods to improve earth grounding.**

### *3.6.5 Using an Induction Set*

Once you know the location of a starting point for the target pipe or cable, you can perform line tracing by placing the transmitter unit directly over and in line with the target. The transmitter and receiver should be at least 30 feet apart to prevent air coupling of the signal between transmitter and receiver.

The stronger the transmitter signal, the further from its location you will be able to detect it with the receiver. Begin tracing the precise path of the target by walking along the supposed path and moving the receiver back and forth across the area. Peak signal strength will be present directly over the target, allowing you to accurately mark its location.

Continue marking the line in this manner until the signal is no longer clear. Move the transmitter down the line (referred to as leapfrogging) and resume the process.

Advantages gained from using this process include the speed and ease of operation. The process is often the only alternative.

Remember to use passive modes in the receiver to help get a starting point. Signals that can be located in the passive mode include:

- 60 Hz AC electrical power
- Cathodic protection signals
- VLF Radio broadcast signals
- Cable TV signals at 31 kHz

Disadvantages encountered while using this process may stem from:

- Very little of the transmitted signal being detectable (only 10-50%)
- Signal couples on all lines, preventing identification of a specific target line

### *3.6.6 Using Inductive Couplers/Clamps*

Clamping is a safe, convenient method of locating and tracing. A clamp or coupler is either standard equipment or an option with most locators.

An advantage to this approach is that clamping safely applies an easy-to-read signal to a live cable without interrupting the supply. A distinct disadvantage is that clamp-applied signals do not travel far when compared to direct-connected signals. This approach can be used on direct buried cables (energized or de-energized) and non-metallic conduit with cables inside.

Surround the cable or non-metallic conduit with the coupler. Apply the signal as close to the dirt as possible. Remember that the signal travels away from the closest good ground, so you may have to go to the "other end" and apply the signal to get it to run.

> **NOTE**
> The cables are in metal conduit, so connect directly to the metal and locate it just like a pipe. The use of a button magnet with a bolt or vise grips can be helpful when connecting the transmitter lead to the metal conduit/pipe.

### *3.6.7 Using Direct Connection*

Direct connection provides you with the strongest and cleanest signals for the receiver to detect. Clip directly to test leads, valves, fittings or couplers.

> **NOTE**
> Be aware that if an electrical insulator kit has been installed, it will stop the signal and prevent you from continuing down the line.

Advantages of this method include:

- *Frequency choice* – Lower frequencies are better because they are more easily detected. Check for the best current measurement frequency or signal strength for the location and conditions.
- *Signal power settings* – Using the correct power level is important. Excess power might create a stronger signal that can be detected at a greater distance, but it also can spread onto other lines, creating possible confusion in detecting and interpreting the signal.

The ground cable and rod should be run out 10–15 feet perpendicular to the side of the line to be located without crossing any other lines. The transmitter signal will couple onto any lines you cross. Look for existing temporary line markers, permanent line markers, paint, flags, stakes, or signs that might indicate the location of the target line. In the absence of any of these indicators, use the two-person sweep as detailed in this document to locate the target line.

Often the limiting factors for signal connection quality or signal distance include improper connections and/or improper grounding. To correct improper connections, make sure the connection lead is attached to shiny metal, not a painted surface, rusty surface, or mud covered area. On larger lines, a "C" clamp, vise grips, or a magnet fitted with a bolt can be used to connect the transmitter lead to the pipe. To improve earth grounding, try one or more of the following:

- Use a longer ground rod. Never use rebar because it is a poor conductor.



*Instructor's Notes:*

- Run water down the ground stake hole.
- Drive the ground stake in at an angle.
- Use multiple ground stakes jumpered, or bonded, together.
- Use an isolated anode as a ground.
- Use a totally isolated pipe as a ground.

#### *3.6.8 Other Locating Techniques*

To find a side leg, tee, or intersection, use the transmitted signal on the line or tune the receiver to one of the passive modes (60 Hz power or radio, CP signal, or CATV). Put the receiver on dual horizontal peak antennas. Then move 4–6 feet off the line and turn the receiver antenna blades parallel to the route of the line. Walk down the line and when you get to a tee or intersection, the receiver will indicate signal and noise to pinpoint the line.

There are several methods or features for accurately pinpointing lines in highly congested areas:

- *Double-ended connection* – The use of double-ended connections is invaluable in locating bonded, plumbed, and common lines in plants, stations, and yards. The transmitter is connected as usual to the pipe or cable. Instead of attaching the ground lead on a ground stake 10–15 feet perpendicular to the line route, you use an extra long extension ground wire spool (30–1,000 feet). Run the extension ground parallel to the line and 10–15 feet off the suspected route of the line. Attach it to the end of the same line. The transmitter is now connected, or hardwired, to both ends of the line.
- *Current measurement* – If the receiver has a current measurement of the transmitter signal, it can be used to pinpoint a specific line. Activate the current measurement function on the receiver, and the line with the highest current measurement will usually be the line the transmitter is connected to or clamped around.
- *Current direction* – Some locators have a help mode called current direction. The theory is that the current will run in one direction on the line the transmitter is connected to. Signal current will run in the opposite direction on all other lines, even if bonded or connected. Regardless of the number of lines in the ROW, the current on the target line will run in the opposite direction from all other lines.
- *Two transmitters on two different frequencies* – Finally, two transmitters can be used at the same time to pinpoint difficult close lines. Connect one transmitter at a low frequency to one of the lines. Using a different frequency, connect a second transmitter to the opposite end of the line. Use the receiver to toggle back and forth between the two frequencies to pinpoint each of the two lines. Accurate depth is usually impossible to determine in these congested situations. Use current measurement if a frequency is on both lines. The higher current will be measured on the line with the transmitter tuned to that frequency. For example, with 8 kHz on both lines, the higher current measurement should be detected on the line connected to the transmitter sending the 8 kHz signal.

#### *3.6.9 Underwater Locating*

When locating in shallow water, the water is invisible to the signal and has no serious impact on signal strength. However, the signal will dissipate over distance. Three methods are available to employ a submersible antenna for underwater locating. They are:

- Wading with antenna
- Antenna taped on to a plastic pole and suspended from a boat
- Antenna carried by a diver

### 3.7.0 Other Methods

The oldest technique, though not a technology, is dowsing, or divining. This method has often been used in the field to locate non-metallic lines without tracer wire or tracer tape. It works much better for some individuals than others.

Wires, welding rods, or other variants are used as locators in the performance of this obscure art. This alternate method of locating pipes and cables will not hold up in court or satisfy federal or state One-Call requirements. While it is difficult to place a value on this method, it is cheap and easy to implement, but does not work equally well for everyone. Many companies will not allow dowsing.

> **WARNING!**
> This method must always be physically verified using another method such as probing, pot holing, or vacuum excavating.

## 4.0.0 ◆ MARKING PIPELINES

Permanent marking of pipelines is required by the Department of Transportation (DOT) regulations. Signs are required at all road crossings, railroad crossings, water crossings, and intervals along the length of the line. Permanent markers are also required at aboveground facilities and pipe sections. Temporary marking of the pipeline is done for any excavators or others performing work in the area that might put the pipeline at risk.



**Audiovisual**

**Show Transparency 8. Review other locating techniques, including underwater locating and dowsing.**

**Safety**

**Stress the importance of physically verifying dowsing.**

**Performance Testing**

**Have each trainee demonstrate Performance Task 1 to your satisfaction. Fill out a Performance Profile Sheet for each trainee.**

**Classroom**

**Ask trainees if they have any questions about the material covered so far.**

**Homework**

**Have trainees review Sections 4.0.0–6.2.0 for the next class session.**

**Materials**

**Ensure that you have everything required for testing during this session.**

**Classroom**

**Ask trainees for examples of places where the DOT requires pipeline signs. Record answers on the board.**

**Classroom**

**Refer trainees back to Figure 1 or bring in a sample of a permanent marker. Discuss places in which permanent markers are installed.**

**State a number of color codes and ask trainees what each color indicates.**

**Performance Testing**

**Have each trainee demonstrate Performance Tasks 2 and 3 to your satisfaction. Fill out a Performance Profile Sheet for each task for each trainee.**

**Audiovisual**

**Show Transparency 9. List instances when permits are required and the people who may supply them.**

**Classroom**

**Introduce the methods for exposing pipeline. Stress that only a competent person may make safety decisions.**

**Review approved procedures for locating, probing, and excavating to expose a pipeline. Stress the importance of following company polices and procedures.**

### 4.1.0 Permanent Markers

Permanent markers are generally colored signs with large lettering that are legible to anyone in the vicinity. In areas where more than one language is common, both may be used on the sign. The marker also includes the company responsible for it and the company emergency number to be contacted for excavations or for observed problems. One such sign is seen in *Figure 1*.

Permanent line markers are installed and maintained as close as practical to the pipeline at all required crossing locations. Otherwise line markers are installed directly over the buried pipelines with the faces approximately perpendicular to the line route. Aboveground pipelines have markers installed to the side. Markers should be visible from all directions.

Where the pipeline crosses a fence, a marker is generally installed over the line. Permanent markers can be mounted on fences, posts, or standards made from scrap pipe. It is preferable they are mounted on posts painted with high visibility paint.

### 4.2.0 Temporary Markers

Temporary pipeline and right-of-way markers must be installed where other persons are engaging or preparing to engage in activities that could damage the pipeline or interfere with operational requirements. Temporary markers may also be installed at the request of individuals or groups.

Temporary markers are provided when needed to more precisely locate the pipeline between permanent markers. Temporary markers can be wire, flags, stakes, or paint applied to the route. The choice may depend on circumstances and the length of line to be marked. The temporary markers should be durable enough to last throughout the proposed work and should also comply with the standard color code (*Table 4*).

**Table 4** Color Code

| Marking Color | Utility Lines |
|---|---|
| Red | Electric |
| Yellow | Gas-Oil |
| Orange | Communication/CATV |
| Blue | Water |
| Purple | Reclaimed Water |
| Green | Sewer |
| Pink | Temporary Survey Markings |
| White | Proposed Excavation |

107T04.EPS

## 5.0.0 ◆ PERMITS

Permits may be required before excavating any sections of the pipeline. The number and types of permits may vary with the location, but in general may include permission from the following:

- Property owner
- DOT
- State
- City/local authorities
- Company permits
- Railroads
- Corps of engineers

It is not uncommon for more than one permit to be required.

## 6.0.0 ◆ EXPOSING PIPELINE

When a portion of the pipeline must be exposed for repairs or other reasons, maps and location devices are used to accurately locate the pipeline and its depth. It is critical that the location is accurately known before the excavation begins. Digging in the vicinity of any underground structure requires special care. The company's procedures will provide specific details on how this should be done, but the following gives a general description of the pipeline locating and exposing process. This includes:

- Excavating
- Pressure reduction
- Danger awareness
- Pipe supports/plugs

All decisions affecting safety on excavation job sites must be made by a competent person.

### 6.1.0 Excavating

Always follow company policies and procedures for the company whose line is being exposed. Excavation prior to line exposure is preferably done by hand where practical. Stripping soil from buried pipeline with mechanical excavating equipment is only done after the pipeline has been located, staked, and probed or exposed by hand. Probing is done across the width of the pipe at intervals less than one-half the pipe diameter. When the pipeline is located, probes are stuck in the ground along the length of the pipe as a guide to excavators.

If the pipe is too deep to reach with the probe, one method of locating it is to machine-excavate to one-half the length of the probe. At that point, the line is reprobed until located and confirmed.



Instructor's Notes:

Digging may then continue by hand or with a backhoe running parallel to the line. After the line is exposed, excavation must proceed parallel to the line so that the operator can keep it in view. The backhoe bucket should never be closer to the pipeline than 2 feet. Final stripping is done by hand.

The cover is removed carefully to avoid damaging the pipe, and spaces are dug at specified intervals under the pipe for the insertion of dirt, plugs, skids, or approved support mediums. Topsoil is removed and kept separate from the rest of the fill for replacement in its original position.

If the soil around the line contains large rock, rock hammers, jack hammers, or rock splitters may be needed to expose the pipeline. If so, the pipe can be shielded with oak planks, blasting mats, oversized pipe, or sleeves.

### 6.1.1 Excavating and Shoring

Working in and around excavations is one of the most hazardous duties performed on the job site. It is essential that excavating and shoring be done correctly to protect the workers and the pipeline when excavations must be made. Excavating is defined as the making of a ditch around the pipeline section. The earth that is excavated can weigh 3,000 pounds for every cubic yard, and the walls that are left after digging must be adequately supported. *OSHA 29 CFR Part 1926 P* defines acceptable trench protective devices and how they are to be used. Excavations of 5 feet or deeper must use shielding and shoring devices or sloping to prevent cave-ins. When the pipe is first exposed and as supporting material is removed from under the pipe, skids or other supports must be placed beneath the pipe about every 20 feet.

In general, the trench is no wider than it is deep. Any excavation deeper than 20 feet requires an engineer to design it and must be approved by a competent person. The shielding or shoring system uses timber, steel, other material, or support structure as a temporary prop for the walls of the trench. This structure may be sloping, vertical, or horizontal. Some factors in excavating and shoring correctly include:

- Soil properties
- Sloping requirements
- Support systems
- Dewatering

### 6.1.2 Soil Properties

The properties of the soil in which excavations are being made are very important when planning trenching and shoring techniques. Properties of the soil will include the grain size, the soil gradation, and the grain shape. Grains are the soil particles, which have been produced from rocks by weathering and chemical processes.

Grain sizes are determined by measuring with calibrated sieves or screens. The sieves are graded by the size of the holes in their mesh. The grains of soil are poured into a series of different size sieves until most of the grains will not pass through the holes in the mesh. The grains caught by each size or grade sieve are classified by that size. *Figure 8* shows a soil classification chart.

In most soils, all of the grains will not be the same size. When passing the samples through a series of sieves, some proportion will be caught in different size meshes. The proportion of each size will identify the soil's gradation. The soil is first classified as coarse-grained or fine-grained. The coarse-grained soil is further classified for gradation according to the following criteria:

- *Well-graded* – Shows a good representation of all particle sizes from largest to smallest
- *Poorly-graded* – Has an excess or deficiency of a certain particle size
- *Uniform gradation* – Most particles are about the same size
- *Gap or skip gradation* – A mixture of two or more uniform soils

| COARSE GRAIN | | | | | | | FINE GRAIN | |
|---|---|---|---|---|---|---|---|---|
| 12" | 3" | | 0.25" | | | | 0.002" | |
| BOULDERS | COBBLES | GRAVEL | | SAND | | | FINES | |
| | | COARSE | FINE | COARSE | MEDIUM | FINE | SILT SIZES | CLAY SIZES |

U.S. STANDARD SIEVES
3" 2" 1" 3/4" 3/8" 4 6 10 20 40 60 100 200

107F08.EPS

*Figure 8* ◆ Soil classification chart.



**Classroom**

**Stress the importance of properly shoring trench walls to protect workers and the pipeline.**

**Refer trainees to Figure 8. Explain how soil's grain size, gradation, and shape affect how it behaves in a trench.**

**Audiovisual**

**Show Transparency 10 (Figure 9). Discuss the different soil behaviors. Stress that water in any soil is a critical factor.**

**Audiovisual**

**Show Transparency 11 (Figure 10). Discuss the maximum allowable slopes for the four OSHA-classified soil types.**

**Show Transparency 12 (Figure 11). Describe benching in Type A soil. Ask trainees if they have any questions.**

The first two may be determined in the field with experience. The others may need to be done in the lab.

The grain shape has an important effect on the properties of the soil. The shape can determine how the grains will stick together, support each other, or slide over each other in the unsupported wall. They will also affect the compaction requirements when backfilling the trench.

The most common shapes are bulky grains, flaky grains, and elongated grains and fibers. Bulky grains are rounded, subrounded, subangular, and angular and usually make up most of the coarse-grained particles. Flaky grains are present in many fine-grained soils. These particles slide against each other and can be compacted well to form a tight mass.

All of these properties determine the types of soils that may be encountered in excavating. The average soil found on most sites is a mixture of sand, silt, clay, and gravel. The amount of each component in the mixture, along with any water, air, gases, chemicals or other materials in the soil, will affect its workability and stability.

Sand is not very stable and does not hold water unless some other material is present. Sand excavations require more support than those in other soils. Silt has fine particles that do hold water, and it is soft when wet. Dry silt is easily broken and powdery. It is not very stable unless it has been compressed and hardened. Clay is also made up of fine particles, but is plastic when wet. Dried clay is hard and difficult to break up. Excavations in clay can stand up well unless water is present, which can cause landslides. Gravel is rock with a range of sizes and, if compacted well, is stable and provides water drainage. The presence of water is a critical factor in any type of soil. *Figure 9* shows different soil behaviors.

### *6.1.3 Sloping Requirements*

Many excavations are started with a vertical cut. Although some soils will stand to considerable depths when cut vertically, most will not. They will fall to a more stable angle, dumping large amounts of material into the trench. Sloping means to cut the side walls of a trench at an angle that will be more stable. The angle should start at the top of the trench wall and continue to the floor of the trench. The slope is cut at the angle of repose for the particular soil, the angle of repose being the greatest angle above the horizontal plane that the soil will rest without sliding.

OSHA has classified soils into four categories and determined the maximum allowable slope for each. The classifications are based on stability and range from stable rock at the high end to Type C as least stable. *Figure 10* shows the maximum angles for soil types.

Stable rock is natural solid material that remains intact when exposed. It includes solid rock, shale, and cemented sands or gravel. Type A soil is cohesive, with a compressive strength of at least 1.5 tons per square foot. These include clay, silty clay, sandy clay, and clay loam. However, no soil can be Type A if the soil is fissured, subject to vibration, has been previously disturbed, or has any factors that decrease stability.

Type B soil is cohesive, with a compression strength of 0.5 to 1.5 tons per square foot. These include granular soils like angular gravel, crushed rock, silt, sandy loam, unstable rock, and any unstable Type A soils. Type C soil is unstable, cohesive soil with a compressive strength less than 0.5 tons per square foot. This includes gravel, loamy soil, sand, submerged soil, soil with water seepage, and unstable submerged rock.

Rules for sloping include the following:

- The angle of repose must be flattened when water is present or when there are silty conditions, loose boulders, and areas where erosion and deep frost are possible.
- The excavated soil, or spoil, must be banked at least two feet away from the edge of the trench, as must any tools and other materials.
- All excavations deeper than 20 feet must also have some type of pre-engineered, certified support system installed.
- Special precautions must be taken in sloping the sides of an excavation adjacent to a previously filled excavation or fill.
- End plates, sheet piling, or other methods of retaining soil should be used at the ditch ends if they cannot be sloped properly.

Benching is allowed in cohesive soils. Benching is excavating the walls in steps, which must rise on the approximate allowable angle of repose. It is also called step back excavation. The same rules apply for benching as for slopes. *Figure 11* shows benching in Type A soil.



Instructor's Notes:

*Figure 9* ◆ Different soil behaviors.

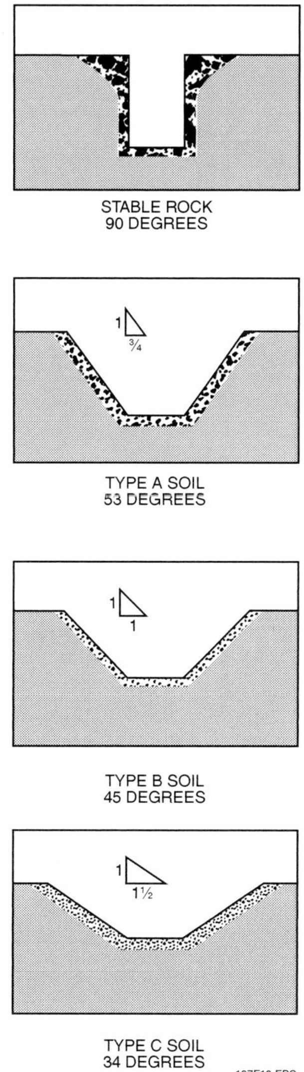


*Figure 10* ◆ Maximum angles for soil types.

**Audiovisual**

**Show Transparency 13 (Figure 12). Ask a trainee to define shoring. Explain the components of the shoring systems in Figure 12.**

**Show Transparency 14 (Figure 13). Describe the two sheet pile restraints. Ask trainees if they have any questions.**

**Show Transparency 15 (Figure 14). Describe trench box placement in relation to the pipe.**

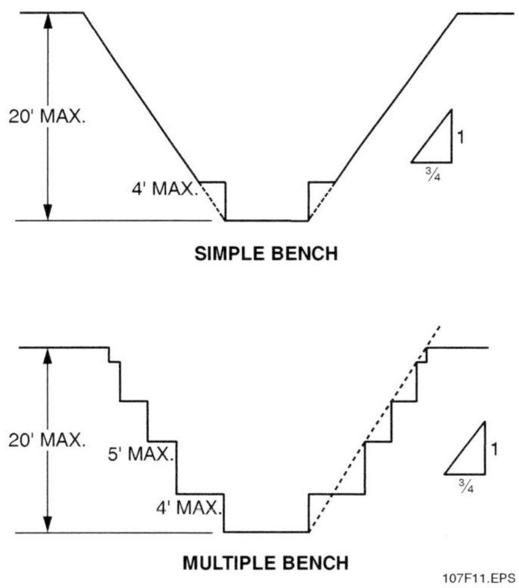


*Figure 11* ◈ Benching in Type A soil.

### *6.1.4 Support Systems*

There are several different types of support systems for trenches. These include shoring, sheet piling, and trench boxes or shields. The type used depends on the following factors:

- Depth of the trench
- Type of soil
- Moisture content of the soil
- Possible weather effects
- Loading of nearby equipment
- Possibility of vibration

Shoring provides a framework to support excavation walls and prevent their movement or collapse. These are installed as the excavation is dug and may be built on site or bought premanufactured. OSHA has strict design specifications for shoring systems. Modern shoring systems are engineered to preload trench walls, providing a positive restraint to soil movement. A basic aluminum hydraulic shoring system may include vertical sheets called uprights, horizontal rails called walers, and cross-brace cylinders. They can be used with or without backing, depending on soil conditions. Cylinder extensions can adapt the components to different widths. In addition to providing a safe work environment, shoring keeps soil from shifting under adjacent structures, such as buried utilities, streets, sidewalks, and foundations. *Figure 12* shows two simple shoring systems.

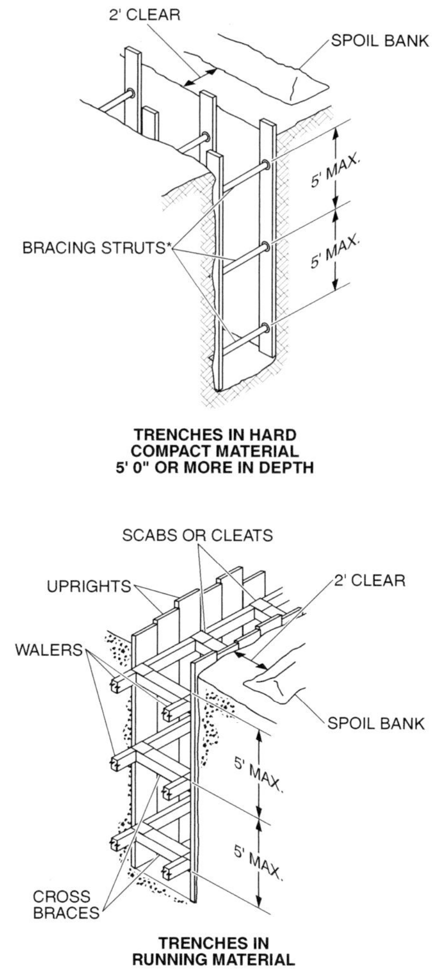


*SCREW JACKS OR TIMBERS NEVER GREATER THAN 5' 0" ON CENTER. (ONE BRACE REQUIRED FOR EACH 4' 0" OF TRENCH DEPTH. NEVER FEWER THAN TWO BRACES.)

107F12.EPS

*Figure 12* ◈ Simple shoring systems.



Instructor's Notes:

Sheet piling is usually constructed of steel with interlocks that tie the sheet piles together. They can also be made of reinforced concrete or tongue-and-groove connected wood. Sheet piles are positioned along the edges of the proposed excavation and driven into the ground before digging. After the sheets are driven, excavation between them can begin. When the hole is one fourth of the final depth, cross braces or tie-back anchors must be installed to restrain the sheet piling (see *Figure 13*).

The trench can then be dug to final depth. After the excavation has been backfilled, the sheet piling is removed by the excavator hooking into the holes at the top and pulling them out.

Trench boxes or shields are large boxes that are lowered into the hole by the excavator and can be moved as the excavation progresses. Trench boxes can be assembled on site, or manufactured boxes can be used. They must be in accordance with *29 CFR 1926.652*, and manufactured boxes must be certified by the manufacturer to the depth and soil conditions in which they are used. Stable trenches allow the box to be installed above the top of the pipe on a wider step trench. In an unstable trench, the box must extend below the pipe, and must be placed carefully to avoid disturbing or damaging the pipe. A suitable backfill material is placed between the trench wall and the box. *Figure 14* shows trench box placement with pipe.

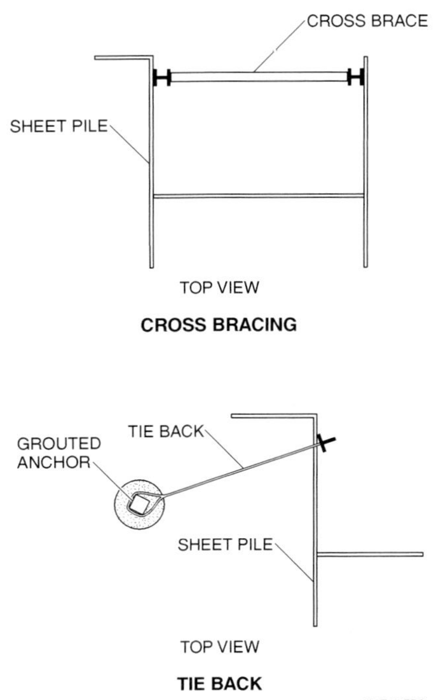


*Figure 13* ◈ Sheet pile restraints.

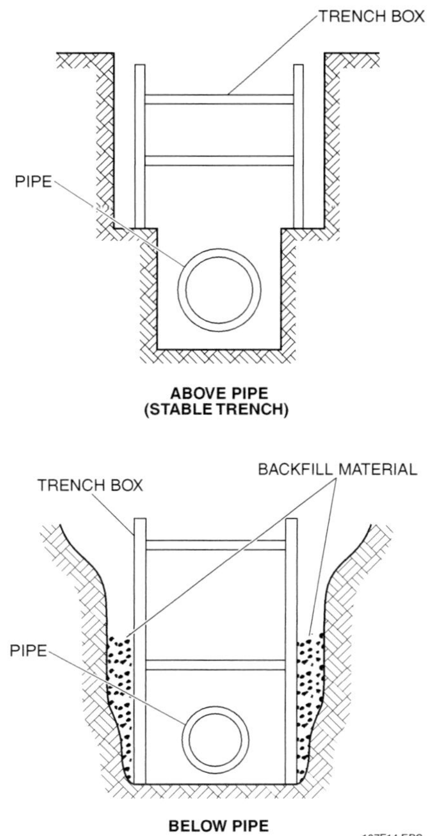


*Figure 14* ◈ Trench box placement with pipe.

**Classroom**

**Illustrate with a simple line drawing how a well point functions.**

**Explain how pressure reduction lessens excavation hazards.**

**Ask trainees if they have any questions about the material covered so far.**

**Homework**

**Have trainees review Sections 6.3.0–6.4.2 for the next class session.**

**Materials**

**Ensure that you have everything required for teaching this session.**

**Teaching Tip**

**See the teaching tip at the end of this module.**

**Audiovisual**

**Show Transparency 16 (Figure 16). Describe unstable trench conditions.**

**Safety**

**Review the safety rules for entering and exiting excavations.**

Trench boxes can be moved along if the excavation is extended. Backfilling can take place behind the box as it is moved. In unstable soils, the excavator can dig inside the box and tamp the box down. The walls must still be correctly sloped above the box, and the box walls must extend at least 18 inches above the point where the walls begin to slope. However, with large diameter pipelines, there may not be room to set a trench box and continue digging. A ditch shield can be temporarily placed and secured in the trench to protect workers exposing the pipeline. When finished digging, the shield can be locked in place.

#### *6.1.5 Dewatering Soil*

When working in waterlogged soils, the excavator may want to install **well points** to dewater the ditch line. Well points can effectively dry out extensive areas along the ditch line by continuously removing underground water.

Each well point functions like a submersible pump. A hollow steel rod with wire mesh on the bottom end is driven into the ground. A flexible hose is attached to its top end and connected to a portable pump. The pump carries the water away from the work site. Many such rods are installed along the line, and the hoses connect to the pump through a pipe along the ground. Dewatering may, in some cases, require additional permits.

Any time pipeline problems or pipeline excavations and repairs occur near a water source, the water must be inspected for any contamination by the pipeline product. Oil and other less refined products produce a blackish sheen on water and will form a consistent sheen on the surface, even after being disturbed. It mostly floats on water because its specific gravity is less than that of water. Gasoline produces a multi-colored sheen on water, representing the refined hydrocarbons and additives. Some components of gasoline may mix with the water molecules and be transported with the water.

### 6.2.0 Pressure Reduction

Some company procedures require pressure reduction in the pipeline before excavation. Reducing the pressure may lessen the hazard during excavation or movement of the line.

### 6.3.0 Danger Awareness

Excavation work presents many dangers. Some of these have been mentioned already. This section further clarifies and emphasizes the dangers involved and how to avoid them. Excavation is considered to be the most hazardous part of construction work. In addition, excavations around petroleum pipelines present the dangers of escaping toxic and/or explosive substances.

These hazards include the following:

- Cave-ins
- Flooding
- Gases
- Electricity
- Falling objects
- Fire or explosion
- Slips, trips, and falls

In all excavations, exits must be provided within 25 feet of each worker. Trenches must be inspected at least daily before work begins, after rainstorms and other weather events, and when there is any sign of change in stability. Any time the excavation is left unattended it must be reinspected. Appropriate personal protective equipment (PPE) must be specified and used.

#### *6.3.1 Cave-Ins*

Cave-ins are the most obvious dangers involved in excavations. When it is stated that one cubic yard of earth can weigh 3,000 pounds, it is important to visualize just how small an amount of earth that is. *Figure 15* shows the dimensions of 1 cubic yard of earth.

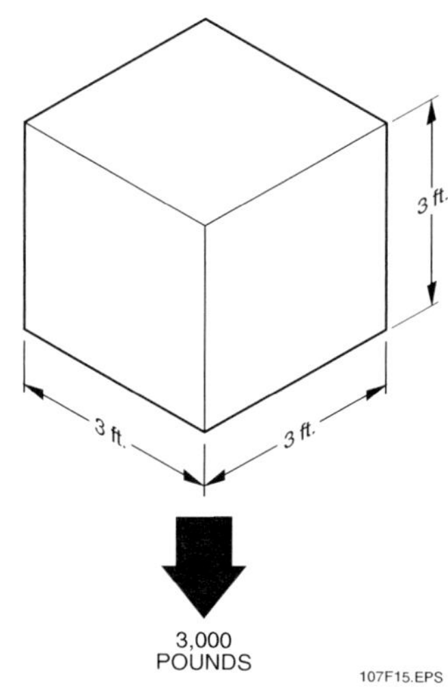


*Figure 15* ◆ Dimensions of 1 cubic yard of earth.



*Instructor's Notes:*

If a worker is buried by collapsing trench walls, the weight alone can be fatal. Even being buried up to the waist can provide enough pressure to result in death. If a worker's face is buried and the weight does not crush him or her to death, the lack of air may do so before he or she can be uncovered. All of this explains why cave-ins are the single most hazardous part of excavation work. The earlier section described how the walls must be sloped and supported. This section describes what the worker must do to maintain their safety against cave-ins. The excavation worker must remember the following:

- No one can predict if an excavation is safe to enter unless the proper support structure or sloping system is provided.
- Excavations in or near backfilled or previously excavated ground are especially dangerous because the soil is loose and does not support itself.
- Saturated soil increases the possibility of a cave-in. The increased pressure of the water exerted on the soil can be the final factor in causing the walls to collapse.
- Clay can be extremely dangerous if dried by the sun. Large chunks of material can break off a trench wall after having been stable and solid for a long period of time.
- It is not safe to assume that frozen earth is stable; frozen ground is not an alternative to shoring.
- Shoring must be adequate to overcome additional pressures from piles of excavated material, adjoining structures, vehicular traffic, and nearby equipment.

A open cut or trench can become more unstable as work is being done. There are some signs to watch for which will indicate an unstable trench (see *Figure 16*). Signs such as these must be reported immediately and work stopped until corrective measures are taken. This is also necessary even if a support system is being used.

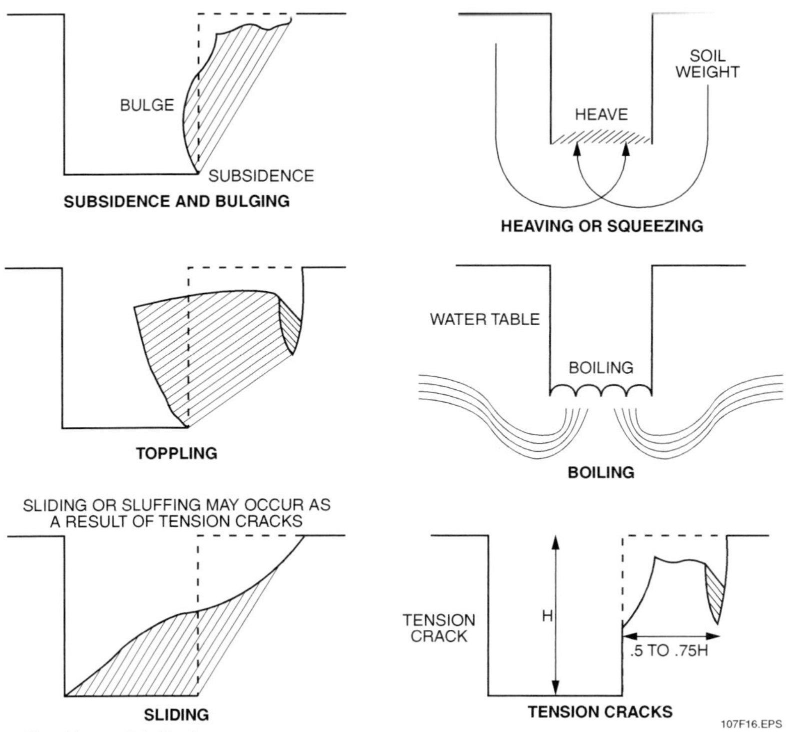


*Figure 16* ◈ Unstable trench indications.

Audiovisual

**Show Transparency 17. Discuss the dangers of flooding, gases, electricity, falling objects, and fire or explosion.**

Classroom

**Refer trainees to Figures 17 and 18 in their Trainee Guides. Explain the different supports and how to distribute stress along the line when pipe is moved.**

These rules must be followed to ensure safety.

- Never enter an excavation to install shoring or trench boxes. These must be in place before entering the trench.
- Check that shoring uprights and cross braces are properly installed.
- Never enter an excavation behind the trench box after it has been moved; this should be backfilled as soon as the box is moved.
- Ensure that an exit is available (within 25 feet) from where any work is to be done. Ladders should have rails that extend 3 feet above the landing.

#### *6.3.2 Flooding*

When working in an excavation, flooding can be a concern. If a water or sewer main is ruptured while excavating, the excavation can quickly fill with water. If dewatering pumps stop, flooding can occur. Just as dangerous are the situations where a natural source of water has been dammed off. Tons of water may be held back, and if the dam fails, it could quickly flood the excavation area. Workers in the trench should make an immediate exit if there is any evidence of gas or liquids in the trench.

#### *6.3.3 Gases*

In some situations, an excavation should be considered a confined space, and appropriate evaluation and controls must be undertaken to ensure against a contaminated atmosphere or toxic substances. Excavations can fill with dangerous gases, or oxygen can be displaced from the bottom of the excavation if located near facilities that store large volumes of chemicals. The liquids or gases, even exhaust fumes from vehicles, could flow or leak into the excavation. When excavating a pipeline for repairs, the liquids or gases could have already leaked into the ground around the pipe. An excavation should never be entered until the atmosphere is determined to be safe by authorized persons and methods.

#### *6.3.4 Electricity*

Electrical shock is a hazard because while excavating, the insulation of an electrical cable could be pierced. When overhead lines are near the excavation, the excavator could come in contact with them, knocking them into the excavation or electrifying the boom. If the boom is electrified, the pipe it is lowering into the excavation will be electrified as well and could impart a deadly shock.

#### *6.3.5 Falling Objects*

Anything around the edges of the excavation could become a dangerous falling object if dropped or knocked over the edge. This includes piles of soil, tools, chunks of dirt from the excavator bucket, parts of the bucket, and other objects. Those workers on the surface must keep such objects away from the edge of the excavation and any excavated dirt at least 2 feet away. Those in the trench should always have on PPE at all times. Do not park vehicles facing toward the trench. Always park vehicles far away from the trench and chock the wheels.

#### *6.3.6 Fire or Explosion*

Gases and flammable liquids can accumulate in a trench from natural sources or from the pipeline being repaired, creating a fire and/or explosion hazard. The trench may have been checked and cleared of such substances before workers entered the trench, but the build-up may occur later. In situations where gas may enter the trench, a constant monitoring system may be needed. When gas is detected, equipment should be immediately shut down. Workers in the trench should make an immediate exit if there is any evidence of gas or liquids in the trench.

### 6.4.0 Pipe Supports/Plugs

A pipeline being excavated will require careful handling and support to avoid damaging the pipe.

#### *6.4.1 Support*

Pipe that is being excavated must be properly secured and supported to lessen the stress on the pipe and the welds. The maximum span between the location of the supports should be determined by an engineer, based on the pipe specifications and other factors.

Skids provide support across the ditch, and skids, blocks, sandbags, airbags, and other materials are used on the bottom of the ditch to support inclined pipe. Other pipeline support methods include cribbing, air bags, earth pillars, and slings (see *Figure 17*).

When supports are being installed or removed, it is important to add or reduce supports in a systematic way to even the stress over the entire line section. Methods of moving pipe include the using two side booms and using one side boom and one backhoe (see *Figure 18*).



Instructor's Notes:

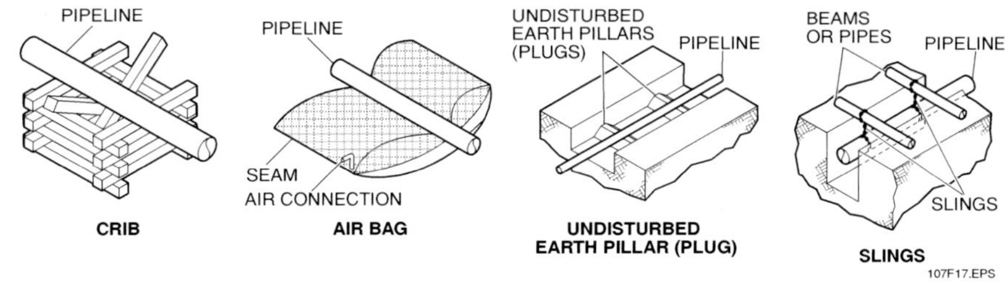


*Figure 17* ◈ Pipeline support methods.

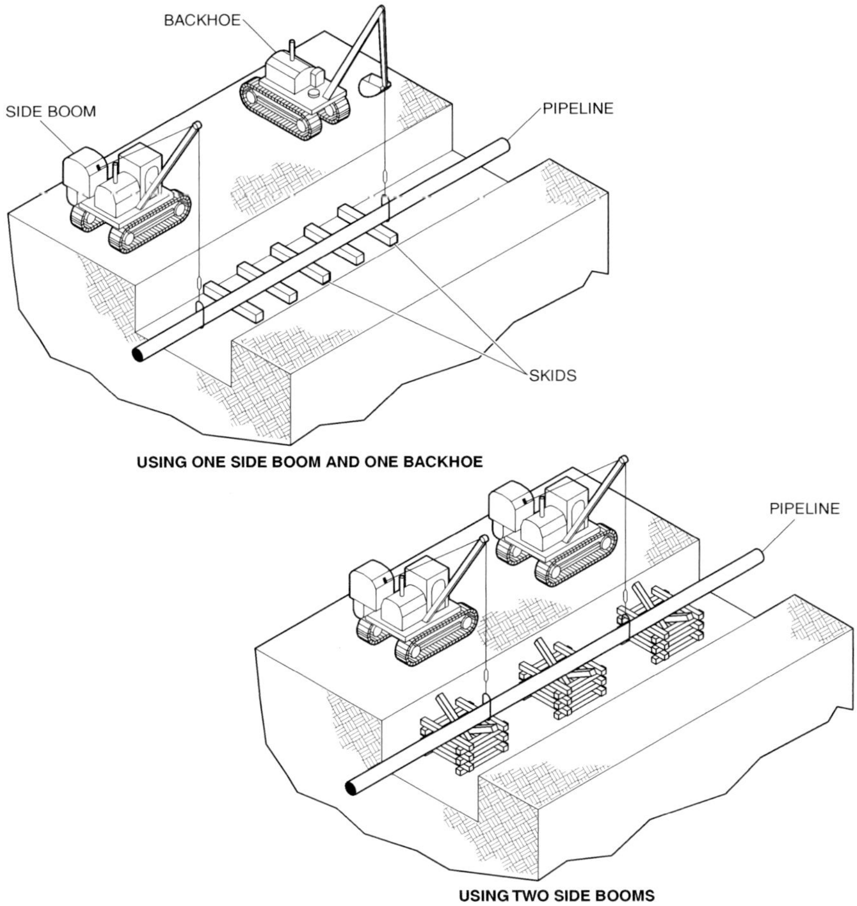


*Figure 18* ◈ Support methods.

**Classroom**

**Describe how earthen plugs are used as pipeline supports.**

**Ask trainees if they have any questions about the material covered so far.**

**Homework**

**Have trainees review Sections 7.0.0–9.0.0 for the next class session.**

**Materials**

**Ensure that you have everything required for testing during these sessions.**

**Audiovisual**

**Show Transparency 18. Discuss dents, gouges, corrosion, coating damage, and pipeline stability.**

**Classroom**

**Ask a trainee to describe the result of leaving cans, tools, or scrap metal in the fill.**

**Ask a trainee to describe the result of leaving rocks in the fill or using frozen earth.**

#### *6.4.2 Plugs*

Earthen plugs can be used as supports. These plugs are left after other supports have been removed. Where possible, earthen plugs should be used to support pipeline stopples and taps.

## 7.0.0 ◆ DAMAGE INSPECTION

Whenever pipe is exposed and before it is backfilled, it must be inspected for damage. Among the types of damage usually encountered are the following:

- Dents and gouges
- Corrosion
- Coating damage
- Pipeline stability
- Welder arcs
- Grinder marks

### 7.1.0 Dents and Gouges

Dents and gouges are usually caused by contact with excavation or moving equipment or by rocks too close to the pipe. The presence of dents and gouges is a danger to the pipeline because they may have reduced the wall thickness of the pipe, creating a weak point, and because they concentrate stress forces at that point. The weakened condition of the pipe wall can be the source of future leaks.

### 7.2.0 Corrosion

Any corrosion on the pipe is an indication that the cathodic protection system is inadequate or that factors exist in the area that have overcome the system. Some factors include organic material touching the pipe wall or release of corrosive material. Corrosion of the pipe will lead to future leaks by weakening and eating away at the pipe wall. The discovery of corrosion on an exposed section of pipe must be followed up by determining the extent of the problem. Adjacent sections of pipe may need to be exposed and inspected.

### 7.3.0 Coating Damage

The coating on the pipe is essential protection for the pipe surface against corrosion and the burial environment effects. The coating may easily be cut, scraped, or otherwise damaged by the excavation and moving equipment or by rocks in the fill.

### 7.4.0 Pipeline Stability

The stability of the pipeline, which was provided during its original burial, may have been disturbed by such natural causes as earth movements, subsidence, or flooding. These could cause the pipe to move horizontally or vertically in its trench, stressing the line. They could also cause a loss of the supporting material in the ditch and fill material. Effects of any such disturbance need to be investigated and remedied before reburial.

## 8.0.0 ◆ REPORTING OBVIOUS DEFECTS

When obvious defects are found on the pipe or its environment, they must be reported on specific forms to management. During pipe inspection, any defects found are listed and described in detail on the pipe inspection form (see *Figure 19*). The information in this report includes the locations, conditions, and descriptions of any damage, as well as actual measurements and sketches.

Repairs of damages and defects are described on a pipe repair report form. This form asks for information, such as type of damage, how it may have happened, where it was located, its description, and how it was repaired. *Figure 20* shows a sample pipe repair report form.

## 9.0.0 ◆ WATER INSPECTIONS

Any time pipeline problems, excavations, or repairs occur near a water source, the water must be inspected for any contamination by the pipeline product. Oil and other less refined products produce a blackish sheen on water and will form a consistent sheen on the surface, even after being disturbed. It mostly floats on water because its specific gravity is less that that of water, representing the refined hydrocarbons and additives. Some components of gasoline may mix with the water molecules and be transported with the water.

## 10.0.0 ◆ BACKFILLING

A single pipeline frequently requires the use of a variety of backfilling procedures to perform the work effectively and economically while complying with specifications regarding pipe and coating protection. Excavators, or trackhoes, and dozers are commonly used to move dirt from the spoil back to the ditch.



*Instructor's Notes:*

**PIPELINE INSPECTION REPORT**

*Frequency: At Time of Inspection*

**PART A — LOCATION**

Reason For Inspection ____________

1. Location: Division ____________ County ____________ State ______ Date of Inspection ____________

A) Mainline
Alignment Sheet ________ Pipeline Size/Segment ________
Beg. M.P. ________ End M.P. ________
Beg. Sta. No. ________ End Sta. No. ________

Note: Station numbers shall be measured from a known landmark (Road Crossing, etc.)

B) Company Facility: Pump Station, Tank Farm, etc.
Location Name ____________
Physical Location of Pipe Inspected ____________
Milepost ____________

**PART B — MATERIAL/CONDITIONS**

2. Piping:
Pipe Size ________ In. O.D. Nom. W.T. ________ In. Actual W.T. ________ Grade ________

3. External Pipe Coating:

A) Coating: ☐ Bare ☐ LAC 450 ☐ Taped ☐ Insulated ☐ Other ________

B) Application: ☐ Mill Coated ☐ Yard Coated ☐ Field Coated ☐ Other

C) Coating Thickness: Factory Applied ____ mile; Field Applied ____ mile

D) Condition of Coating: ☐ Good ☐ Fair ☐ Poor

E) Bond: ☐ Good ☐ Fair ☐ Poor

F) Cause of Coating Failure: ☐ Damage ☐ Disbondment ☐ Defective Application ☐ Decomposition ☐ Other ________

4. Environment: ☐ Clay ☐ Wet ☐ Loam ☐ Dry ☐ Sand ☐ Moist ☐ Gumbo ☐ Caliche ☐ Rock ☐ Other ________
Depth of Cover (Inches) ________
Soil Resistivity Measurement ____________ (ohm-cm) Measurement Method ____________

5. Coating Repaired ☐ No ☐ Yes
Repair System ____________

6. Pipe-to-Soil Potential at Inspected Location (At Grade) ________ (volts) (At Pipe/Soil Interface) ________ (volts)

7. Were any Non-Destructive Tests Completed? ☐ No ☐ Yes Type ________

8. Evidence of: ☐ Corrosion ☐ Defect ☐ Damage ☐ Stress Corrosion Cracking (S.C.C.) ☐ Calcareous Deposits

9. Test Station Added ☐ No ☐ Yes I.D. No. ________

NOTES:
1) If S.C.C. or Corrosion is Noted, Complete Below and Back of Report (External Only, See Note 4)
2) If Defect or Damage is Noted, Complete Back of Report
3) If Repairs are Required, AAPL Form 502 "REPAIR REPORT FOR LEAK DAMAGE, CORROSION OR TEST FAILURE" Must be Attached.
4) AAPL Form 508 "INTERNAL CORROSION INSPECTION REPORT" Should be Completed When Internal Corrosion Is Detected.

SAMPLE

**PART C — EXTERNAL CORROSION**

10. General Corrosion Information:

A. DESCRIPTION
☐ Less than 10% of Nom. Wall Thickness, Complete Front of Report
☐ Greater than 10% of Nom. W.T., Complete Front and Back of Report
☐ Pitting ☐ General ☐ Localized
Max Depth ________ (In.) Longitudinal Length ________ (In.)

B. STATUS: ☐ Active ☐ Inactive ☐ Unknown

C. CAUSE: ☐ Galvanic ☐ Stray Current ☐ Unknown ☐ Other ________

11. Cathodic Protection: ☐ Impressed ☐ Galvanic

12. Pipe-to-Soil Potential:
a) Last Pipe-to-Soil Measurement at Nearest Point Each Side of Corroded Area
Upstream ________ (volts) Downstream ________ (volts)
b) Distance from Corroded Area to Measurement Points:
Upstream ________ (feet) Downstream ________ (feet)
c) Date of Measurements: Upstream ________ Downstream ________

13. pH of Soil Near Corroded Area ________ pH of Moisture Under Coating ________

14. Pipe Surface Temperature if Fluid is Heated ________ (deg F)

Remarks ____________

Distribution: (Original) Corr. Ctrl./Eng (Copy) Division Office (Copy) Other ________

________ Technician or Inspector ___/___/___ Date ________ Supervisor ___/___/___ Date

107F19A.EPS

*Figure 19* ◆ Pipeline inspection report form (1 of 2).



**Classroom**

**Refer trainees to Figures 19 and 20. Discuss the pipe inspection report and the pipe repair report.**

**Performance Testing**

**Have each trainee demonstrate Performance Task 4 to your satisfaction. Fill out a Performance Profile Sheet for each trainee.**

**Classroom**

**Describe how oil looks on water and how gasoline looks in water. Explain why oil floats on water.**

**Ask trainees if they have any questions about the material covered so far.**

**Homework**

**Have trainees review Sections 10.0.0–10.3.3 for the next class session.**

# PIPELINE INSPECTION REPORT

*Frequency: At Time of Inspection*

**PART D — DEFECT, DAMAGE OR CORROSION INVESTIGATION**

Notes:
1) Tabulate all Dents, Gouges, Slivers, Corrosion, S.C.C., etc.
2) List Reference Point - Valve, Tee, Center Road etc., as shown on Alignment Sheet with appropriate Station Number, if available.
3) Indicate Location of Longitudinal Seam. If Pipe is Vertical Substitute Compass direction for o'clock Position. For Spiral Weld Pipe, Note as Such.
4) Defects Such as Dents, Slivers, Gouges, etc., shall be tabulated individually. Corrosion Cracks and/or Corrosion Pits Closely Grouped May be Tabulated as One Defect.
5) Defect Width and Length Will be Measured Along the Girth and Longitudinal Axis Respectively.

Reference Point: ☐ Upstream End (Station No.) ______________ Description ______________________

☐ Downstream End (Station No) ______________ Description ______________________

Upstream End Downstream End

SKETCH DEPICTING PIPE CUT AT BOTTOM AND LAID FLAT

NOTE: Indicate Girth Weld Location If Known

______ O'Clock Position →

← Length of Inspected Area →



**PART E — DEFECT, DAMAGE OR CORROSION MEASUREMENTS AND LOCATION**

| No. | Type | Distance from Girth Weld | Distance from Long Seam | Width (Circumferencial) | Length (Longitudinal) | Maximum Depth | Number of Anomalies |
|---|---|---|---|---|---|---|---|
| 1. | | | | | | | |
| 2. | | | | | | | |
| 3. | | | | | | | |
| 4. | | | | | | | |
| 5. | | | | | | | |
| 6. | | | | | | | |
| 7. | | | | | | | |
| 8. | | | | | | | |
| 9. | | | | | | | |
| 10. | | | | | | | |
| 11. | | | | | | | |
| 12. | | | | | | | |
| 13. | | | | | | | |
| 14. | | | | | | | |

If Repair is Required, send to Division Manager for Approval.

Repair Required ☐ No ☐ Yes Type of Repair ______________________________

Approved by ______________________ Title ______________________

Remarks: ______________________________________________

107F19B.EPS

*Figure 19* ◆ Pipeline inspection report form (2 of 2).



Instructor's Notes:

## REPAIR REPORT FOR LEAK, DAMAGE, OR TEST FAILURE

*Frequency: As Needed*

Report Date ____________

**PART A — GENERAL**

☐ Leak Report ☐ Test Failure Report ☐ New Construction ☐ Existing Facility

Instructions: Complete this side of the form for each incident regardless of cause. Check appropriate box for specific cause of leak, damage, or failure and complete the pertinent part(s) on reverse side.

☐ Corrosion
Part E

☐ Damage by Outside Forces
Part F

☐ Construction Defect or Material Failure
Part G

☐ Other (Describe incident in detail in writing and attach it to this form when parts E, F and G are not applicable)

**PART B — DESCRIPTION OF INCIDENT**

1. Description of Leak or Damage (Rupture, Pinhole, Dimensions, etc.): ____________

2. How was Leak or Damage detected? (Patrol, Landowners, etc.): ____________

3. Location of Leak, Failure or Damage:
   a) Nearest Town ____________ County ____________ State ____________
   b) Facility Name ____________ c) Mile Post ____________ d) Survey Station No. ____________
4. Time of Leak, Failure or Damage:
   a) Detected at: Date: ____________ Time: ____________

5. Estimated pressure at point and time of incident (PSIG) ____________
6. Maximum allowable operating pressure (PSIG) ____________

**PART C — DESCRIPTION OF MATERIAL**

1. Origin of Leak, Failure or Damage:
   ☐ a. Body of Pipe ☐ d. Pump ☐ g. Tap Connection ☐ j. Other (Specify) ____________
   ☐ b. Girth Weld ☐ e. Valve ☐ h. Fabricated Assembly ____________
   ☐ c. Longitudinal Weld ☐ f. Heater ☐ i. Fitting (Type) ____________
2. Material which Leaked, was Damaged or Failed is:
   ☐ a. Steel ☐ b. Plastic ☐ c. Other (Specify) ____________
3. Pipe Description:
   a) Nominal diameter (inches) ____________ b) Nominal wall thickness (inches) ______
   c) Pipe Specification: ____________ d) Grade ______
4. Original Installation Date: ____________
5. Leak or Damage Investigation Information:
   a) X-Ray Findings ____________
   b) Ultrasonic Findings ____________
   c) NDT Report Attached ____________

**PART D — REPAIR**

Method of Repair: ☐ Temporary ☐ Permanent Date: ____________

a) Pipe ☐ Welded Sleeve ☐ Clock Spring ☐ Other *(Specify)* ____________
☐ Replaced Pipe Heat No. ______ Length ______ Pretest Date ______ Grade ______ WPS # ______

b) Component ☐ Replaced ☐ Reconditioned ☐ Other *(Specify)* ____________

c) Manufacturer ____________ Attach MTR (or Reference)

____________ *Technician or Inspector* ___/___/___ *Date* ____________ *Supervisor* ___/___/___ *Date*

Retention: Pipe - Life of System
Other - 1 Year

SAMPLE

107F20A.EPS

*Figure 20* ◈ Pipe repair report form (1 of 2).

## REPAIR REPORT FOR LEAK, DAMAGE, OR TEST FAILURE

*Frequency: As Needed*

Report Date ____________

**PART E — CORROSION**

1. Was there external corrosion? ☐ Yes ☐ No

   If yes, complete AAPL Form No. 501 "Pipeline Inspection Report" and attach

2. Was there internal corrosion? ☐ Yes ☐ No

   If yes, complete AAPL Form No. 508 "Internal Corrosion Inspection Report" and attach.

**PART F — DAMAGE BY OUTSIDE FORCES**

1. Primary cause of incident
   - ☐ a. Damage by equipment by or for operator
   - ☐ c. Damage by earth movement
   - ☐ b. Damage by equipment operated by outside party
   - ☐ d. Other (Specify) ____________
2. Location information for excavating and berming incidents:
   - a) Did the operator receive prior notification that equipment would be used in the area?

     ☐ (1) Yes ☐ (2) No (3) Date ____________ (4) Time ____________
   - b) Was the pipeline marked or identified?

     ☐ (1) Yes ☐ (2) No

     (1) If "Yes," what type of marking or identification was used to advise outside party of location of pipeline?

     ☐ Permanent Markers ☐ Excavation ☐ Other (Specify) ____________

     ☐ Map furnished ☐ On-site observation

     ☐ Temporary stakes ☐ Paint
3. Damage by earth movement:

   ☐ a. Subsidence ☐ c. Landslide ☐ e. Other (Specify) ____________

   ☐ b. Earthquake ☐ d. Washout

   f. Was the earth movement caused by direct or indirect action of others? ☐ (1) Yes ☐ (2) No

   If "Yes" explain ____________



**PART G — CONSTRUCTION DEFECT OR MATERIAL FAILURE**

1. Primary cause of incident: ☐ a. Construction defect ☐ b. Material failure
2. Explain all known facts and related conditions pertaining to the above.

Distribution: (Original) Operations Center
(Copy) Division Office
(Copy) Engineering Dept. (For System Modifications)
(Copy) Other ____________

Attachments: AAPL Form 101-A, B; If incident was associated with a Leak or Accident.
AAPL Form 501; If pipe was exposed and/or external corrosion was found.
AAPL Form 508; If internal corrosion was found.

Retention: Pipe - Life of System

107F20B.EPS

*Figure 20* ◆ Pipe repair report form (2 of 2).



Instructor's Notes:

Adequate fill material must be provided underneath the pipe as well as over it. The spoil bank is the main source for backfill material, but loss through excavation may require more material. The backfill material around the pipe must not damage the pipe or coating, and additional material should meet the same specifications of original construction.

The backfill is laid around the pipe in such a manner that it supports the pipe evenly, reducing stresses that may later cause pipe damage. The bedding, side fill, and cover should provide a uniform, even blanket around the pipe.

When filling is completed, the surface should be restored as closely as possible to its original condition, including the replacement of topsoil.

Backfilling correctly and safely involves the following tasks:

- Preparing the fill
- Positioning skid piles and sandbags
- Placing bedding or slurry
- Performing soil compaction

### 10.1.0 Preparing Fill

Any rocks and foreign objects left in the ditch or in the backfill material are potential hazards to the pipe and its coating. Foreign material such as cans, tools, clamps, or scrap metal, if left in the ditch, can diminish cathodic protection or damage the pipe in that area.

Fill dirt, particularly that used for bedding and close to the pipe, may be screened and sifted to remove rocks, large chunks, and other debris. The spoil from a new ditch made through rock cannot be used as fill because of the amount of rock debris. New fill dirt must be brought in to pad and fill most of the ditch. Sometimes rocky soil may be used in the upper part of the ditch, as long as the rocks are small, well away from the pipe, and only used up to about 1 foot from the surface. The final layer must be clean fill dirt or topsoil. If excavated soil has been contaminated with product from a leak before or during repair, it should not be used as fill.

Spoil from ditches dug in frozen ground can be as dangerous as rocky soil. The frozen bits of earth are irregular and hard as rock. They may be used if they can be broken up sufficiently to not pose a hazard to the pipe. However, frozen soil is difficult to distribute evenly and, when it thaws, may result in areas of unsupported pipe or uneven fill. In some areas, soil that was unfrozen when excavated can freeze at the surface, particularly overnight. This fill would cause the same problems.

### 10.2.0 Removing and Positioning Skid Piles and Sandbags

Before backfilling, skids must be removed. This is only done while other support is being provided under the pipe. In rocky rough soils, **padding** of earth, sand bags, or other approved supports may be laid under the pipe. Synthetic materials are also often used to provide a padded surface on the bottom of the ditch.

### 10.3.0 Placing Bedding

In preparing the ditch for replacement of the pipe, pipeline specifications should be followed. A sound pipe installation begins with a stable foundation. This foundation may include a layer of stabilization material to support the bedding, a layer of bedding material to cushion and support the pipe, and initial backfill to blanket the pipe. These three elements can be critical to the load-bearing ability of the pipeline. *Figure 21* shows this trench bedding arrangement.

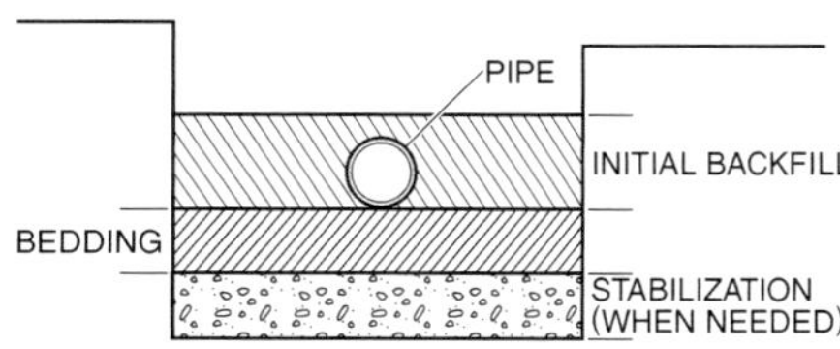


*Figure 21* ◆ Trench bedding arrangement.

#### *10.3.1 Stabilization*

Some soil conditions may require a stabilization layer to support the pipe bedding. Unless the pipe is moved to a different trench, the need for such a layer should be evident in the existing trench. If so, the stabilization layer under the exposed pipe may need to be repaired after excavation.

If the pipeline is laid in a new trench, the bottom should be inspected to determine the need for stabilization. A worker can walk in the trench, shifting weight from one foot to the other and determine if the soil underneath moves, shifts, or gives. If it does not feel solid, stabilization may be needed.

Crushed stone is often used, in sizes ranging from less than an inch to large rip-rap. The material is poured into the trench and spread to a uniform depth.



**Classroom**

**Stress the importance of supporting pipe while removing skids.**

**Audiovisual**

**Show Transparency 19 (Figure 21). Describe the components of a stable foundation.**

**Classroom**

**Describe how workers inspect the trench bottom.**

**Classroom**

**Stress that the type of bedding to be used is spelled out in the job specifications.**

**Ask trainees if they have any questions about the material covered so far.**

**Homework**

**Have trainees review Sections 10.4.0–10.5.0 and answer the Review Questions for the next class session.**

**Materials**

**Ensure that you have everything required for testing during this session.**

**Classroom**

**Describe the steps in soil compaction, including extra compaction, crowning, and the use of ditchbreakers.**

**Teaching Tip**

**See the suggested Teaching Tip at the end of this module.**

#### *10.3.2 Bedding*

Bedding should be sufficient to provide uniform, firm support for the pipe and to maintain pipe grade. The type of bedding support is dictated by specifications. The load-bearing capacity of the pipeline is determined from the strength of the pipe itself and the type of bedding.

Bedding extends from the top of the stabilization layer, if present, to the bottom of the pipe. Some materials used include native soil, sand, stone, or concrete. Specifications commonly require 4 to 6 inches of bedding material.

The material is poured into the trench and leveled to grade along the full length of the trench. Lasers are often used to check grade.

#### *10.3.3 Initial Backfill*

After laying the pipe on the bedding, the initial backfill is added. This is material placed from the bedding or the bottom of the pipe to some specified level above the pipe. The purpose of this initial backfill is to protect the pipe. You must use only clean fill material free of rocks, clumps, foreign objects, or organic materials. Initial backfill may include any of the following:

- **Haunching**, tamping, or filling to the spring line
- Tamping or filling to the top of the pipe
- Tamping or filling to some point above the pipe

Haunching is the process of placing and consolidating material in areas that may remain void or loose when the rest of the backfill is added. Because bedding only supports a narrow strip along the bottom of the pipe, sand or dirt is haunched in along the bottom sides or haunches of the pipe. This material is compacted with a hand tamp or other such tool. *Figure 22* shows some different pipe laying conditions.

### 10.4.0 Performing Soil Compaction

The replaced soil frequently needs to be compacted to avoid a depression caused by later settlement and because loose, uncompacted earth is subject to erosion by rainwater. By consolidating the soil particles in the fill, compaction also strengthens the pipeline cover and increases the protection of the blanket of fill around the pipe.

Compaction can be accomplished by driving a heavy crawler-type tractor over the backfilled ditch. Extra cover is often used at road crossings, ditches, and streams. Extra compaction is allowed at road crossings and other locations where heavy loads are expected to cross the pipeline.

Backfill is normally crowned after compaction to allow for some settlement. **Ditch breakers** may be required on the slope of a ditch, diversion level, or surface to prevent backfill from washing away.

## 10.5.0 ◆ STEPS FOR BACKFILLING

Backfilling a trench following pipeline maintenance is a delicate operation with the goal of protecting the pipeline from damage. Use the following steps as a checklist to properly backfill a trench:

*Step 1* Identify soil types and their suitability to support pipe. Some soil types compact better than others, making them more suitable to support the pipeline.

*Step 2* Perform a visual inspection of backfill material to identify foreign objects that could damage the pipeline system. For example, use a shovel to spot check for large objects or foreign material that could be mixed in. If you see any foreign material or objects, see that they are removed.

*Step 3* Determine when backfill material is unsuitable for backfill around the pipeline. Remember that soft material must be used near the coating. The backfill may need to be replaced if there are too many rocks, roots, brush, or other foreign material.

*Step 4* Use suitable backfill material within six inches of pipe or per company policy.

*Step 5* Fill equally along both sides of the pipe until the pipe is covered. This will help prevent pipe movement and provide additional support for pipe.

*Step 6* Crown the backfill according to company procedures. This is usually performed to compensate for settlement of backfill.

**NOTE**

Use *49 CFR Part 195* to find properly pipeline cover in *195.248(a)*.

**NOTE**

Provide extra cover at road crossings, ditches, and streams. These areas face an increased risk from third-party damage, and extra cover allows some extra protection. In special circumstances like these, you should reference procedural manuals, such as your company manual and *49 CFR Part 195.*



Instructor's Notes:

*Step 7* Compact the soil, as appropriate for proper pipe support, size of support, and spacing of support. Use a tamping tool, backhoe, of other appropriate tool. Be sure that you know when the soil is properly compacted.

> **NOTE**
> Perform extra compaction at road crossings or other locations where heavy loads may cross pipeline. This provides extra protection for pipe. Less settlement also means that there's less of a chance of third-party damage.

| LAYING CONDITION | DESCRIPTION |
|---|---|
|  TYPE 1 | FLAT-BOTTOM TRENCH. LOOSE BACKFILL. |
| 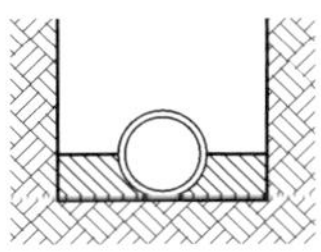 TYPE 2 | FLAT-BOTTOM TRENCH. BACKFILL LIGHTLY. CONSOLIDATED TO CENTER LINE OF PIPE. |
| 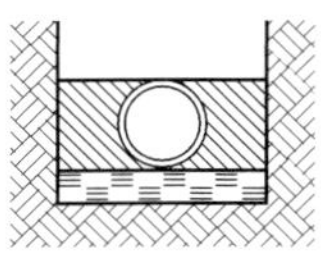 TYPE 3 | PIPE BEDDED IN 4" MINIMUM LOOSE SOIL OR BEDDING MATERIAL. BACKFILL LIGHTLY CONSOLIDATED TO TOP OF PIPE. |
| 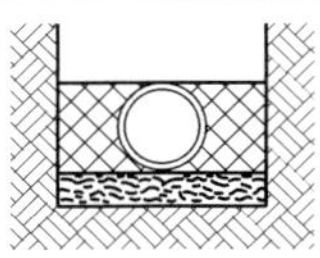 TYPE 4 | PIPE BEDDED IN SAND, GRAVEL, OR CRUSHED STONE TO DEPTH OF ⅛ PIPE DIAMETER, 4" MINIMUM. BACKFILL COMPACTED TO TOP OF PIPE. |
| 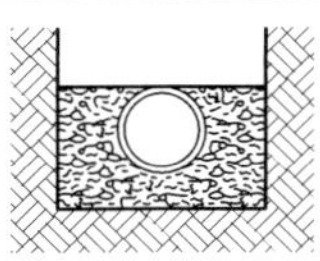 TYPE 5 | PIPE BEDDED TO ITS CENTER LINE IN COMPACTED GRANULAR MATERIAL, 4" MINIMUM UNDER PIPE. COMPACTED GRANULAR OR SELECT MATERIAL TO TOP OF PIPE. |

NOTE: These options vary with job specifications and locations.

107F22.EPS

*Figure 22* ◆ Different pipe laying conditions.



## Audiovisual

**Show Transparency 20 (Figure 22). Review the different pipe laying conditions. Ask a trainee to explain what is meant by haunching.**

**Show Transparency 21. Review the steps for backfilling. Stress the importance of following the company manual and *49 CFR Part 195* or *Part 192*.**

## Performance Testing

**Have each trainee demonstrate Performance Task 5 to your satisfaction. Fill out a Performance Profile Sheet for each trainee.**

**Classroom**

**Go over the Review Questions. Encourage trainees to ask questions. Have trainees prepare for the Module Examination.**

**Examination**

**Administer the Module Examination. Be sure to record the results of the Exam on Craft Training Report Form 200 and submit the results to the Training Program Sponsor.**

**Performance Testing**

**Ensure that all Performance Tasks have been completed and Performance Profile Sheets for each trainee are filled out. Be sure to record the results of the testing on Craft Training Report 200 and submit the results to the Training Program Sponsor.**

## Summary

Excavating and backfilling are necessary tasks associated with the maintenance of a buried pipeline. Although original excavation and backfill were done during construction, there are times during the lifetime of the pipeline that it must be exposed and/or brought to the surface. Although modern materials, methods of protection, and maintenance practices have lessened the frequency of problems and the needs for repairs, it is still required from time to time.

The first step is always locating the pipeline and any other buried structures correctly. Until this is done, no excavation work should be started. The One-Call system is used to notify all parties who have structures in the vicinity, while markers and maps can give their general locations. Location devices and probes pinpoint the actual presence of the pipeline.

Excavation can be a dangerous procedure if not done properly. Procedures are regulated by OSHA. It can also be dangerous to the pipeline itself, the public, and environment. For this reason, excavation and backfill procedures are monitored and regulated by the DOT, EPA, and other state and local entities.

For the safety of workers, it is important to ensure that trenches are safe places in which to do the necessary work. It is important also to know how to deal with hazardous product when encountered. Using heavy equipment to dig and backfill can be dangerous to workers and to the pipeline itself. Specific procedures are followed to avoid accidents and damage from excavators and backhoes.

Once a pipeline is exposed, examine and inspect how burial and operation are affecting its condition. This is in addition to performing the work that caused the excavation to be done. Reports are made on the inspection, and any repairs are performed. Ensuring that the pipeline is in good condition when reburied can help alleviate future problems.

The surrounding environment, especially any nearby water sources, is examined for contamination when the work is completed. Any contamination, if found, could be the result of a leak for which the pipeline was dug up or a spill during repairs. If found, contamination must be reported and procedures followed for cleaning it up right away.

Backfilling the trench over the pipeline involves more than just shoveling the dirt back in the hole. Specific procedures for preparing the trench with stabilization material, bedding, and initial backfill are followed to give support and protection for the pipeline. Completing the backfill must also meet the standards and regulations required for providing a durable, protective cover for the pipeline as well as a surface that meets the environmental and local usage needs



Instructor's Notes:

**Review Questions**

1. Accidents during _____ are among the leading causes of injuries and deaths in the construction industry.
   a. excavation
   b. welding
   c. backfilling
   d. pipelifting

2. Any plan to excavate must be preceded with alerting _____.
   a. the pipeline company
   b. supervisors
   c. the One-Call system
   d. local authorities

3. _____ should be used to help locate the pipeline with relation to surface features.
   a. Local sources
   b. Probes
   c. Road maps
   d. Alignment maps

4. Signs are required at all _____ crossings.
   a. public
   b. highway and street
   c. river and stream
   d. rail, road, and water

5. _____ may be required before excavating any sections of the pipeline.
   a. Shields
   b. Permits
   c. Surveys
   d. Pipe supports

6. Excavations of _____ feet or deeper must use shielding and shoring devices or sloping to prevent cave-ins.
   a. 12
   b. 8
   c. 6
   d. 5

7. _____-graded shows a good representation of all particle sizes from largest to smallest.
   a. Gap
   b. Well
   c. Poorly
   d. Uniformly

8. Soil classifications are based on stability and range from stable rock at the high end to _____ at least stable.
   a. Type D
   b. Type C
   c. Type B
   d. Type A

9. When working in waterlogged soils, the excavator may want to install _____ to dewater the ditch line.
   a. well points
   b. pumps
   c. drains
   d. hoses

10. _____ soil increases the possibility of a cave-in.
    a. Frozen
    b. Clay
    c. Saturated
    d. Dry

11. A pipeline being excavated will require careful handling and _____ to avoid damaging the pipe.
    a. support
    b. cover
    c. shielding
    d. insulation

12. Repairs of damages and defects are described on a(n) _____ report form.
    a. pipeline inspection
    b. excavation
    c. maintenance
    d. pipe repair

13. A sound pipe installation begins with a _____.
    a. clean ditch
    b. soil compaction
    c. stable foundation
    d. layer of gravel

14. The type of bedding support is dictated by _____.
    a. local supplies
    b. depth of the trench
    c. geologic features
    d. specifications

15. The replaced soil frequently needs to be _____ to avoid a depression caused by later settlement.
    a. crowned
    b. compacted
    c. tamped
    d. terraced

GLOSSARY

# Trade Terms Introduced in This Module

*Backfilling:* The technique for covering a completed pipeline so that adequate fill material is underneath and over the pipe to protect it from loose rocks, abrasion, shifting, and washouts.

*Competent person:* An OSHA term for a person who is capable of identifying existing and predictable hazards in the surroundings or working conditions that are unsanitary, hazardous, or dangerous to employees and who has authorization to take prompt corrective measures to eliminate them.

*Ditch breaker:* A device that divides a ditch into sections to block water movement. Used for pipelines in areas where washouts are a threat.

*Gouge:* Surface damage that reduces the pipe wall thickness by displacement or removal of metal.

*Haunching:* The process of providing support material for the bottom sides, or haunches, of the pipe on either side of where the pipe lays on the bedding.

*Padding:* Screened or sifted dirt, clean gravel, or foam placed in a ditch to protect pipe from damage caused by sharp rocks or rough soil.

*Pot-holing:* A procedure used to locate underground installations by hand-digging a small hole in the ground.

*Probe:* A sensing element that extends into the air, gas, or liquid stream to measure pressure, temperature, or velocity. In the pipeline industry, a T-shaped hand tool, 4 to 6-feet long, used to locate underground installations by pushing it into the soil repeatedly at regular intervals.

*Well point:* A device installed in waterlogged soil to dry out areas along the pipe ditch line. Hundreds may be needed in some areas.



Instructor's Notes:

# MODULE 62107-02 — TEACHING TIPS

The following are suggested activities or instructional methods to help you teach the material in this AIG.

## General

When you call on someone to answer a question, the rest of the class relaxes or even tunes out because they expect that the question and answer will take place only between you and the trainee you called on. Instead, use this technique to involve more trainees in answering questions and to keep them on their toes.

1. Ask trainees to define a term or explain a concept.
2. After one trainee has answered, ask a trainee seated nearby if the answer is right. Then ask whether a trainee in the back of the room agrees.
3. Ask trainees to explain why they think an answer is right or wrong.
4. Use the session to clear up incorrect ideas and encourage trainees to learn from their mistakes.

## Section 2.0.0 *OSHA Requirements*

Administer this quick quiz to help trainees focus on OSHA requirements. Note that while some answers may appear to be correct, only those answers as given in the OSHA requirements are correct in this quiz. To give this quiz, follow these steps:

1. Photocopy and hand out copies of the quiz.
2. Give trainees time to complete the quiz (5 to 10 minutes).
3. Have trainees exchange papers.
4. Read the correct answers, tell trainees to write down the correct answer next to incorrect answers, and have trainees exchange papers again.
5. Ask trainees if they have questions about OSHA requirements.

### *Answers to Quick Quiz*

1. c. a competent person
2. a. exposure to harmful levels of atmospheric contaminants
3. a. accumulated water
4. b. support systems such as shoring, bracing, or underpinning shall be provided
5. c. walkways or bridges with standard guardrails shall be provided

**Quick Quiz** ***OSHA Requirements***

Complete the following sentences by circling the correct response:

**1.** Structural ramps that are used solely by employees as a means of access to and egress from excavations or ramps for equipment shall be designed by _____.

a. the site supervisor

b. an OSHA inspector

c. a competent person

d. an OSHA inspector and DOT supervisor

**2.** Tests and controls shall be used to prevent _____.

a. exposure to harmful levels of atmospheric contaminants

b. exposure to standing water

c. exposure to hazardous levels of loose rock, shale, or loosened earth

d. exposure to hazardous levels of sound frequencies

**3.** Employees shall not work in excavations in which there is/are _____.

a. accumulated water

b. workers from other trades

c. heavy equipment

d. rocks

**4.** Where the stability of adjoining buildings, walls, or other structures is endangered by excavation operations, _____.

a. the central excavation site shall be relocated to a more stable site

b. support systems such as shoring, bracing, or underpinning shall be provided

c. no explosive materials shall be used at any location on the site

d. additional insurance shall be provided by the general contractor

**5.** Where employees or equipment are required or permitted to cross over excavations, _____.

a. warning lights and signs with standard symbols shall be provided

b. safety nets of OSHA-approved gauge shall be provided

c. walkways or bridges with standard guardrails shall be provided

d. chain-link fences at all entrances to the excavation shall be provided

## Section 6.3.1 *Cave-Ins*

This simple exercise will help trainees visualize just how heavy one cubic yard of earth really is. You will need an empty 3' × 3' × 3' cardboard box and a tape measure. Trainees will need pencils and paper to take notes. Allow 30 minutes for this exercise.

1. Ask a trainee to measure the box and to guess how much it would weigh if filled with earth. State that the box, if filled with earth, might weigh as much as 3,000 pounds.
2. Ask trainees for examples of objects or animals that weigh 3,000 pounds.
3. Note that an adult African elephant can weigh up to 14,000 pounds; a baby elephant starts life at around 350 pounds. Ask trainees to figure out how many 350-pound baby elephants it would take to equal 3,000 pounds of earth in the box. (About nine.)
4. Tell trainees to consider how much a seemingly small amount of earth weighs. Use this information to discuss the importance of safety and why cave-ins are the single most hazardous part of excavation work.

## Section 10.4.0 *Soil Compaction*

This simple exercise will give trainees a close-up look at different soils and how they compact. Advance preparation is necessary. The number of buckets and garden trowels or spoons you will need depends on the number of soil types you can collect. You will need the following items:

- Plastic buckets
- Two-liter bottles of water (if a water source is not available) or empty bottles to fill with water
- Small bags containing various types of soil available to you; for example, sand, potting soil, clay, fill dirt, dirt that contains rocks or twigs
- Garden trowels or large spoons

Trainees will need pencils and paper to take notes for later discussion.

1. Explain that soil that appears to be compacted may actually have void spaces in it that will settle later on. Emphasize the importance of properly compacting soil.
2. Assign trainees to a bucket and soil type. Tell them to fill the buckets to within 2 inches of the top and to signal when they feel their soil is properly compacted.
3. Have trainees carefully pour water into their buckets and observe how much their soil settles.
4. Point out that different soils settle at different rates. Ask trainees which soils settled the most and which soils settled the least.
5. Note that how soil behaves is an important component of pipeline excavation and backfilling.

## MODULE 62107-02 — ANSWERS TO REVIEW QUESTIONS

| Answer | Section |
|---|---|
| **1.** a | 2.0.0 |
| **2.** c | 3.0.0 |
| **3.** d | 3.2.0 |
| **4.** d | 4.0.0 |
| **5.** b | 5.0.0 |
| **6.** d | 6.1.1 |
| **7.** b | 6.1.2 |
| **8.** b | 6.1.3 |
| **9.** a | 6.1.5 |
| **10.** c | 6.3.1 |
| **11.** a | 6.4.0 |
| **12.** d | 8.0.0 |
| **13.** c | 10.3.1 |
| **14.** d | 10.3.2 |
| **15.** b | 10.4.0 |

## CONTREN™ LEARNING SERIES — USER UPDATES

The NCCER makes every effort to keep these textbooks up-to-date and free of technical errors. We appreciate your help in this process. If you have an idea for improving this textbook, or if you find an error, a typographical mistake, or an inaccuracy in NCCER's Contren™ textbooks, please write us, using this form or a photocopy. Be sure to include the exact module number, page number, a detailed description, and the correction, if applicable. Your input will be brought to the attention of the Technical Review Committee. Thank you for your assistance.

*Instructors* – If you found that additional materials were necessary in order to teach this module effectively, please let us know so that we may include them in the Equipment/Materials list in the Instructor's Guide.


**Write:** Curriculum Revision and Development Department
National Center for Construction Education and Research
P.O. Box 141104, Gainesville, FL  32614-1104

**Fax:** 352-334-0932

**E-mail:** curriculum@nccer.org


Craft ____________ Module Name ____________

Copyright Date ____________ Module Number ____________ Page Number(s) ____________

Description ____________

____________

____________

____________

(Optional) Correction ____________

____________

____________

(Optional) Your Name and Address ____________

____________

____________

# Index

# Index

# *Index*

# Index